Bambara groundnut: Utilization and Future Prospects

Victoria A. Jideani • Afam I. O. Jideani

Bambara groundnut: Utilization and Future Prospects

Springer

Victoria A. Jideani
Department of Food Science & Technology
Cape Peninsula University of Technology
Bellville, South Africa

Afam I. O. Jideani
Department of Food Science and Technology, Faculty of Science
Engineering and Agriculture
University of Venda
Thohoyandou, South Africa

ISBN 978-3-030-76076-2 ISBN 978-3-030-76077-9 (eBook)
https://doi.org/10.1007/978-3-030-76077-9

This Springer imprint is published by the registered company Springer Nature Switzerland AG
The registered company address is: Gewerbestrasse 11, 6330 Cham, Switzerland

This book is dedicated to enthusiasts of Bambara groundnut.

Foreword

Bambara groundnut: Utilization and Future Prospects

Forming the heart of my research focus the past decade, I have seen the interest in Bambara groundnut shifting from merely tapping into the potential of an "underutilised and neglected species", to recognition of the crop as an important "future food" resource. Considering this important shift, this new book *Bambara Groundnut: Utilization and Future Prospects* is rather timely and can boast as being the first reference book with a sole, comprehensive focus on Bambara groundnut.

Various topics of Bambara groundnut research are touched upon, which highlight how this crop can be utilised in an interdisciplinary manner. As a climate-smart pulse crop, Bambara groundnut is especially relevant in our current climate of global transitions. Hereto Chap. 3 provides an apt overview of the resource efficiency of Bambara groundnut linked to environmental conditions and agronomic factors, whilst also highlighting how this aspect can be beneficially applied towards food and nutrition security. This provides insight to the following chapters, which address aspects of Bambara groundnut ingredient functionality and application in various traditional, model and commercial food systems. In addition, Chap. 14 highlights the current barriers and opportunities of the Bambara groundnut value chain, which has the potential to become a sustainable food system engaging all stakeholders, especially at the small-scale farming level. The final chapter concludes with the current and future research directions envisioned for Bambara groundnut, which not only cements the "future food" status of the crop, but also reinforces the opportunities for fruitful interdisciplinary research amongst scientists in the food, agriculture and biotechnology domains.

The publication of such a comprehensive book on Bambara groundnut as food source is highly welcomed. Professors Victoria Jideani and Afam Jideani are commended for undertaking such an endeavour, which is strengthened by their scientific rigour and active vision for value addition of Africa's indigenous crops.

Wageningen University
Wageningen, the Netherlands

Claudine F. Diedericks

Preface

Bambara groundnut [*Vigna subterranea*] is an extremely hardy grain legume that produces reasonable yields even under conditions of drought and low soil fertility. Consequently, it is one of the climate-smart crops. The crop, previously underutilized, has received a growing interest among researchers and consumers for its balanced nutritional profile. Ethno-medicinal benefits have been claimed by the indigenous consumers of Bambara groundnut (BGN). Such knowledge passed down from generation to generation face the risk of being lost especially as the indigenous knowledge of BGN lie with the elderly, as the young generation do not seem to care due to urbanization and changing lifestyle. Hence, the book provides in-depth information on the past and recent research on its breeding, food and feed utilization, indigenous knowledge, and prospects.

Topics covered on BGN include production, genetic diversity, breeding, and climate-smart crop; physical and engineering properties; postharvest handling and technofunctionality; functional food and ingredients potential; BGN protein, starch, and fibre; ethnonutritional, ethnomedical, nutraceutical indigenous knowledge; phytonutrients and antioxidant activity; BGN global research; marketing and commerce; and future BGN research directions. A significant number of illustrations focus on these topics.

This book is a complete reference for food scientists, technologists, engineers, and farmers working in the bioresource technology field, and is of particular interest to academics and professionals working in the food industry, food processing, and biotechnology. A search of the world of literature using text mining and network analysis of some database did not reveal any published book on BGN, hence this vacuum is now filled. This book is therefore distinct, being the only reference book on BGN, written by scientists that have conducted scientific research on BGN, considering the growing interest in plant proteins by food professionals around the globe.

Department of Food Science & Technology, Cape Peninsula
University of Technology, Bellville, South Africa — Victoria A. Jideani
Department of Food Science and Technology,
University of Venda, Thohoyandou, South Africa — Afam I. O. Jideani

Contents

About the Authors

Victoria A. Jideani Professor Victoria Jideani possesses a BSc honours degree in food science and technology (1983) from the University of Ife (now Obafemi Awolowo University), Ile-Ife, Nigeria; an MSc degree in food science (1993) from the University of Leeds, UK; a diploma in computer applications (1994) from the Women Technology Centre, Leeds, UK; and a PhD in food microbiology (2001) from the Abubakar Tafawa Balewa University (ATBU), Bauchi, Nigeria. She is an IBM Certified Specialist, SPSS Statistics (2013). Professor Jideani is founder and leader of the Cereals and Legumes Biopolymers Research for Food Security (CLBRFS) group at the Cape Peninsula University of Technology (CPUT), South Africa. Its focus is on value addition to African indigenous climate-smart grains and seeds, a means for Africa to achieve food and nutrition security. Professor Jideani has two registered patents and over 90 publications in both international and regional peer-reviewed journals, conference proceedings, and books/book chapters. Her patent on Bambara groundnut value-added products is the object of a proposed CPUT spin-off company. She is internationally recognized for her research contribution in cereal science, food processing, food product development, optimization, and modelling of food processes. She is an established researcher according to the National Research Foundation of South Africa rating for scientists and a professional member of (1) Professional Natural Scientist, South Africa from 2010; (2) South African Association of Food Technologists (SAAFoST); (3) Institute of Food Technologists (IFT); (4) American Association of Cereal Chemists; (5) European Association for Integrating Food Science and Engineering Knowledge Into the Food Chain (ISEKI-Food Association); and (5) Nigerian Institute of Food Science and Technology (NIFST). She received the TW Kambule-NSTF Award: Researcher, 2019/2020 National Science and Technology Forum (NSTF)-South32 Awards Finalist in recognition of her outstanding contribution to science, engineering, technology (SET), and innovation, through research and its outputs (by an individual from 6 years up to 15 years of research work from the commencement of the research career, predominantly in South Africa).

Afam I. O. Jideani Afam I. O. Jideani is a professor of food science and technology at the University of Venda (Univen), South Africa, and the director of the Postharvest-Handling Special Interest Group of ISEKI Food Association (IFA) in Vienna, Austria. He obtained his PhD from the University of Leeds, UK. He was the director of research and development at the National Food Technology Research Centre, Botswana, and briefly chaired the Food Legislation and Advisory Committee of the National (Botswana) Food Control Board. He pioneered a postgraduate degree programme in food science and technology at Univen. He is a National Research Foundation (South Africa) rated researcher and has over 160 publications in scientific journals, books, and book chapters. He is a member of the Education Committee as well as the Food Safety Working Group – Scientific Communications of the International Union of Food Science and Technology (IUFoST). He is a professional member of the following scientific bodies: (i) South African Association of Food Technologists (SAAFoST); (ii) Institute of Food Technologists (IFT); (iii) IFA, where he is the current South African country representative, IFA Academy member, and an advisory board member of *International Journal of Food Studies of IFA*; and (iv) Nigeria Institute of Food Science and Technology (NIFST). He is a fellow of the International Academy of Food Science and Technology (IAFoST) and NIFST.

Chapter 1
Introduction

The chapter gives an overview on Bambara groundnut (*Vigna subterranea*) in relation to the world of food legumes, importance in the increasing growth in world population, the socio-economic, nutrition and cultural significance to the people, the reason for its growth in popularity especially in the achievement of sustainable healthy diets and environment, the outlook as an underutilized and under-researched smart food legume crops, the increasing attention and scientific investigation received in all areas including physical properties, chemical and physical modifications, breeding, genotype, applications, ingredients, and techno-functionality.

The world food legumes comprise all legumes grown for human food. These food legume species include soybean [*Glycine max* (L.) Merr.], groundnut (*Arachis hypogaea* L.) (both sometimes regarded as oilseeds), pea (*Pisum sativum* L), faba bean (*Vicia faba* L.), common bean (*Phaseolus vulgaris* L), mung bean (*Vigna radiata* Wilczek), adzuki bean [*Vigna angularis* (Willd) Ohwi & Ohashi], cowpea (*Vigna unguiculate* L. Walp), Bambara groundnut (*Vigna subterranea*) as well as runner bean (*Phaseolus multiflorus* Willd), lima bean (*Phaseolus lunatus* L), chickpea (*Cicer arietinum* L), lentil (*Lens culinaris* Medik), grass pea (*Lathyrus sativus* L), lupine(*Lupinus* L), rice bean [*Vigna umbellata* (Thumb.) Ohwi and Ohashi], black gram [*Vigna mungo* (L.) Hepper], hyacinth bean [*Lablab purpureus* (L.) Sweet], pigeon pea [*Cajanus cajan* (L.) Millsp], velvet bean [*Mucuna pruriens* (L.) DC.], winged bean [*Psophocarpus tetragonolobus* (L.) DC], guar bean [*Cyamopsis tetragonoloba* (L.) taubert], sword bean [*Canavalia gladiata* (Jacq.) DC.], and jack bean [*Canavalia ensiformis* (L.) DC.] (Ashagbon, 2017; Li et al., 2017; Nyau et al., 2015).

Legume importance is increasing due to the growing world population, projected to double by 2050. The Bambara groundnut (BGN) (*Vigna subterranea/Voandzeia substerranea* (L) Thouars), a legume of the family Fabaceae (Duke, 1981), is a nutritious legume crop with growing global appeal as plant food, grown as subsistence crop mostly by women farmers and plays a crucial role in both socio-economic, nutrition and the culture of the people (NRC, 2006; Nadembega, 2016; Atoyebi et al., 2018; Dje et al., 2005; Mkandawire, 2007; Misheck et al., 2013; Omonhinmin & Ogunbodede, 2013; Ndidi et al., 2014; Muhammad et al., 2020). The following

V. A. Jideani, A. I. O. Jideani, *Bambara groundnut: Utilization and Future Prospects*, https://doi.org/10.1007/978-3-030-76077-9_1

are the synonyms: *Glycine subterranea* L. and *Voandzeia subterranea* (L.) Thouars ex DC. (USDA, 2009). Whole BGNs are the edible seed of the BGN plant. It is described as a complete diet (Quedraogo et al., 2012), and a low-fat source of protein with a carbohydrate and protein content of approximately 65% and 18% respectively and other vital nutrients and antinutrients (Atoyebi et al., 2018). The seed contains about 49–63.5% carbohydrate, 15–25% protein, 4.5–7.4% fat, 5.2–6.4% fiber, 3.2–4.4% ash and 2% mineral (Amarteifio & Moholo, 1998). Compared with other legumes, it is high in protein, low in fat, high in phenolics with antioxidant properties (Kone et al., 2011; Baoua et al., 2014; Murevanhema & Jideani, 2013), with starch (Afolabi et al., 2017) and characterised as gluten-free (Taylor & Awika, 2017).

Bambara groundnut (BGN) is described as an indigenous African legume with a long history of cultivation mainly for human food and animal feed due to its high protein content (FAO, 2001; Adeleke et al., 2017) and having a tremendous nutritional, economic and social potential. In sub-Saharan Africa, BGN crop is predominant and ranks third among cultivated legumes after peanuts (*Arachis hypogaea*) and cowpea (*Vigna unguiculata*) (Hampson et al., 2000; Omonhinmin & Ogunbodede, 2013; Muhammad et al., 2020; FAO, 2001, 2018; Berchie et al., 2010; Ndidi et al., 2014). It is also in some parts of Asia and the Americas (Duodu & Apea-Bah, 2017; Atoyebi et al., 2018) and similar to other popular legumes such as cowpea, soybean, pea, etc. (Wang & Castonguay, 2014; Odongo et al., 2015; Ashagbon, 2017). The fruits are in the soil, thereby protecting them from insect and bird damage, ravaging other crops like cowpea, common bean and soybean, whose tastiest parts are above the ground. However, soybean (*Glycine max* (L.) Merr.) and groundnut (*Arachis hypogaea* L.) are not treated as food legumes in China but are defined as oil crops (Zheng et al., 1997). Although China is one of the leading world producers of legume crops, there is no production of BGN. In China, the six major and well-known food legume species are pea, faba bean, common bean, mung bean, adzuki bean, and cowpea. Those with more local distributions include runner bean, lima bean, chickpea, lentil, grass pea, lupine, rice bean, black gram, hyacinth bean, pigeon pea, velvet bean, winged bean, guar bean, sword bean, and jack bean (Li et al., 2017). China has remained the world's leading producer of peas, faba beans, mung beans, and adzuki beans, as documented by FAO statistics and China Agriculture Statistical Reports (Li et al., 2017). Legumes are an essential part of the human diet (Amarowicz, 2020), and their demand as a healthy food will markedly increase in China, India and the rest of the world; hence BGN has potential. In the study of Cook (2017), the conclusion was with recommendations to help improve small-scale farmers' production of BGN to alleviate poverty and improve food and economic security.

Diedericks (2020) in the study titled "Beyond a subsistence crop: Structure and functionality of Bambara groundnut proteins" concluded that (i) BGN protein isolates showed potential as partial animal protein replacers, based on the change in the rheological responses when combined with whey protein isolates and (ii) new knowledge from the study gave a better understanding of the structure-function behavior of BGN proteins, which can be used to strengthen the position and steer the application of BGN as an alternative plant protein source.

Bambara groundnut is an indigenous African crop grown as food across the continent (Atiku & Mohammed, 2004) and Indonesia, Malaysia and Sri Lanka (Eltayeb et al., 2011). It is among the most edible plants in whole grain/seed/nut, fruits, leaves, and tuber/roots. The seeds are used as feed for humans, poultry and livestock (Brink & Belay, 2006; Ijarotimi & Keshinro, 2012). It is used to prepare many dishes (Adebowale et al., 2005; Jideani, 2015; Mashau et al., 2020). Yet, it is classified as underutilized and not very common or unknown to consumers, national and international food baskets (Jideani, 2015). Value-added products from this legume, such as beverages and fibers, indicate that BGN can break the cycle of poverty and malnutrition in sub-Saharan Africa (Jideani, 2015). Bamshaiye et al. (2011) stated that BGN is used to alleviate nutritional problems, especially for the rural population. The State of the World (SoW) (2011) made the following observation about BGN: "*... this little bean, indigenous to tropical Africa, is highly overlooked by scientists, development agencies, and humanitarian programs, even though it packs a lot of nutrition. The bean is growing in popularity because it is a hardy plant, able to withstand high temperatures and dry conditions. ...The Bambara bean is high in protein, particularly methionine, which makes the protein more complete than that in other beans*". In Zimbabwe (Chandiposha et al., 2013) and Nigeria, BGN has found various food uses: it can be fried or boiled and eaten as a snack, pounded into flour and used in the preparation of soup or porridge, fried or steamed food products such as *akara, moin-moin* and *okpa*, and the preparation of local food drink, *kunnu* and stiff porridge, *tuwo*. In Zambia, BGN flour is used in making bread (Adebowale et al., 2005) and milk prepared from BGN gave a preferred flavor when compared with milk from soybean, cowpea, and pigeon pea (Piyarat, 2008).

Bambara groundnut seed utilization in human health shows that the plant has a significant impact on human nutrition by improving the quality of nutrition of many rural communities (Omoikhoje, 2008; Kone et al., 2011). The Lancet Commissions review on sustainable food systems (Willett et al., 2019) recommended to move toward healthy diets derived from whole grains, nuts, fruits, vegetables, and legumes, a low to moderate amount of seafood and poultry, low or no inclusion of red meat, processed meat, added sugar, refined grains, and starchy vegetables to achieve sustainable healthy diets and environment. BGN seeds contain sufficient quantities of proteins, carbohydrates, lipids and the highest concentration of soluble fiber among the beans.

The addition of BGN into cereal-based products improved nutritional, functional and sensory quality of such products (Oluwatofunmi et al., 2015; Adebayo-Oyetoro et al., 2017; Mashau et al., 2020). Hence, in developing economies, BGN is a suitable legume for cereal-legume-based diet formulation, especially during rapid growth and complementary feeding in infants. Adebayo-Oyetoro et al. (2017) showed that millet-based products serve as alternative protein and energy sources for infants and young children and sensory qualities were significantly different with BGN flour. At 90:10% millet-based *fura* and BGN, Adebayo-Oyetoro et al. (2017) stated that the *fura* product was generally acceptable and that pretreatment of BGN before usage improved colour, texture and water absorption capacity. This report and others make BGN a good protein source, making it desirable to complement other cereal foods.

Bambara groundnut has various uses in the confectionery industry and has medicinal uses (Atiku & Mohammed, 2004; Brink & Belay, 2006). It is a good source of fiber, calcium, iron and potassium and unusually high in methionine, an essential sulphur-containing amino acid. BGN essential amino acids are comparable to soybeans in lysine, methionine and cysteine (Omoikhoje, 2008). Indigenous knowledge associated with BGN includes leaf preparations applied to abscesses and infected wounds; sap from leaves applied to the eyes to treat epilepsy; roots sometimes taken as an aphrodisiac, and pounded seeds mixed with water and used to treat cataracts. BGN seeds have the highest concentration of soluble fiber compared to other beans. Some reported BGN phytochemicals, including anthocyanin and flavonoids, impart many beneficial health effects and reduce the risk of many chronic illnesses. Considerable studies have been done on the nutraceutical properties of sword bean, common beans, lentils, peas, soybeans, pigeon pea, BGN, kidney beans, lima beans and jack beans (Ashagbon, 2017; Nyau et al., 2015) legumes with remarkable adaptation to tropical conditions (Ashagbon, 2017; Li et al., 2017). Unfortunately, consumers are unaware of BGN's rich nutritional benefits (Jideani, 2015).

The numerous positive attributes notwithstanding, BGN is a severely underutilized and under-researched legume. Nonetheless, BGN is a popular crop because of its resistance to drought and pests and its ability to produce on poor soils. Hence, it is among the most adaptable plants with the comparative capability to withstand harsh climatic conditions and yields well on poor soils (Collinson et al., 1996, 1997, 1999; Tweneboah, 2000; Baryeh, 2001; Azam-Ali et al., 2001; Hillocks et al., 2012; Mabhaudhi & Modi, 2013; Ndidi et al., 2014; Odongo et al., 2015; Ho et al., 2017; Tsoata et al., 2016, 2017a; Bonny et al., 2019), generally on abandoned or fallow land (Mohamadou et al., 2018), a drought-tolerant and easy-to-cultivate legume (Sirivongpaisal, 2008; Tsoata et al., 2016, 2017b) in drier areas with short, inconsistent rainfall (FAO, 2001; Berchie et al., 2010; Ndidi et al., 2014), and useful in climate change adaptation strategies (Hillocks et al., 2012), grown as landraces (Zeven, 1998) in rotation or intercropped with cereal, root and tuber crops (Bonny et al., 2019). BGN (*Voandzeia subterranean* (L) Thouars) is a species of African origin (Stephens, 2003) used locally as a vegetable and first found in West Africa (Anonymous: List-based Rec./soil conserve. Serve, USA Database of the USDA, Belts Ville; 1986. Available: http://www.tropical.or/reference/11.)

Bambara groundnut seeds, grain, flour and starch have been receiving increasing attention and scientific investigation in all areas including physical properties, chemical and physical modifications, breeding, genotype, applications, ingredients, and techno-functionality (Eknayake et al., 1999; Adebowale & Lawal, 2002a, b, 2004; Mkandawire, 2007, Afolabi, 2012; Murevanhema & Jideani, 2013; Uarrota et al., 2013, Kaptso et al., 2015; Ademiluyi et al., 2015; Oyeyinka et al., 2015, 2016; Ogundele et al., 2017). An area receiving attention is the seed's hard-to-cook phenomenon, a common challenge with some cereal and legume grains (Teixeira et al., 2016). Other technological attributes affecting usage as food include strong beany flavour, presence of antinutrients and poor dehulling and milling characteristics (Oluwatofunmi et al., 2015).

Consumers are demanding new food products that have health benefits and the increasing need for convenience food (Coltram et al., 2000; Ja'afar-Furo et al., 2011). Some underutilized climate-smart food crops contain healthy nutritional components (Kone et al., 2011; Jideani & Murevanhema, 2013; Jideani & Diedericks, 2014). One such crop is BGN because of its resistance to drought and pests and its ability to produce a good harvest when grown on poor soils. It has long storage life (Amarteifio & Moholo, 1998), a much-desired property of a smart crop in the face of global warming. BGN's rich nutritional profile is "a native solution to Africa's Food Crisis" (Stone et al., 2011). There are seven BGN varieties in terms of color (Fig. 1.1), namely black, red, cream/black-eye, cream/brown-eye, cream/no-eye, speckled/flecked/spotted and brown (DPP, 2011). Usually, the seeds are mixed when purchased. In various research reports, BGN seeds are separated into as many assessions or landraces based on the seed and eye color (Figs. 1.1 and 1.2).

Fig. 1.1 Varieties of Bambara groundnut (*Vigna subterranea*)

Fig. 1.2 Seed morphological genetic diversity among different landraces of Bambara groundnut (Note: A = Bidi, B = Maiki, C = Bidilalle, D = Karu, F = Maizane, G = Doii, H = Ex-Sokoto, I = Jatau, J = Bidiyashi, K = Hawayenzaki, L = Zabuwa, M = Yar-Gombe, N = Maibargo, O = Giwa, P-Dunabaki, Q = Cancaraki) (Muhammad et al., 2020)

References

Adebayo-Oyetoro, A. O., Shotunde, A. B., Adeyeye, S. A. O., & Ogundipe, O. O. (2017). Quality evaluation of millet-based fura powder supplemented with Bambara groundnut. *International Journal of Food Science, Nutrition and Dietetics, 6*(3), 358–362. https://doi.org/10.19070/2326-3350-1700063

Adebowale, K. O., & Lawal, O. S. (2002a). Effect of annealing and heat moisture conditioning on the physicochemical characteristics of Bambara groundnut (*Voandzeia subterranea*) starch and flour. *Food Chemistry, 78*, 305–311.

Adebowale, K. O., & Lawal, O. S. (2002b). Effect of annealing and heat-moisture conditioning on the physicochemical characteristics of Bambara groundnut (*Voandzeia subterranea*) starch. *Nahrung/Food, 46*, 311–316.

Adebowale, K. O., & Lawal, O. S. (2004). Comparative study of the functional properties of Bambara groundnut (*Voandzeia subterranea*n), jack bean (*Canavalia ensiformins*) and mucuna bean (*Mucuna pruriens*) flours. *Food Research International, 37*, 355–365.

Adebowale, Y. A., Adeyemi, I. A., & Oshodi, A. A. (2005). Functional and physicochemical properties of flours of six Mucuna species. *African Journal of Biotechnology, 4*, 416–468.

Adeleke, O. R., Adiamo, O. Q., & Fawale, O. S. (2017). Nutritional, physicochemical, and functional properties of protein concentrate and isolate of newly-developed Bambara groundnut (*Vigna subterrenea* L.) cultivars. *Food Science & Nutrition, 6*(1), 229–242. https://doi.org/10.1002/fsn3.552

Ademiluyi, A. O., Oboh, G., Boligon, A. A., & Athayde, M. L. L. (2015). Dietary supplementation with fermented legumes modulate hyperglycemia and acetylcholinesterase activities in Streptozotocin-induced diabetes. *Pathophysiology, 22*, 195–201.

Afolabi, T. A. (2012). Synthesis and physicochemical properties of carboxymethylated Bambara groundnut (*Voandzeia subterranean*) starch. *International Journal of Food Science and Technology, 47*, 445–451. https://doi.org/10.1111/j.1365-2621.2011.02860.x

Afolabi, T. A., Opara, A. O., Kareem, S. O., & Oladoyinbo, F. O. (2017). In vitro digestibility of hydrothermally modified Bambara groundnut (*Vigna subterranean* L.) starch and flour. *Food Science & Nutrition, 6*, 36–46. https://doi.org/10.1002/fsn3.510

Amarowicz, R. (2020). Legume seeds as an important component of human diet. *Food, 9*, 1812. https://doi.org/10.3390/foods9121812. www.mdpi.com/journal/foods

Amarteifio, J. O., & Moholo, D. (1998). The chemical composition of four legumes consumed in Botswana. *Journal of Food Composition and Analysis, 11*(4), 329–333.

Anonymous: List-based Rec/soil conserve. Serve, USA database of the USDA, Belts Ville; 1986. Available: http://www.tropical.or/reference/11

Ashogbon, A. O. (2017). Evaluation of compositional and some physicochemical properties of Bambara groundnut and cocoyam starch blends for potential industrial applications. *American Journal of Food and Nutrition, 5*(2), 62–68. https://doi.org/10.12691/ajfn-5-2-3

Atiku, S., & Mohammed, K. (2004). *Cultivation of Bambara groundnut in Nigeria*. ABU, Zaria: Report of a field study.

Atoyebi, J. O., Osilesi, O., Abberton, M., Adebawo, O., & Oyatomi, O. (2018). Quantification of selected anti–nutrients and bioactive compounds in African Bambara Groundnut (*Vigna subterranea* (L.) Verdc.). *American Journal of Food and Nutrition, 6*(3), 88–95. https://doi.org/10.12691/ajfn-6-3-5

Azam-Ali, S. N., Sesay, A., Karikari, S. K., Massawe, F. J., Aguilar-Manjarrez, J., Brennan, M., & Hampson, K. J. (2001). Assessing the potential of an underutilized crop – A case study using Bambara groundnut. *Experimental Agriculture, 37*, 433–472.

Bamishaiye, O. M., Adegbola, J. A., & Bamishaiye, E. I. (2011). Bambara groundnut: An underutilized nut in Africa. *Advances in Agricultural Biotechnology, 1*, 42–51. (60-72).

Baoua, I. B., Amadou, L., Baributs, D., & Murdock, L. L. (2014). Triple bag hermetic technology for post-harvest preservation of Bambara groundnut (*Vigna subterranean*). *Journal of Stored Products Research, 58*, 48–52.

Baryeh, E. A. (2001). Physical properties of Bambara groundnuts. *Journal of Food Engineering, 47*(4), 321–326.

Berchie, J. N., Sarkodie, A. J., Adu-Dapaah, H., Agyemang, A., Addy, S., Asare, E., & Donkor, J. (2010). Yield evaluation of three early maturing Bambara groundnut (*Vigna subterranea* L. Verdc.). *Journal of Agronomy, 9*, 175–179.

Bonny, A., Ngantcha, M., Scholtz, W., Chin, A., Nel, G., Anzouan-Kacou, J.-B., Karaye, K., Damasceno, A., & Crawford, T. C. (2019). Cardiac Arrhythmias in Africa: Epidemiology, management challenges, and perspectives. *Journal of the American College of Cardiology, 73*(1), 100–109.

Brink, M., & Belay, G. (Eds.). (2006). *Plant resources of tropical Africa. Cereals and pulses* (p. 214). PROTA Foundation/Backhuys Publishers/CTA.

Chandiposha, M., Chagonda, I., & Makuvaro, V. (2013). Utilisation of common grain crops in Zimbabwe. *African Journal of Food Science, 7*(9), 253–257. https://doi.org/10.5897/AJFS2013.1011

Collinson, S. T., Azam-Ali, S. N., Chavula, K. M., & Hodson, D. A. (1996). Growth development and yield of Bambara groundnut (*Vigna subterranea*) in response to soil moisture. *The Journal of Agricultural Science, 126*, 307–318.

Collinson, S. T., Clawson, E. J., Azam-Ali, S. N., & Black, C. R. (1997). Effects of soil moisture deficits on the water relations of Bambara groundnut (*Vigna subterranean* L. Verdc.). *Journal of Experimental Botany, 48*, 877–884.

Collinson, S. T., Berchie, J., & Azam-Ali, S. N. (1999). The effect of soil moisture on light interception and the conversion coefficient for three landraces of Bambara groundnut (*Vigna subterranea*). *Journal of Agricultural Science, 133*, 151–157.

Coltrain, D., Barton, D., & Boland, M. (2000). *Value added: Opportunities and strategies*. Department of Agricultural Economics, Kansas State University.

Cook, D. 2017. *Small scale farmers utilization and perceptions of Bambara groundnut production in South Africa: A case study in a semi-arid region of Limpopo*. Masters dissertation, University of Cape Town. http://hdl.handle.net/11427/27485 https://open.uct.ac.za/handle/11427/27485

Diedericks, C. F. (2020). *Beyond a subsistence crop: Structure and functionality of Bambara groundnut proteins*. PhD dissertation, Wageningen University, Wageningen. ISBN 9789463953177–159, https://library.wur.nl/WebQuery/wurpubs/563553

Djè, Y., Bonny, B. S., & Zoro Bi, A. I. (2005). Observations préliminaires de la variabilité entre quelques morphotypes de voandzou (*Vigna subterranea* L. Verdc., *Fabaceae*) de Côte d'Ivoire. *Biotechnologie, Agronomie, Société et Environnement/Biotechnology, Agronomy, Society and Environment, 9*(4), 249–258.

DPP. (2011). *Production guidelines for Bambara groundnuts* (pp. 1–20). Directorate Plant Production, Department of Agriculture, Forestry and Fisheries.

Duke, J. A. (1981). *Voandzeia subterranea* (L). Thouars. In *Handbook of legumes of world economic importance* (pp. 307–310). Plenum Press.

Duodu, K. G., & Apea-Bah, F. B. (2017). Chapter 9: African legumes: Nutritional and health-promoting attributes, pp. 223–269. In J. Taylor & J. Awika (Eds.), *Gluten-free ancient grains: Cereals, pseudocereals, and legumes: Sustainable, nutritious, and health-promoting foods for the 21st century* (1st ed., pp. 356). Woodhead Publishing. eBook ISBN: 9780081008911 https://www.elsevier.com/search-results?query=Bambara%20groundnut&labels=books&page=1

Eknayake, S., Jansz, E. R., & Nair, B. M. (1999). Proximate composition, mineral and amino acid content of mature *Canavalia gladiata* seeds. *Food Chemistry, 66*, 115–119.

Eltayeb, A. R. S. M., Ali, A. O., Abou-Arab, A. A., & Abu-Salem, F. M. (2011). Chemical composition and functional properties of flour and protein isolate extracted from Bambara groundnut (*Vigna subterranean* (L.) Verdc.). *African Journal of Food Science, 5*(2), 82–90.

FAO. (2001). *Global mapping system for Bambara groundnuts production* (FAO agricultural information management series No. 1). Viale delle Terme di Caracalla, 00100 Rome, Italy.

FAO. (2018). *Statistic of production (FAOSTAT)*. Food and Agriculture Organization of the United Nations.

Hampson, K. J., Azam-Ali, S. H., Sesay, A., Mukwaya, S., & Azam-Ali, S. N. (2000). *Assessing opportunities for increased utilisation of Bambara groundnut in Southern Africa* (52p). Tropical Crops Research Unit, University of Nottingham. Internal Report.

Hillocks, R. J., Bennett, C., & Mponda, O. M. (2012). Bambara nut: A review of utilisation, market potential and crop improvement. *African Crop Science Journal, 20*(1), 1–16.

Ho, W. K., Chai, H. H., Kendabie, P., Ahmad, N. S., Jani, J., Massawe, F., Kilian, A., & Mayes, S. (2017). Integrating genetic maps in Bambara groundnut [*Vigna subterranea* (L) Verdc.] and their syntenic relationships among closely related legumes. *BMC Genomics, 18*, 192. https://doi.org/10.1186/s12864-016-3393-8

Ijarotimi, O. S., & Keshinro, O. O. (2012). Formulation and nutritional quality of infant formula produced from germinated popcorn, Bambara groundnut and African locust bean flour. *Journal of Microbiology, Biotechnology and Food Sciences, 1*(6), 1358–1388.

Ja'afar-Furo, M. R., Bello, K., & Sulaiman, A. (2011). Assessment of the prospects of value addition among small-scale rural enterprises in Nigeria: Evidence from North-Eastern Adamawa State. *Journal of Development and Agricultural Economics, 3*(3), 144–149.

Jideani, V. (2015). *Should the Bambara groundnut remain underutilized? FST Magazine August 2015.* A Publication of the South African Association of Food Science and Technology, South Africa.

Jideani, V. A., & Diedericks, C. F. (2014). Nutritional, therapeutic, and prophylactic properties of *Vigna subterranea* and *Moringa oleifera*. In O. Oguntibeju (Ed.), *Antioxidant-antidiabetic agents and human health* (pp. 187–207). https://doi.org/10.5772/57338

Jideani, V. A., & Murevanhema, Y. Y. (2013). *A Bambara groundnut milk beverage and probiotic yoghurt produced therefrom and a method for the production thereof South Africa-Complete Patent.*

Kaptso, G. K., Njintang, Y. N., Nguemtchouin, M. M. G., Scher, J., Hounhouigan, J., & Mbofung, C. M. (2015). Physicochemical and micro-structural properties of flours, starch and proteins from two varieties of legumes: Bambara groundnut (*Vigna subterranea*). *Journal of Food Science and Technology, 52*, 4915–4924.

Koné, M., Paice, A. G., & Touré, Y. (2011). Chapter 22: Bambara groundnut [*Vigna subterranea* (L.) Verdc. (Fabaceae)] usage in human health. In P. V. R. Watson RR & V. B. Patel (Eds.), *Nuts and seeds in health and disease prevention* (1st ed., pp. 189–196). Academic, p. 1226. https://doi.org/10.1016/C2009-0-01960-6. Hardcover ISBN: 9780123756886, eBook ISBN: 9780123756893.

Li, L., Yanga, T., Liua, R., Redden, B., Maaloufd, F., & Zonga, X. (2017). Food legume production in China. In Advances in crop science: Innovation and sustainability (Y. Xu, Ed.). *The Crop Journal, 5*(2), 115–126. https://doi.org/10.1016/j.cj.2016.06.001

Mabhaudhi, T., & Modi, A. T. (2013). Growth, phenological and yield responses of a Bambara groundnut (*Vigna subterranea* (L.) Verdc.) landrace to imposed water stress under field conditions. *South African Journal of Plant and Soil, 30*(2), 69–79. https://doi.org/10.1080/02571862.2013

Mashau, E. M., Mabodze, T., Tshiakhatho, O. J., Silungwe, H., & Ramashia, S. E. (2020). Evaluation of the content of polyphenols, antioxidant activity and physicochemical properties of tortillas added with Bambara groundnut flour. *Molecules, 25*, 3035.

Misheck, C., Chagonda, I., & Makuvaro, V. (2013). Utilisation of common grain crops in Zimbabwe. *African Journal of Food Science, 7*(9), 253–257. https://doi.org/10.5897/AJFS2013.1011

Mkandawire, C. H. (2007). Review of Bambara groundnut production in sub-Saharan Africa. *Agricultural Journal, 2*(4), 464–470.

Mohamadou, M., Ngakou, A., & Tchuenguem F-N, F. (2018). Influence de l'interaction mycorhizes-insectes floricoles sur les rendements de *Vigna subterranea* variété rouge (*Fabaceae*) à Dang (Ngaoundéré, Cameroun). *International Journal of Biological and Chemical Sciences, 12*(4), 1897–1913. https://doi.org/10.4314/ijbcs.v12i4.30. Available online at http://www.ifgdg.org

Muhammad, I., Rafii, M. Y., Ramlee, S. I., Nazli, M. H., Harun, A. R., Oladosu, Y., Ibrahim Musa, I., Arolu, F., Chukwu, S. C., Haliru, B. S., Akos, I. S., Halidu, J., & Arolu, I. W. (2020).

Exploration of Bambara groundnut (*Vigna subterranea* (L.) Verdc.), an underutilized crop, to aid global food security: Varietal improvement, genetic diversity and processing. *Agronomy, 10*, 766. https://doi.org/10.3390/agronomy10060766. www.mdpi.com/journal/agronomy

Murevanhema, Y. Y., & Jideani, V. A. (2013). Potential of Bambara groundnut (*Vigna subterranea* (L.) Verdc) milk as a probiotic beverage. A review. *Critical Review of Food Science and Nutrition, 53*, 954–967.

Nadembega, S. (2016). *Food production and food security in Burkina Faso: Case of the Bambara groundnut in three communes of the three agro-ecological zones*. Diploma of Master 2, Université Catholique de l'Afrique de l'Ouest, Bobo Dioulasso, 90 p.

Ndidi, U. S., Ndidi, C. U., Aimola, I. A., Bassa, O. Y., Mankilik, M., & Adamu, Z. (2014). Effects of processing (boiling and roasting) on the nutritional and antinutritional properties of Bambara groundnuts (*Vigna subterranea* [L.] Verdc.) from Southern Kaduna, Nigeria. *Journal of Food Processing*. https://doi.org/10.1155/2014/472129

NRC. (2006). *Lost crops of Africa: Volume II: Vegetables*. Chapter 11. Locust bean. Washington, DC: National Research Council, 2006. The National Academies Press.

Nyau, V., Prakash, S., Rodrigues, J., & Farrant, J. (2015). Antioxidant activities of Bambara groundnuts as assessed by FRAP and DPPH assays. *American Journal of Food and Nutrition, 3*(1), 7–11. https://doi.org/10.12691/ajfn-3-1-2

Odongo, F. O., Oyoo, M. E., Wasike, V., Owuoche, J. O., Karanja, L., & Korir, P. (2015). Genetic diversity of Bambara groundnut (*Vigna subterranean* (L) Verde.) landraces in Kenya using microsatellite markers. *African Journal of Biotechnology, 14*(4), 283–291.

Ogundele, O. M., Minaar, A., & Emmambux, M. N. (2017). Effects of micronisation and dehulling of pre-soaked Bambara groundnut seeds on microstructure and functionality of the resulting flours. *Food Chemistry, 214*, 655–663.

Oluwatofunmi, E. G., Ishola, I. S., & Bamidele, F. J. (2015). Formulation and nutritional evaluation of maize, Bambara groundnut and cowpea seeds blends complementary food. *American Journal of Food and Nutrition, 3*(4), 101–105. https://doi.org/10.12691/ajfn-3-4-2

Omoikhoje, S. O., 2008. Assessment of the nutritive value of Bambara groundnut as influenced by cooking time. *Livestock Research for Rural Development, 20* (4), Article #55. Retrieved February 11, 2021, from http://www.lrrd.org/lrrd20/4/omoi20055.htm

Omonhinmin, C. A., & Ogunbodede, O. O. (2013). Genetic diversity, taxonomy and legumins implications of seed storage protein profiling in Fabaceae. *African Journal of Biotechnology, 12*(17), 2157–2163.

Oyeyinka, S. A., Singh, S., Patrick, O. A., Gerrano, A. S., & Amonsou, E. O. (2015). Physicochemical properties of starches with variable amylose contents extracted from Bambara groundnut genotypes. *Carbohydrate Polymers, 133*, 171–178.

Oyeyinka, S. A., Singh, S., Ma, Y., & Amonsou, E. O. (2016). Influence of high-pressure homogenization on the physicochemical properties of Bambara starch complexed with lysophosphatidylcholine. *LWT- Food Science and Technology, 74*, 120–127.

Piyarat, S. (2008). Structure and functional properties of starch and flour from Bambara groundnut. *Songklanakarin Journal of Science and Technology, 30*, 51–56.

Quedraogo, M., Zagre M'bi, B., Jorgensen, S. T., & Liu, F. (2012). Effect of mounding times on yield of Bambara groundnut (*Vigna subterranea* (L.) Verdc.) landraces in Sahel-Burkina Faso. *African Journal of Agricultural Research, 7*(32), 4505–4511.

Sirivongpaisal, P. (2008). Structure and functional properties of starch and flour of Bambara Groundnut. *Songklanakarin Journal of Science and Technology, 30*(1), 51–56.

SoW. (2011). *African indigenous crops. Innovations that nourish the planet*. The State of the World.

Stephens, J. M. (2003). *Bambara groundnut (Voandezeia subterranean (L) Thouars)*. Institute of Food and Agriculture Sciences (IFAS).

Stone, G. W., Maehara, A., Lansky, A. J., de Bruyne, B., Cristea, E., Mintz, G. S., Mehran, R., McPherson, J., Farhat, N., Marso, S. P., Parise, H., Templin, B., et al. (2011). A prospective natural-history study of coronary atherosclerosis. *The New England Journal of Medicine, 364*,

226–235. https://doi.org/10.1056/NEJMoa1002358. https://www.nejm.org/doi/full/10.1056/nejmoa1002358#article_citing_articles

Taylor, J., & Awika, J. (2017). Gluten-free ancient grains: Cereals, pseudocereals, and legumes: Sustainable. In *Nutritious, and health-promoting foods for the 21st century* (1st ed., p. 356). Woodhead Publishing. eBook ISBN: 9780081008911.

Teixeira, N. C., Queiroz, V. A. V., Rocha, M. C., Amorim, C. P., Soares, T. O., Monteiro, M. A. M., & Junqueira, R. G. (2016). Resistant starch content among several sorghum (*Sorghum bicolor*) genotypes and the effect of heat treatment on resistant starch retention in two genotypes. *Food Chemistry, 197*, 291–296.

Tsoata, E., Njock, S. R., Youmbi, E., & Nwaga, D. (2016). Early effects of water stress on some biochemical and mineral parameters of mycorrhizal *Vigna subterranea* (L.) Verdc. (*Fabaceae*) cultivated in Cameroon. *International Journal of Agronomy and Agricultural Research, 7*(2), 21–35.

Tsoata, E., Temegne, N. C., & Youmbi, E. (2017a). Analysis of early biochemical criterion to screen four Fabaceae plants for their tolerance to drought stress. *International Journal of Current Research, 9*(1), 22568–44575.

Tsoata, E., Temegne, N. C., & Youmbi, E. (2017b). Analysis of early growth criterion to screen four *Fabaceae* plants for their tolerance to drought stress. *Research Journal of Life Sciences, Bioinformatics, Pharmaceutical and Chemical Sciences, 2*(5), 94–109.

Tweneboah, C. K. (2000). *Modern agriculture in the tropics, food crops*. Co-wood Publishers.

Uarrota, V. G., Amante, E. R., Demiate, I. M., Vieira, F., Delgadillo, I., & Maraschin, M. (2013). Physicochemical, thermal, and pasting properties of flours and starches of eight Brazilian maize landraces (*Zea mays* L.). *Food Hydrocolloids, 30*, 614–624.

USDA. (2009). *GRIN – Germplasm Resources Information Network*. National Germplasm Resources Laboratory, Beltsville, Maryland. https://www.feedipedia.org/node/1640 Viewed06-01-2021. http://www.ars-grin.gov

Wang, N., & Castonguay, G. (2014). Effect of maturity on physicochemical and cooking characteristics in yellow peas (*Pisum sativum*). *Canadian Journal of Plant Science, 94*(3), 565–571. https://doi.org/10.4141/cjps2012-331

Willett, W., Rockström, J., Loken, B., Springmann, M., Lang, T., Vermeulen, S., Garnett, T., Tilman, D., DeClerck, F., Wood, A., Jonell, M., Clark, M., Gordon, L. J., Fanzo, J., Hawkes, C., Zurayk, R., Rivera, J. A., De Vries, W., Majele Sibanda, L., Afshin, A., Chaudhary, A., Herrero, M., Agustina, R., Branca, F., Lartey, A., Fan, S., Crona, B., Fox, E., Bignet, V., Troell, M., Lindahl, T., Singh, S., Cornell, S. E., Srinath Reddy, K., Narain, S., Nishtar, S., & Murray, C. J. L. (2019). Food in the Anthropocene: The EAT-Lancet Commission on healthy diets from sustainable food systems. *Lancet, 393*(10170), 447–492. https://doi.org/10.1016/S0140-6736(18)31788-4

Zeven, A. (1998). Landraces: A review of definitions and classifications. *Euphytica, 104*, 127–139. https://doi.org/10.1023/A:1018683119237

Zheng, Z. J., Wang, S. M., & Zong, X. X. (1997). *Monograph on food legumes in China* (pp. 3–24). China Agriculture Press. (53–365, in Chinese).

Chapter 2
World Production, Ecology and Postharvest Handling of Bambara Groundnut

2.1 Introduction

Bambara groundnut (BGN) originated from sub-Saharan Africa. The accuracy of its production estimate is difficult, considering the general case with underutilized orphan crops. There has been a significant focus on agronomy over post-harvest processing (Adeleke et al., 2017). The science and technology of BGN production covers the areas of genetics (Olukolu et al. 2012; Somta et al. 2011), physiology, meteorology, and soil requirements (Tsoata et al. 2015). The genus *Vigna* (Family *Leguminosae*) is a vital legume taxon with about 90 species with seven cultivated as economic crops. BGN is the third most crucial food legume after peanut (*Arachis hypogaea*) and cowpea (*Vigna unguiculata*) (Baryeh, 2001; Odongo et al., 2015; Omonhinmin & Ogunbodede, 2013; Siegel & Fawcett, 1976). Newly developed BGN (*Vigna subterrenea* L.) cultivars were reported by Adeleke et al. (2017).

2.2 Origin and Ecology of Bambara Groundnut

The origin and current distribution of BGN, its classical and vernacular nomenclature, appropriate climatic conditions for its growth, soil requirements, adaptability of the crop to climate change, ecology and intrinsic properties have received considerable attention (Basu et al., 2007a; IPGRI, IITA, BAMBET, 2000; Jensen et al., 2011; Lengwati, 2020; Nordin & Singh, 2015; Sambo, 2014). BGN (*Vigna subterranea* (L.) Verdc.) originated from the African continent (Dalziel, 1937; Ho et al., 2017; Goli et al., 1997; Hepper, 1963; Jacques-Félix, 1946; Mkandawire, 2007; Rassel, 1960; Temegne et al., 2018; Verdcourt, 1980). It is a crop that can adapt to various agro-ecological conditions (Agnolucci et al., 2020). The ecology of BGN based on literature shows its distribution from West to southern Africa via Central

V. A. Jideani, A. I. O. Jideani, *Bambara groundnut: Utilization and Future Prospects*, https://doi.org/10.1007/978-3-030-76077-9_2

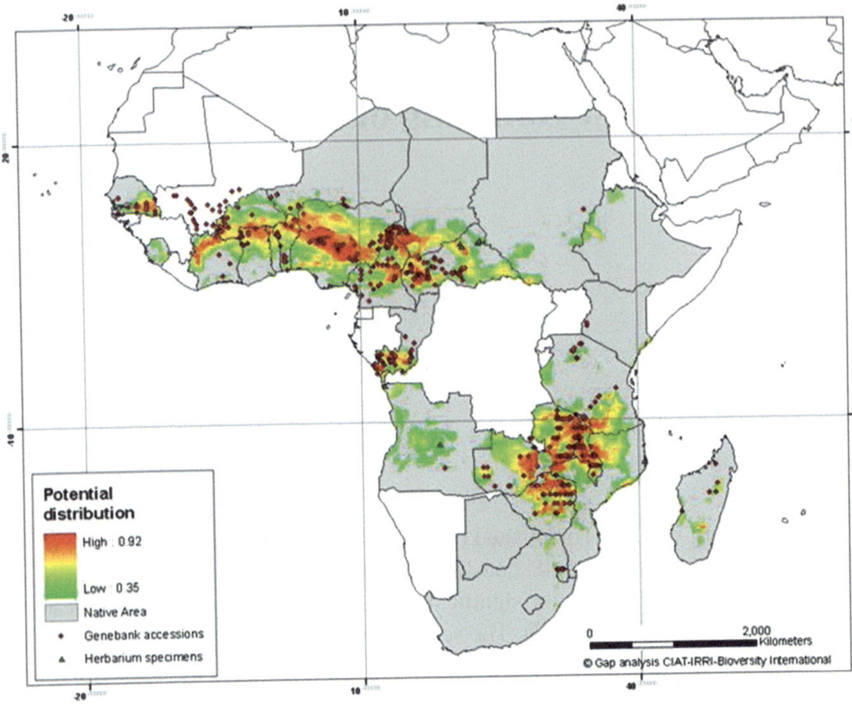

Fig. 2.1 Distribution of Bambara groundnut (*Vigna subterranea* (L.) Verdc.) in Africa (CIAT-IRRI-Biodiversity international) (Heller et al., 1997)

Africa (Fig. 2.1). Bambara groundnut is a dicotyledonous angiosperm (Fig. 2.2) with its Angiosperm Phylogeny Group IV (APG IV, 2016) classification detailed in Table 2.1.

Bambara groundnut is the common name derived from a tribe named Bambara, whose members live today mainly in Mali (Verdcourt, 1980). The exact area of BGN origin in Africa has been a subject of debate. Dalziel found the wild state in 1901 in Yola Nigeria (Dalziel, 1937; Hepper, 1963) and reported that Ledermann saw the wild plant near Garoua in northern Cameroon. Hepper confirmed Dalziel conclusion in 1957 (Heller et al., 1997). Therefore, BGN is originated from sub-Saharan (SSA) Africa, more specifically from the north-eastern region of Nigeria and northern Cameroon, both in West Africa. However, the real centre of diversity or origin of BGN remains controversial. The South-East Africa region could be a secondary centre of domestication for BGN. Presently cultivated throughout tropical Africa, Indonesia, Malaysia, India, Sri Lanka, Philippines, South Pacific, parts of northern Australia, Papua New Guinea, Central and extensively in the Americas (Alhassan & Egbe, 2013; Amadou et al., 2001; Basu et al., 2007a, b Duke, 1981; Duodu & Apea-Bah, 2017; PROTA, 2006).

Detailed in Table 2.2 are some BGN common names based on language (Alhassan & Egbe, 2014; DAFF, 2016; Heuzé et al., 2016). In French, it is called voandzou,

Fig. 2.2 (**a**) Bambara groundnut plant, a dicotyledonous angiosperm (APG, IV, 2016); (**b**) Eurema eximia taking nectar from a flower of *Vigna subterranea*; (**c**) Paragus borbonicus collecting pollen from a flower of *Vigna subterranean* (Mohamadou et al., 2018)

Table 2.1 Bambara groundnut angiosperm phylogeny group

Kingdom:	*Plantae*
Sub Kingdom:	*Tracheobionta*
Class:	*Magnoliopsida* (Dicotyledon)
Sub Class:	*Rosidae*
Order:	*Fabales*
Family:	*Fabaceae*
Sub Family:	*Fabaceae Papilionoideae*
Tribe:	*Phaseoleae*
Sub Tribe:	*Phaseolinae*
Genus:	*Vigna*
Species:	*Vigna subterranea* (L.) Verdc.

Table 2.2 Some Bambara groundnut common names

Language	Common name
English	Congo goober; earth pea Bambara bean, Bambara groundnut, ground bean, hog-peanut, Congo groundnut, Madaascar groundnut
Spanish	Guisante de tierra; mani africano; mani de Bambara
French	Pois Bambara; pois souterrain; voandzou
Portuguese	Mandubi-d'Angola
German	Angolaerbse; Erderbse
Italy	Pisello di terra
Indonesian	Kacang bogor
Swahili	Mnjugu-mawe
South Africa	Jugoboon, ditloo-marapo, izindlubu, indlubu, jugo bean, nduhu, phonda, ndlowu, njugo, tindlubu, inhlowa
Nigeria	Okpa, Gurjiya, Epa-roro, Karo, Okirikiri, Adikpo, Kparuru, Ikpeyiole and Carol
Ghana	Aboboi akyii
Zambia	Ntoyo cibemba
Zimbabwe	Nyimo

Heuzé et al. (2016), DAFF (2016) and Alhassan and Egbe (2014)

"pois Bambara" or "pois de terre" or "pois arachide". In English, it is called Bambara groundnut, Bambara pea, Bambara nut, Bambara-bean, Congo goober, earth pea, ground-bean, or hog-peanut (DAFF, 2016; FAO, 1994; Heuzé et al., 2016; USDA, 2017; Verdcourt, 1980).

In Cameroon, several vernacular names exist depending on languages: Matop/ Wondenkana (Bassa'a), Matob (Douala), Moto/Mɔto/Matoba (Fe'efe'e), ngalaa-ji (Fulfulde), Motobo (Bafia), Debbi (Foulbe), Mogran (Toupouri), Souonchié (Moundang), Atob/Metop (Ewondo), Deppi, Jwə/Djué or matobɔ (Ghomala), ndzu/ Ndən (Mankon), Kezo Ke Zon (Nugunu), mendzo wah'a sah (Bagam), etc. (Temegne et al., 2015; Temegne, 2018). In Southern Guinea Savanna region of Nigeria, some of the popular landraces include, Karo, Okirikiri, Adikpo, Kparuru, Ikpeyiole and Carol (Alhassan & Egbe, 2014).

2.3 World Production of Bambara Groundnut

Production of BGN is facing a decline in areas of production in Africa. With the focus or priority of many farmers on major crops such as cassava, peanut (groundnut), maize, soybean, cowpea, etc., BGN is gradually becoming an endangered crop species (Ahoussou et al., 1995). However, in a study, 77.5% of the non-producers indicated that, if supported, they would be willing to go into the production showing there is a potential to increase the current level of production of BGN crop (Adzawla et al., 2016).

Bambara groundnut production estimate accuracy is difficult, considering the general case with underutilized crops, due to its widespread use in subsistence

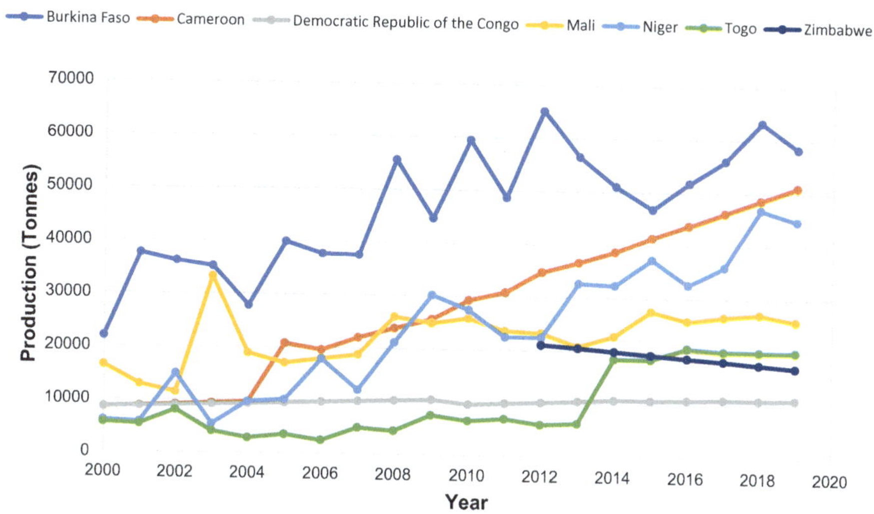

Fig. 2.3 World production of Bambara groundnut from 2000 to 2019 (FAOSTAT, 2020)

Table 2.3 Rate of production in sub-Saharan Africa by country

Country	Regression parameters		
	Rate of production (year^{-1})	Constant	R^2
Burkina Faso	0.017	−29.73	0.648
Cameroon	0.045	−85.47	0.920
Democratic Republic of Congo	0.006	−7.27	0.900
Mali	0.013	−21.44	0.391
Niger	0.047	−89.39	0.833
Togo	0.041	−78.11	0.612
Zimbabwe	−0.013	30.02	0.999

farming systems for which there are no reliable data. However, some sources estimate its annual world production at about 330,000 MT, of which West Africa produces 45–50% (Alhassan & Egbe, 2013; PROTA, 2006). Figure 2.3 details the production data for the largest producers of BGN from the year 2000 to 2019 in sub-Saharan Africa and confirms the difficulty of obtaining reliable production data.

The world production of BGN comes from Africa led by Burkina Faso, Cameroon, the Democratic Republic of Congo, Mali, Niger, Togo and Zimbabwe. The production range from 5701 tonnes in 2000 for Togo to 58,432 tonnes in 2019 for Burkina Faso. The rate of production was estimated using Eq. 2.1.

$$\log_{10} \text{production}(\text{tonnes}) = mx + c \tag{2.1}$$

Where m = rate of production and c = constant.

In terms of production rate, Cameroon, Niger and Togo seem to have a faster and consistent growth rate than Burkina Faso (Table 2.3). Researchers' recent efforts are getting attention by the renewed interest in Bambara groundnut production (DPP, 2011). There is a steady increase in output of BGN except for Zimbabwe, where the rate of production took a dip. The production figures for Zimbabwe were only available from 2012. Surprisingly, there are no available data for Nigeria and Ghana for BGN production. These two countries have BGN as their staple and are well cultivated. Recently, entrepreneurs from Singapore and USA have partnered with farmers in Ghana to supply BGN (Gan, 2020).

2.3.1 *Production, Yield, Intercropping & Mycorrhizal Inoculations in Sub-Saharan Africa*

Ajayi and Lale (2001) estimate that grain yields of BGN are between 650 and 850 kg ha^{-1} with significant variations between different sub-Saharan Africa (SSA) countries. Mkandawire (2007) reviewed the production of BGN (*Vigna subterranea* (L.)) in SSA. On yield characteristics, several researchers have identified important crop yield characters as the number of flowers per plant, the number of seeds/pods, seed size, spikelet number/spike, 100-seed weight, fruit weight/plant (Agnolucci et al., 2020; Arshad, et al 2005, Atoyebi, et al., 2017; Jonah, 2011; Manggoel et al., 2012; Oyiga & Uguru, 2011; Touré et al., 2012). Maunde et al. (2015) using simple correlation and path coefficient analysis determined the yield characteristics in BGN and reported that (i) a strong positive correlation between the number of pods/plants (r = 0.74), 100-grain weight (r = 0.35) and harvest index (r = 0.83) with seed yield, and (ii) the essential yield determinant as the number of pods/plant and harvest index. Table 2.4 details seven common BGN varieties and their characteristics and Fig. 2.4 shows some BGN landraces.

Most often, production involves BGN/maize intercropping. To determine some landraces' suitability, Alhassan and Egbe (2014) investigated the optimal planting density of BGN intercropped with maize. Their outcomes include (i) intercropping decreased the number of pods and grain yields of BGN component, (ii) number of

Table 2.4 The Bambara groundnut (BGN) cultivated varieties

BGN variety	Characteristics
Black	Mainly one-seeded, early maturing, usually small to medium-sized kernel.
Red	Large kernels, late maturing, good yielder but prone to rotting onsite.
Cream/black-eye	A large kernel with a good yield.
Cream/brown-eye	A moderate kernel with a good yield.
Cream/no eye	Very small pods and kernels produce one seed and a lower yield.
Speckled/flecked/spotted	Predominantly purple colour, small kernels and one-seeded pods.
Brown	Variations between light and dark brown, medium to large-sized kernels.

DAFF (2016); Bonny et al., (2019)

Fig. 2.4 Different morphotypes of Bambara groundnut landraces. Unit 1: 14 accessions; Unit 2: 16 accessions; Unit 3: 13 accessions; Unit 4: 13 accessions; Unit 5: 13 accessions; Unit 6: 13 accessions; Unit 7: 13 accessions; Unit 8: 13 accessions; Unit 9: 16 accessions; Unit 10: 13 accessions; Unit 11: 13 accessions (Khan et al., 2020)

pods and grain yields increased with increased planting density, (iii) BGN/maize intercropping was productive, with maize as the dominant component and (iv) the marginal rate of returns for the best combinations was 116.13%, suggesting profitability of the intercropping systems.

Oyejobi and Ajiboye (2015) studied the efficacy of mycorrhizal fungal inoculations on seed germination and plant growth of BGM (TVsu 283) and observed that (i) the fungal strains (*Glomus mosseae, Aspergillus niger, A. fumigatus and Malbranchea gypsea*) significantly increased plant growth parameters, (ii) the highest rate of germination (increased number of leaves – 54.2%, leaf length – 23.8%, leaf surface area – 49.1%, stem girth – 20.0%, and the number of shoots – 40.0%) at 42nd day observed in *A. fumigatus* inoculant, and (iii) mycorhiza symbiosis helped BGN plant nutrient requirements making it possible to have improved yield using bioinoculant/biofertiliizer. A similar study was done by Mohamadou et al. (2018) on the influence of mycorrhical-floricultural insect interaction on the yields of BGN red variety.

Phosphorus fertilization on arbuscular mycorrhizal fungi in the BGN rhizosphere is an area of interest in solving production challenges. BGN diseases and pests contribute to the low production and poor soil condition, particularly phosphorus (P) deficiency (Temegne et al., 2015, 2017). A high level of phosphate fertilizer (200 kg. ha^{-1} P_2O_5) significantly reduces the intensity of mycorrhization. Moreover, tropical soils are highly leached, abundantly clayey and strongly acid, resulting in low mineral availability, especially P; hence plants in some regions must establish symbiotic relationships to strengthen their hydromineral nutrition (Mohale et al., 2014; Temegne et al., 2017). In the study of Temegne et al. (2017) on the effect of P fertilizer application on the diversity of arbuscular mycorrhizal fungi (AMF) under the rhizosphere of three BGN landraces, plants were grown in a farm under different simple superphosphate (SSP) levels (0, 50, 100, 150 and 200 kg.ha^{-1} P_2O_5) resulting in (i) BGN landraces were the host of several AMF species and were highly colonized with all showing hyphae, vesicles and even spores, (ii) high level of SSP (200 kg.ha^{-1} P_2O_5) significantly ($p < 0.001$) reduced the intensity of mycorrhization, as well as the diversity indices compared to the control, (iii) a density of AMF spores reaching 1930 was obtained in 100 g of trapping substrate, (iv) 16 AMF species belonging to nine genera (Glomus, Acaulospora, Gigaspora, Racocetra, Rhizophagus, Funneliformis, Septoglomus, Diversispora and Claroideoglomous), were isolated and identified, and (v) *Rhizophagus intraradices* and *Septoglomus constrictum* were the most abundant. AMF species identified in BGN rhizosphere can be multiplied and used as biological fertilizer to increase the yield of the legume.

2.3.2 *Disease Detection and Control- Viral & Fungal Diseases*

Plant crop diseases do not spare BGN. Worldwide, plant crop production is affected by numerous diseases and pests; hence the need for novel and rapid methods for the timely detection of pests and diseases to develop control measures with greater efficiency (Siricharoen et al., 2016; Wiesner-Hanks et al., 2018). An artificial intelligence (AI)-based BGN disease and pest detection system (Selvaraj et al., 2019; Howard et al., 2017), involving AI, deep learning (Mohanty et al., 2016), transfer learning, convolutional neural networks, and mobile application will support BGN farmers. These technologies are discussed further in Sect. 12.4. Moreover, different state-of-the-art models for digital crop disease detection are reported. In developing countries with limited human infrastructure capacity, traditional pest and disease identification approaches rely on agricultural extension specialists. These rely on empirical knowledge, which is less effective in overcoming farming challenges. Digital detection systems (AI, deep learning, transfer learning, etc.) for early identification of a crop disease or pest leading to digital image processing and predictive analytics, with greater accuracy, can result in faster interventions with reduced impacts on food supply chains (Brahimi et al., 2018; Kamilaris, 2018; Selvaraj et al., 2019). Mohanty et al. (2016) predicted the crop-disease from every

Fig. 2.5 Example of leaf images from the PlantVillage dataset, representing every crop-disease pair used. (**1**) Apple Scab, Venturia inaequalis (**2**) Apple Black Rot, Botryosphaeria obtusa (**3**) Apple Cedar Rust, Gymnosporangium juniperi-virginianae (**4**) Apple healthy (**5**) Blueberry healthy (**6**) Cherry healthy (**7**) Cherry Powdery Mildew, Podoshaera clandestine (**8**) Corn Gray Leaf Spot, Cercospora zeae-maydis (**9**) Corn Common Rust, Puccinia sorghi (**10**) Corn healthy (**11**) Corn Northern Leaf Blight, Exserohilum turcicum (**12**) Grape Black Rot, Guignardia bidwellii, (**13**) Grape Black Measles (Esca), Phaeomoniella aleophilum, Phaeomoniella chlamydospora (**14**) Grape Healthy (**15**) Grape Leaf Blight, Pseudocercospora vitis (**16**) Orange Huanglongbing (Citrus Greening), Candidatus liberibacter spp. (**17**) Peach Bacterial Spot, Xanthomonas campestris (**18**) Peach healthy (**19**) Bell Pepper Bacterial Spot, Xanthomonas campestris (**20**) Bell Pepper healthy (**21**) Potato Early Blight, Alternaria solani (**22**) Potato healthy (**23**) Potato Late Blight, Phytophthora infestans (**24**) Raspberry healthy (**25**) Soybean healthy (**26**) Squash Powdery Mildew, Erysiphe cichoracearum (**27**) Strawberry Healthy (**28**) Strawberry Leaf Scorch, Diplocarpon earlianum (**29**) Tomato Bacterial Spot, Xanthomonas campestris pv.vesicatoria (**30**) Tomato Early Blight, Alternaria solani (**31**) Tomato Late Blight, Phytophthora infestans (**32**) Tomato Leaf Mold, Passalora fulva (**33**) Tomato Septoria Leaf Spot, Septoria lycopersici (**34**) Tomato Two Spotted Spider Mite, Tetranychus urticae (**35**) Tomato Target Spot, Corynespora cassiicola (**36**) Tomato Mosaic Virus (**37**) Tomato Yellow Leaf Curl Virus (**38**) Tomato healthy (Mohanty et al., 2016)

crop-disease from their data set (Fig. 2.5). However, deep learning detection is not intended to replace existing methods for disease diagnosis; it rather supplements laboratory tests as a viable additional method to help prevent yield loss.

Bambara groundnut production is seriously threatened by root-knot nematodes (RKNs; *Meloidogyne spp.*) and cultural methods have not been very effective in controlling RKN. Jada et al. (2011) evaluated carbofuran in preventing *M. javanica* on three BGN cultivars (*Kwachanjiwa, Kwaheuma,* and *Kwatolotolo*) using three application timings (at planting, 3 and 6 weeks after planting, and none). They reported that applying carbofuran at planting provided the most significant reduction in *M. javanica* population levels, leading to increased yields in BGNs compared to the other two application timings. Both *Kwachanjiwa* and *Kwatolotolo* provided similar high yields compared to *Kwaheuma,* which was most likely related to the *M. javanica* tolerance in these cultivars.

2.3.2.1 Viral Diseases of Bambara Groundnut

As a plant crop, BGN is not spared by viral diseases affecting the yield (Thottappilly & Rossel, 1997). BGN and cowpea are two legume crops cultivated under the same climatic conditions and sometimes in association (Konate et al., 2017). Hence factors affecting their breeding, cultivation, virus infection and production could be similar (Barro et al., 2018). Four viruses have been reported in BGN from Burkina Faso using serological and biological tests. They include cowpea aphid-borne mosaic virus (CABMV) (Fig. 2.6), bean typical mosaic virus strain-blackeye cowpea mosaic (BCMV-BlCM), peanut mottle virus (PeMoV) and cowpea mottle virus (CPMoV) (Barro et al., 2018; Sérémé, 1989; Drabo et al., 1997; Néya, 2011; Konate et al., 2019). An estimate of 12 viruses infect the crop and among these include CABMV, BCMV-BlCM, PeMoV, and CPMoV.

Konate et al. (2017, 2019), being the first to report the serological and molecular characterization of viruses infecting BGN (Fig. 2.7), (i) identified and characterized the main viruses occurring in BGN from Burkina Faso (Table 2.5) as Potyvirus CABMV and BCMV-BlCM in 8.57% of the samples with double antibody sandwich enzyme-linked immunosorbent assay (DAS-ELISA) test and 14.29% of the samples when reverse transcription-polymerase chain reaction (RT-PCR) test, (ii)

Fig. 2.6 Cowpea (Vigna unguiculata) and Bambara groundnut (Vigna subterranea) plants showing CABMV symptoms on their leaves. (**a**) Cowpea plant infected by CABMV showing severe mosaic symptoms on leaves; (**b**) Bambara groundnut plant showing leaf curling, decreased size and stoppage of growth induced by CABMV infection; (**c**) Bambara groundnut healthy plant

Fig. 2.7 Bambara groundnut (Vigna subterranea) landraces with their phenotypic characteristics (Konate et al., 2019)

obtained a phylogenetic tree based on 476 nucleotides in coat protein showed that all Burkina Faso CABMV isolates clustered together with Uganda isolate (KT726938), and (iii) distinguished two groups within these isolates.

2.3.2.2 Fungal Disease of BGN Seed

Aspergillus flavus is a facultative pathogen that derives energy as a saprophyte on plant debris rich in carbohydrates like BGN. Like other fungi, *A. flavus* does synthesize and release a plethora of secondary metabolites such as aflatoxin B1 and aflatoxin B2, aspergillic acid, aflam, nitropropionic acid, and kojic acid. Colonization of *A. flavus* and the succeeding aflatoxin contamination have been found in agriculturally important crops like maize, legumes, nuts, and their products (Yu, 2012). The aflatoxins production is a significant source of food security threat in agricultural produce (Okayo et al., 2020), exceptionally high nutrient legume like BGN. Reports of incidences of *A. flavus* on groundnuts (*Arachis hypogea*) call vigilance and surveillance of other foodstuffs like BGN.

Exposure to high aflatoxin doses that are greater than 1000 ppb leads to acute aflatoxicosis, characterized by liver damage, hepatitis, and, in some cases, death (Williams et al., 2004). The International Agency for Research on Cancer (IARC) documented aflatoxin B1 as the most lethal and classified it as a probable human carcinogen (IARC, 1993). However, only 40–50% of *A. flavus* strains can produce

Table 2.5 Percentage of nucleotide identity between Burkina Faso isolates of cowpea aphid-borne mosaic virus (CABMV) and blackeye cowpea mosaic virus strain of bean common mosaic virus (BCMV-BICM) and Genbank related sequences based on 476 nucleotides in the coat protein (Konate et al., 2017)

Accession no.	Country	E1	E2	E3	E4	E5	E6	E10	E18	E113	E138	E139	E142
BCMV (AJ312438)	China	67.8	97.3	66.9	68.1	66.9	66.9	66.9	68	66.9	97.3	64.3	97.7
BICMV (AY575773)	Taiwan	66.7	94.1	64.5	68.3	64.5	64.5	64.5	68.1	64.5	94.1	65.5	94.3
BCMV (DQ666332)	Colombia	67	84	66.7	67	66.7	66.7	66.7	66.8	66.7	84	66.6	84
WMV (EU660586)	France	67.1	68.7	65.5	66.9	65.5	65.5	65.5	68.5	65.5	68.7	66.7	68.5
ZYMV (AB458595)	Syria	66.7	67.7	66	64.5	65.8	66	65.8	67.3	65.8	67.7	66.8	67.7
CABMV (KT726938)	Uganda	93.1	68.5	97.5	93.1	97.7	97.5	97.7	93.3	97.7	68.5	85.1	68.5
CABMV (Y17822)	Monguno	84.2	69.3	84.2	84.5	84.2	84.2	84.5	84	84.5	69.3	78.4	69.3
CABMV (AJ132414)	Nigeria	80.3	67.7	81.8	80.1	82	81.8	82	80.8	82	67.7	81.2	67.7
CABMV (Y18634)	Morocco	80.3	67	81.6	80.3	81.8	81.6	81.8	80.8	81.8	67	81.6	67
CABMV (KF725712)	Brazil	77	69.1	75.2	77.2	75	74.8	75	77.2	75	69.1	77.1	69.1
CABMV (AF348210)	Zimbabwe	77.5	67.4	76.1	77.7	76.3	76.1	76.3	77.9	76.3	67.4	76.1	67.4

E1, MF277031 E10, MF277037
E2, MF277032 E18, MF277038
E3, MF277033 E113, MF277039
E4, MF277034 E138, MF277040
E5, MF277035 E139, MF277041
E6, MF277036 E142, MF277042

the toxins (Davari et al., 2015), and hence there is a need to distinguish the toxic from the nontoxic ones. Accurate identification of aflatoxigenic fungi is paramount to develop mitigation measures against fungal infections and mycotoxin production.

Although not much is in the literature on the fungal infestation of BGN, the pathogenesis of *A. flavus* on essential agricultural products is a crucial concern on human health due to the synthesis and secretion of the hazardous secondary metabolite, aflatoxin. Therefore, morphological classification using molecular tools such as polymerase chain reaction (PCR)-based methods that target and amplify the specific DNA regions are better for sensitive and rapid diagnosis (Norlia et al., 2018; Sweeney et al., 2000). Identification of fungi within the *Aspergillus* genera can be challenging, requiring an integrated approach to attain a reliable identification and characterization of isolates capable of synthesizing aflatoxins (Okayo et al., 2020).

The use of chemical fungicides poses major environmental and food safety issues. Jiang et al. (2020) reported on eco-friendly biocontrol alternatives rapamycin-producing *Streptomyces hygroscopicus* LYJ637 carried in a carboxylmethylcellulose/poly (vinyl alcohol) [CMC/PVA] blend that prevented *Botryosphaeria canker* (a fungus) effectively on Chinese hickory, which provides an alternative approach to chemical control strategies.

2.4 Traditional Postharvest Handling of Bambara Groundnut

There has been a significant focus on BGN agronomy over post-harvest processing (Adeleke et al., 2017). The postharvest handling of BGN includes harvesting, drying, shelling, winnowing and storage (Aviara et al., 2013). Like peanut, BGN pods are produced underground and attached to the root of the plant. Bambara groundnut is harvested at about 4 months from planting by manually pulling up the plant from the soil, or using the hoe to dig up the pods or animal-drawn cultivator. The pods are then manually detached from the roots and collected. The harvesting method is labour-intensive and increases losses (Aviara et al., 2013; DAFF, 2016).

The traditional technologies involved in shelling include sun-drying followed by pounding using mortar and pestle. After harvesting, the plants are left for a day or two and stacked in windrows to dry and ready for shelling when the pods rattle upon shaking (DAFF, 2016). Due to temperature variation, there is the risk of over-drying of the pods with resulting low shelling output. Mortar and pestle are used for crushing the dry pods. Sometimes for small quantities, the dry pods are poured into jute or hessian bags and beaten with sticks on flat surfaces or poured on the ground and trodden with feet to have them shelled. Other methods include the use of stone to crack the pods on a flat surface or the beating of jute or hessian bag containing the pods against the wall of a building (Aviara et al., 2013). All the methods are laborious and could result in losses due to seed damage. The cracking process may result in less damage to the seeds, but it is time and labour intensive.

After, cracking the pods, the seeds are separated from the seed-pod mixture by winnowing. The method involves the dropping of the seed-pod mixture from a height against a prevailing wind. The lighter materials are blown off by the wind, while the heavier seeds are collected in a container placed on the ground below the height (Aviara et al., 2013). The winnowing process increases losses.

Bambara groundnut is stored in both shelled and unshelled conditions. When the quantity is small, the unshelled BGN can be stored in pots and when large they are stored in granaries. When shelled, the seeds are treated with agrochemicals before storing in bags, drums and other containers (Aviara et al., 2013). BGN seeds keep well in pods and shelling should only occur when the seeds are required for food or sale because the seeds are susceptible to weevil damage (DAFF, 2016).

2.5 Conclusion

From the origin, ecology, world production and traditional postharvest handling of Bambara groundnut (BGN) it originated from sub-Saharan Africa more specifically from the north-eastern region of Nigeria and northern Cameroon, both in West Africa. Presently, cultivated throughout tropical Africa, Indonesia, Malaysia, India, Sri Lanka, Philippines, South Pacific, parts of northern Australia, Papua New Guinea, Central and extensively in the Americas. BGN production estimate accuracy is difficult, considering the general case with underutilized crops, due to its widespread use in subsistence farming systems for which there are no reliable data. Researchers' recent efforts are getting attention by the renewed interest in BGN production. The traditional postharvest handling of BGN includes harvesting, drying, shelling, winnowing and storage. Like peanut, BGN pods are produced underground and attached to the root of the plant. Bambara groundnut is harvested at about 4 months from planting by manually pulling up the plant from the soil, or using the hoe to dig up the pods or animal-drawn cultivator. The pods are then manually detached from the roots and collected. The harvesting method is labour-intensive and increases losses. BGN seeds keep well in pods and shelling should only occur when the seeds are required for food or sale because the seeds are susceptible to weevil damage.

References

Adeleke, O. R., Adiamo, O. Q., & Fawale, O. S. (2017). Nutritional, physicochemical, and functional properties of protein concentrate and isolate of newly-developed Bambara groundnut (*Vigna subterrenea* L.) cultivars. *Food Science & Nutrition, 6*(1), 229–242. https://doi.org/10.1002/fsn3.552

Adzawla, W., Donkoh, S. A., Nyarko, G., O'Reilly, P. J., Olayide, O. E., Mayes, S., Feldman, A., & Azman, H. R. (2016). Adoption of Bambara groundnut production and its effects on farmers' welfare in Northern Ghana. *African Journal of Agricultural Research, 11*(7), 583–594. https://doi.org/10.5897/AJAR2015.10568

Agnolucci, P., Rapti, C., Alexander, P., De Lipsis, V., Holland, R. A., Eigenbrod, F., & Ekins, P. (2020). Impacts of rising temperatures and farm management practices on global yields of 18 crops. *Nature Food, 1*, 562–571. https://doi.org/10.1038/s43016-020-00148-x

Ahoussou, N., Koffi, G., & Sangaré, A. (1995). *Rapport de pays pour la conférence technique internationale de la FAO sur les ressources phytogénétiques* (75 p). FAO. http://www.fao.org/fileadmin/templates/agphome/documents/PGR/SoW1/africa/COTEDIVO.pdf

Ajayi, F. A., & Lale, N. E. S. (2001). Seed coat texture, host species and time of application affect the efficacy of essential oils applied for the control of *Callosobruchus maculatus* (F.) (Coleoptera: Bruchidae) in stored pulses. *International Journal of Pest Management, 47*(3), 161–166. https://doi.org/10.1080/09670870010004495

Alege, G. O., Abu Ngozi, E., & Sunday, C. E. (2014). Seed protein electrophoresis of some members of the family fabaceae. *African Journal of Biotechnology, 13*(36), 3730–3735.

Alhassan, G. A., & Egbe, M. O. (2013). Participatory rural appraisal of Bambara groundnut (*Vigna subterranea* (L.) *Verdc.)* production in Southern Guinea Savannah of Nigeria. *Journal of Agricultural Science, 1*, 18–31.

Alhassan, G. A., & Egbe, M. O. (2014). Bambara groundnut/maize intercropping: Effects of planting densities in Southern guinea savanna of Nigeria. *African Journal of Agricultural Research, 9*(4), 479–486. https://doi.org/10.5897/AJAR2013.7955

Amadou, H. I., Bebeli, P. J., & Kailtsikes, P. J. (2001). Genetic diversity in Bambara groundnut (*Vigna subterranean* L) germplasm revealed by RAPD markers. *Genome, 44*, 995–999.

APG IV. (2016). Angiosperm Phylogeny Group. "An update of the Angiosperm Phylogeny Group classification for the orders and families of flowering plants: APG IV" (PDF). *Botanical Journal of the Linnean Society, 181*(1), 1–20.

Arshad, M., Ghafoor, A., & Qureshi, A. S. (2005). Inheritance of qualitative traits and their linkage in blackgram [*Vigna mungo* (L) Hepper]. *Pakistan Journal of Botany, 37*(1), 41–46.

Atoyebi, J. O., Oyatomi, O., Osilesi, O., Adebawo, O., & Abberton, M. (2017). Morphological characterization of selected African accessions of Bambara groundnut (*Vigna subterranea* (L.) Verdc.). *International Journal of Plant Research (SAP Series), 7*(2), 29–35.

Aviara, N. A., Lawal, A. A., Atiku, A. A., & Haque, M. A. (2013). Bambara groundnut processing, storage and utilization in North East Nigeria. *Continental Journal of Engineering Sciences, 8*, 28–36.

Barro, A., Batieno, T. B. J., Sawadogo, N., De Tignegre, J. B., Nanema, K. R., Traore, R. E., & Sawadogo, M. (2018). Resistance evaluation of five varieties of cowpea and their F1 descendants from a diallel crossing to cowpea aphid-borne mosaic virus in Burkina Faso. *International Journal of Biological and Chemical Sciences, 12*(2), 668–675. https://doi.org/10.4314/ijbcs.v12i2.5

Baryeh, E. A. (2001). Physical properties of Bambara groundnuts. *Journal of Food Engineering, 47*(4), 321–326.

Basu, S., Roberts, J. A., Azam-Ali, S. N., & Mayes, S. (2007a). Bambara groundnut. In C. M. Kole (Ed.), *Genome mapping and molecular breeding in plants – Pulses, sugar and tuber* (pp. 159–173). Springer.

Basu, S. M., Mayes, S., Davey, M., Roberts, J. A., Azam-Ali, S. N., Mithen, R., & Pasquet, R. S. (2007b). Inheritance of 'domestication' traits in Bambara groundnut (*Vigna subterranea* (L.) Verdc.). *Euphytica, 157*, 59–68. https://doi.org/10.1007/s10681-007-9396-4

Bonny, B. S., Seka, D., Adjoumani, K., Koffi, K. G., Kouonon, L. C., & Sie, R. S. (2019). Evaluation of the diversity in qualitative traits of Bambara groundnut germplasm (*Vigna subterranea* (L.) Verdc.) of Côte d'Ivoire. *African Journal of Biotechnology, 18*(1), 23–36. https://doi.org/10.5897/AJB2018.16691. ISSN: 1684-5315.

Brahimi, M., Arsenovic, M., Laraba, S., Sladojevic, S., Boukhalfa, K., & Moussaoui, A. (2018). Deep learning for plant diseases: Detection and saliency map visualization. In *Human and machine learning* (pp. 93–117). Springer.

CABI *Vigna subterranea* (Bambara groundnut) in Invasive species compendium https://www.cabi.org/isc/datasheet/56585. Viewed 09-01-2021.

DAFF. (2016). *Bambara groundnuts (Vigna subterranea)- Production guideline*. DAFF.

Dalziel, J. M. (1937). *Voandzeia Thou. The useful plants of West Tropical Africa* (pp. 269–271). Crown Agents.

Davari, E., Mohsenzadeh, M., Mohammadi, G., & Rezaeian-Doloei, R. (2015). Characterization of aflatoxigenic *Aspergillus flavus* and *Aspergillus parasiticus* strain isolates from animal feedstuffs in Northeastern Iran. *Iranian Journal of Veterinary Research, 16*(2), 150–155.

DPP. (2011). *Production guidelines for Bambara groundnuts* (pp. 1–20). Directorate Plant Production, Department of Agriculture, Forestry and Fisheries.

Drabo, I., Sérémé, P., & Dabiré, C. (1997). Country reports. In J. Heller, F. Begemann, & J. Mushonga (Eds.), *Conservation and improvement of Bambara groundnut (Vigna subterranea (L.) Verdc.)* (pp. 19–26). BP 550 Hararé, Zimbabwé.

Duke, J. A. (1981). *Voandzeia subterranea* (L). Thouars. In *Handbook of legumes of world economic importance* (pp. 307–310). Plenum Press.

Duodu, K. G., & Apea-Bah, F. B. (2017). Chapter 9: African legumes: Nutritional and health-promoting attributes, pp. 223–269. In J. Taylor, & J. Awika (Eds.), *Gluten-free ancient grains: Cereals, pseudocereals, and legumes: Sustainable, nutritious, and health-promoting foods for the 21st century* (1st ed., pp. 356). Woodhead Publishing, eBook ISBN: 9780081008911. https://www.elsevier.com/search-results?query=Bambara%20groundnut&labels=books&page=1

FAO. (1994). *Definition and classification of commodities (draft): 4. Pulses and derived products.* Food and Agriculture Organization. Retrieved 13 Dec 2017.

FAOSTAT. (2020). *Food and Agriculture data.* Food and Agriculture Organisation of the United Nations. http://www.fao.org/faostat/en/#home. Accessed 25 Jan 2021.

Gan, E. (2020). "We fell in love with the Bambara groundnut": How a start-up is improving livelihoods of small farmers in Africa. *Eco-Business*, 1–11.

Goli, A. E., Begemann, F., & Ng, N. Q. (1997). Characterization and evaluation of IITA's Bambara groundnut collection. In J. Heller, F. Begemann, & J. Mushonga (Eds.), *Conservation and improvement of Bambara groundnut (Vigna subterranea (L.) Verdc.)* (pp. 101–118). BP 550. Hararé, Zimbabwé.

Heller, J., Begemann, F., & Mushonga, J. (1997). Bambara groundnut (*Vigna subterranea* (L.) Verdc.). Promoting the conservation and use of underutilized and neglected crops. In *Proceedings of the workshop on Conservation and Improvement of Bambara groundnut (Vigna subterranea (L.) Verdc.)*, 14–16 November 1995, Harare (Zimbabwe), Institute of Plant Genetics and Crop Plant Research, Gatersleben/Department of Research & Specialist Services, Rome (Italy).

Hepper, F. N. (1963). The Bambara groundnut (*Voandzeia subterranea*) and Kersting's groundnut (*Kerstingiella geocarpa*) wild in West Africa. *Kew Bulletin, 16*, 395–407.

Heuzé, V., Tran, G., & Lebas, F. (2016). *Bambara groundnut (Vigna subterranea) seeds.* Feedipedia, a programme by INRAE, CIRAD, AFZ and FAO. https://www.feedipedia.org/node/530

Ho, W. K., Chai, H. H., Kendabie, P., Ahmad, N. S., Jani, J., Massawe, F., Kilian, A., & Mayes, S. (2017). Integrating genetic maps in Bambara groundnut [*Vigna subterranea* (L) Verdc.] and their syntenic relationships among closely related legumes. *BMC Genomics, 18*, 192. https://doi.org/10.1186/s12864-016-3393-8

Howard, A. G., Zhu, M., Chen, B., Kalenichenko, D., Wang, W., Weyand, T., Andreetto, M., & Adam, H. (2017). Mobilenets: Efficient convolutional neural networks for mobile vision applications. *arXiv preprint arXiv:1704.04861*.

IARC. (1993). Some naturally occurring substances: Food items and constituents, heterocyclic aromatic amines and mycotoxins. *International Agency for Research on Cancer (IARC) Monograph on the Evaluation of Carcinogenic Risk to Humans, 56*, 82–87, IARC, Lyon, France.

IPGRI, IITA, BAMNET. (2000). *Descriptors for Bambara Groundnut (Vigna subterranea (L.) Verdc.).* International Plant Genetic Resources Institute, Rome, Italy; International Institute of Tropical Agriculture, Ibadan, Nigeria; The International Bambara Groundnut Network, Germany, 48 p. https://www.bioversityinternational.org/fileadmin/_migrated/uploads/tx_news/Descriptors_for_Bambara_groundnut__Vigna_subterranea__324.pdf

Jacques-Félix, H. (1946). Remarques sur l'origine et la geocarpie du *Voandzeia subterranea* Thou. *Bulletin de la Société Botanique de France, 93*(9), 260–362.

Jada, M. Y., Gungula, D. T., & Jacob, I. (2011). Efficacy of carbofuran in controlling root-knot nematode (*Meloidogyne javanica* Whitehead, 1949) on cultivars of Bambara groundnut (*Vigna subterranea* (L.) Verdc.) in Yola, Nigeria. *International Journal of Agronomy*. https://doi.org/10.1155/2011/358213

Jensen, E. S., Peoples, M. S., Boddey, R. M., Gresshoff, P., Hauggaard-Nielsen, H., Alves, B. J. R., & Morrison, M. J. (2011). Legumes for mitigation of climate change and the provision of feedstock for biofuels and biorefineries. A review. *Agronomy for Sustainable Development, 32*(2), 329–364.

Jiang, H., Rao, Y., Mei, L., & Wang, Y. (2020). Antifungal activity of rapamycin on *Botryosphaeria dothidea* and its effect against Chinese hickory canker. *Pest Management Science*. https://doi.org/10.1002/ps.6035

Jonah, P. M. (2011). Phenotypic and genotypic correlation in Bambara ground nut (*Vigna subterranean* L. Verdc.) in Mubi, Adamawa State, Nigeria. *World Journal of Agricultural Science, 7*(3), 298.

Kamilaris, A. (2018). Prenafeta-Boldu FX. Deep learning in agriculture: A survey. *Computers and Electronics in Agriculture, 147*, 70–90.

Khan, M. H., Md. Rafii, M. Y., Ramlee, S. I., Jusoh, M., & Mamun, A. l. (2020). Genetic variability, heritability, and clustering pattern exploration of Bambara groundnut (*Vigna subterranea* L. Verdc) accessions for the perfection of yield and yield-related traits. *BioMed Research International*, Article ID 2195797, 31 pages. https://doi.org/10.1155/2020/2195797

Konate, M. N., Ouedraogo, M., Neya, B. J., Bangratz, M., Palanga, E., Nandkangre, H., Ouoba, A., Nanema, R., Sawadogo, N., & Sawadogo, M. (2017). Molecular characterization of virus isolates from genus Potyvirus infecting *Vigna subterranea* in Burkina Faso. *African Journal of Biotechnology, 16*(39), 953–1961. https://doi.org/10.5897/AJB2017.16168

Konate, M. N., Ouoba, A., Nandkangre, H., Palanga, E., Ouedraogo, T., Ouedraogo, M., & Sawadogo, M. (2019). Assessment of bambara groundnut landraces for resistance to Cowpea aphid borne mosaic virus. *International Journal of Biological and Chemical Sciences, 13*(6), 2704–2713. https://doi.org/10.4314/ijbcs.v13i6.22

Lengwati, M. D. (2020). The development of analytical methodologies and advances in plant physiology, measurable amounts of elements in nutrients contribution and carbon assimilation by five grain legumes, Bambara groundnut; groundnut, cowpeas, blackgram and mungbeans planted in Mpumalanga province, South Africa during 2011/2012 season. *Acta Scientific Agriculture, 4*(7), 67–75.

Manggoel, W., Uguru, M. I., Ndam, O. N., & Dashbak, M. A. (2012). Genetic variability, correlation and path coefficient analysis of some yield components of ten cowpea (*Vigna unguiculate* L. Walp) accessions. *Journal of Plant Breeding and Crop Science, 4*(5), 80–86.

Maunde, S. M., Tanimu, B., & Mahmud, M. (2015). Correlation and path coefficient analysis of yield characteristics of Bambara (*Vigna subterranea* L. Verde). *African Journal of Environmental Science and Technology, 9*(1), 12–15.

Mkandawire, C. H. (2007). Review of Bambara groundnut (*Vigna subterranean* (L.) Verdc). Production in sub-Sahara Africa. *Agricultural Journal, 2*(4), 465–470.

Mohale, K. C., Belane, A. K., & Dakora, F. D. (2014). Symbiotic N nutrition, C assimilation, and plant water use efficiency in Bambara groundnut (*Vigna subterranea* L. Verdc) grown in farmers' fields in South Africa, measured using ^{15}N and ^{13}C natural abundance. *Biology and Fertility of Soils, 50*(2), 307–319. https://doi.org/10.1007/s00374-013-0841-3

Mohamadou, M., Ngakou, A., & Tchuenguem F-N, F. (2018). Influence de l'interaction mycorhizes-insectes floricoles sur les rendements de *Vigna subterranea* variété rouge (*Fabaceae)* à Dang (Ngaoundéré, Cameroun). *International Journal of Biological and Chemical Sciences, 12*(4), 1897–1913. https://doi.org/10.4314/ijbcs.v12i4.30. Available online at http://www.ifgdg.org

Mohanty, S. P., Hughes, D. P., & Salathe, M. (2016). Using deep learning for image-based plant disease detection. *Frontires in Plant Science, 7*, 1419. https://doi.org/10.3389/fpls.2016.01419

Néya, B. J. (2011). *Sérologie, pathogénie, épidémiologie et contrôle de la mosaïque Cowpea aphid-borne mosaic virus (CABMV) du niébé (Vigna unguiculata (L.) Walp.) transmise par des pucerons (Aphis craccivora, A. gossypii) au Burkina Faso*. Thèse de doctorat, Université de Ouagadougou, p. 153.

Nordin, N., & Singh, A. (2015). Seedling emergence of Bambara groundnut (*Vigna subterranea* (L.) Verdc.) landraces at various temperature conditions. *Procedia Environmental Sciences, 29*, 201–202.

Norlia, M., Jinap, S., Nor-Khaizura, M. A. R., Son, R., Chin, C. K., & Sardjono. (2018). Polyphasic approach to the identification and characterization of aflatoxigenic strains of *Aspergillus* section *Flavi* isolated from peanuts and peanut-based products marketed in Malaysia. *International Journal of Food Microbiology, 282*, 9–15.

Odongo, F. O., Oyoo, M. E., Wasike, V., Owuoche, J. O., Karanja, L., & Korir, P. (2015). Genetic diversity of Bambara groundnut (*Vigna subterranean* (L) Verde.) landraces in Kenya using microsatellite markers. *African Journal of Biotechnology, 14*(4), 283–291.

Okayo, R. O., Andika, D. O., Dida, M. M., K'Otuto, G. O., & Gichimu, B. M. (2020). Morphological and molecular characterization of toxigenic *Aspergillus flavus* from groundnut kernels in Kenya. *International Journal of Microbiology*. https://doi.org/10.1155/2020/8854718

Olukolu, B. A., Mayes, S., Stadler, F., Ng, N. Q., Fawole, I., Dominique, D., Azam-Ali, S. N., Abbott, A. G., & Kole, C. (2012). Genetic diversity in Bambara groundnut (*Vigna subterranea* (L.) Verdc.) as revealed by phenotypic descriptors and DArT marker analysis. *Genetic Resources and Crop Evolution, 59*(3), 347–358.

Omonhinmin, C. A., & Ogunbodede, O. O. (2013). Genetic diversit, taxonomy and legumins implications of seed storage protein profiling in Fabaceae. *African Journal of Biological Sciences, 12*(17), 2157–2163.

Oyejobi, G. K., & Ajiboye, A. (2015). Efficacy of mycorrhizal inoculations on seed germination and plant growth of Bambara groundnut *Vigna subterranean* (TVsu 283). *African Journal of Agricultural Research, 10*(9), 1022–1030.

Oyiga, B. C., & Uguru, M. I. (2011). Inter-relationships among pod and seed yield traits in banmara ground nut (*Vigna subterranean* L. Verdc.) in the derived Savanah agro-ecology of South Eastern Nigeria under two planting dates. *International Journal of Plant Breeding, 5*(2), 106–111.

PROTA (Plant Resources of Tropical Africa). (2006). In M. Brink, G. Belay (Eds.), *Cereals and pulses* (pp. 213–217). PROTA Foundation, Netherlands.

Rassel, A. (1960). Voandzou, Voandzeia subterranean Thouars, and its cultivation in Kwango. *Bulletin Agricole du Congo Belge, 51*, 1–26.

Sambo, B. E. (2014). Endangered, neglected, indigenous resilient crops: A potential against climate change impact for sustainable crop productivity and food security. *IOSR JAVS, 7*(2), 34–41.

Selvaraj, M. G., Vergara, A., Ruiz, H., Safari, N., Elayabalan, S., Ocimati, W., & Blomme, G. (2019). AI-powered banana diseases and pest detection. *Plant Methods, 15*, 92. https://doi.org/10.1186/s13007-019-0475-z

Sérémé, P. (1989). *Seed-borne pathogens of Bambara groundnut (Vigna subterranea (L.) Verdc.) in Burkina Faso and their significance*. A report of research conducted from September–December, 1982 at the Danish Government Institute of Seed Pathology for Developing Countries, Copenhagen, Denmark, 49 p.

Siegel, A., & Fawcett, B. (1976). *Food legume processing and utilization: With special emphasis on application in developing countries* (88 p). International Development Research Centre.

Siricharoen, P., Scotney, B., Morrow, P., & Parr, G. (2016). A lightweight mobile system for crop disease diagnosis. In *International conference on image analysis and recognition* (pp. 783–791). Springer.

Somta, P., Chankaew, S., Rungnoi, O., & Srinives, P. (2011). Genetic diversity of the Bambara groundnut (*Vigna subterranean* (L.) verdc.) as assessed by SSR markers. *Genome, 54*, 898–910.

Sweeney, M. J., Pamies, P., & Dobson, A. D. W. (2000). The use of reverse transcription-polymerase chain reaction (RT-PCR) for monitoring aflatoxin production in *Aspergillus parasiticus* 439. *International Journal of Food Microbiology, 56*(1), 97–103.

Temegne, N. C. (2018). *Improvement in the performances of Voandzou (Vigna subterranea (L.) Verdc.) in response to phosphate deficiency through chemical and biological fertilization*. PhD thesis, Fac Sci Univ Yaounde I.

Temegne, N. C., Mbogne, T. J., Nbendah, P., Youmbi, E., Taffouo, V. D., & Ntsomboh-Ntsefong, G. (2015). Effect of phosphate deficiency on growth and phosphorus content of three Voandzou (*Vigna subterranea* (L.) Verdc.) varieties. *IOSR Journal of Agriculture and Veterinary Science, 8*(9), 52–59.

Temegne, N. C., Wakem, G.-A., Taffouo, V. D., Mbogne, T. J., Onguene, A. N., Youmbi, E., & Ntsomboh-Ntsefong, G. (2017). Effect of phosphorus fertilization on arbuscular mycorrhizal fungi in the Bambara groundnut rhizosphere. *African Journal of Microbiology Research, 11*(37), 1399–1410. https://doi.org/10.5897/AJMR2017.8680. ISSN 1996-0808.

Temegne, N. C., Gouertoumbo, W. F., Wakam, G. A., Nkou, F. T. D., Youmbi, E., & Ntsomboh-Ntsefonng, G. (2018). Origin and ecology of Bambara groundnut (*Vigna subterranea* (L.) Verdc): A review. *Journal of Ecology & Natural Resources, 2*(4), 000140.

Thottappilly, G., & Rossel, H. W. (1997). Identification and characterization of viruses infecting bambara groundnut (*Vigna subterranea*) in Nigeria. *International Journal of Pest Management, 43*(3), 177–185.

Touré, Y., Koné, M., Tanoh, H. K., & Koné, D. (2012). Agromorphological and phenological variability of 10 Bambara groundnut [*Vigna subterranea* (L.) Verdc. (Fabaceae)] landraces cultivated in the Ivory Coast. *Tropicultura, 30*(4), 216–221.

Tsoata, E., Njock, S. R., Youmbi, E., & Nwaga, D. (2015). Early effects of water stress on some biochemical and mineral parameters of mycorrhizal *Vigna subterranean* (L.) Verdc. (*Fabaceae*) cultivated in Cameroon. *International Journal of Agronomy and Agricultural Research, 7*(2), 21–35.

USDA. (2017). *Taxon: Vigna subterranea (L.) Verdc*. USDA.

Verdcourt, B. (1980). The correct name for the Bambara groundnut. *Kew Bulletin, 35*(3), 474.

Wiesner-Hanks, T., Stewart, E. L., Kaczmar, N., DeChant, C., Wu, H., Nelson, R. J., Lipson, H., & Gore, M. A. (2018). Image set for deep learning: Field images of maize annotated with disease symptoms. *BMC Research Notes, 11*(1), 440.

Williams, J. H., Phillips, T. D., Jolly, P. E., Stiles, J. K., Jolly, C. M., & Aggarwal, D. (2004). Human aflatoxicosis in developing countries: A review of toxicology, exposure, potential health consequences, and interventions. *The American Journal of Clinical Nutrition, 80*(5), 1106–1122.

Yu, J. (2012). Current understanding on aflatoxin biosynthesis and future perspective in reducing aflatoxin contamination. *Toxins, 4*(11), 1024–1057.

Chapter 3
Global-View of Bambara Groundnut as Climate Smart Food Crop for Food Security and Bioeconomy

3.1 Introduction

Climate change brings warmer temperatures and increasingly erratic and volatile weather and rainfall patterns with consequent global food production (Wong & Schuchard, 2012). The challenges climate change possess include (1) Food production has relied on a stable supply of raw materials sensitive to the climate and vulnerable to changes in the natural environment with multiple unpredictability levels in availability, quality and price; (2) The demand on farmers ability to track, understand and respond to changes in weather with climate change producing unpredictable and unfamiliar conditions challenging to manage and (3) The pressure to meet the nutritional needs of a growing world population requires double the amount of food currently produced. Simultaneously, these challenges have a solution in "climate-smart" agriculture that combines enhanced crop resilience, productivity and food security with reduced environmental impact (Wong & Schuchard, 2012). Bambara groundnut (BGN) as a climate-smart crop (CSC) and a drought-resistant, nitrogen-fixing legume will play a role in food security. However, the hard-to-cook and photoperiod sensitivity to pod filling traits needs to be solved to make it a significant contribution to human well-being (Mayes et al., 2019). This chapter aims to assess the role of BGN in food security and bioeconomy while highlighting efforts towards reaching the potential of this crop in the face of climate change.

3.2 Genetic Diversity of Bambara Groundnut Related to Climate Change

Bambara groundnut is an African legume crop also grown in Southeast Asia in Indonesia and Thailand (Sean et al., 2019). It is a geocarpic plant producing and ripening the seeds under the ground like peanuts (*Arachis hypogaea* L.). Still,

V. A. Jideani, A. I. O. Jideani, *Bambara groundnut: Utilization and Future Prospects*, https://doi.org/10.1007/978-3-030-76077-9_3

compositionally, the seed is closer to chickpea [*Cicer arietinum*] (Halimi et al., 2019). Reduced and erratic rainfall patterns associated with climate change require crops and agricultural systems with more resilience (Sean et al., 2019). One of the most significant challenges facing humanity is the global food security threat due to climate change [drought, desertification, salinization, flooding and increased temperature], population expansion, and urbanization (Siise et al., 2014). BGN has the agronomic characteristics of (1) high tolerance to drought and adaptation to marginal soils; (2) useful nutritional features (Siise et al., 2014; Suhairi et al., 2018) and (3) legume-rhizobia symbiosis and nitrogen-fixation potentials for improved soil fertility (Aliyu et al., 2016; Berchie et al., 2013; Jørgensen et al., 2009; Siise et al., 2014). Therefore, it is suitable for incorporation into resource-poor, low input sustainable farming systems with an advantage as a crop of choice for climate-smart agriculture (Uko, 2017). Hence, it has the potential to contribute to food security (Berchie et al., 2013). BGN is reported to have a centre of origin in north-eastern Nigeria and northern Cameroon and exists in the wild from central Nigeria eastwards to southern Sudan and the landraces presently cultivated throughout tropical Africa (Sean et al., 2019) with few varieties developed through controlled breeding. Several organizations hold various landraces and wild-type accessions of BGN, developing improved genotypes with desired phenotypic traits that meet the demand of multiple actors, especially farmers could be very useful (Aliyu et al., 2016; Khan et al., 2020). About 4500 accessions of BGN conserved globally are kept in 16 countries and three regional or international gene banks. For instance, the Genetic Resources Centre of the International Institute of Tropical Agriculture (IITA) has 1910 accessions collected from various countries across sub-Saharan Africa as detailed in Fig. 3.1 (FAO, 2018; Sean et al., 2019).

Drought tolerance is a plant's ability to grow, flower and displays economic yield under a suboptimal water supply. Plants respond and adapt to and survive under drought stress as depicted in Fig. 3.2 through morphological, biochemical and physiological mechanisms (Farooq et al., 2009).

3.2.1 Morphological Mechanisms of Drought Tolerance

Escape This is attained through a shortened life-cycle or growing season, allowing the plant to reproduce before the environment becomes dry. Drought escape occurs when phenological development is successfully matched with periods of soil moisture availability, the growing season is shorter and terminal drought stress predominates. In general, plants may escape drought stress by cutting short their growth duration and avoiding stress with the maintenance of high tissue water potential either by reducing water loss or improved water uptake, or both. Some plants are reported to reduce their surface area either by leaf shedding or production of smaller leaves (Farooq et al., 2009). Developing short-duration varieties is an effective strategy for minimizing yield loss from terminal drought, as early maturity helps the crops to avoid the period of stress (Kumar & Abbo, 2001). However, the yield is

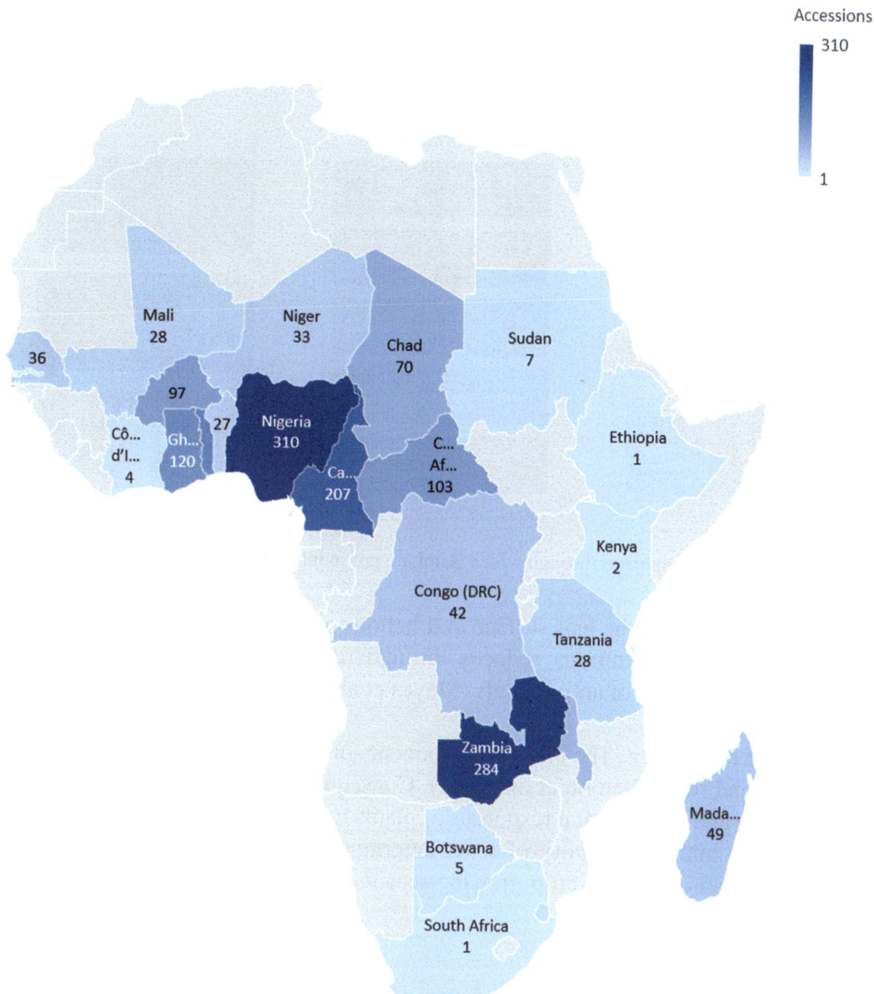

Fig. 3.1 Origin and distribution of BGN accessions held by the International Institute of Tropical Agriculture [IITA]. (Adapted from Sean et al., 2019)

generally correlated with the length of crop duration under favourable growing conditions, and any decline in crop duration below the optimum will reduce yield (Farooq et al., 2009).

Avoidance This is the ability of plants to maintain normal physiological processes under mild, moderate and intermittent drought stress through enhancing soil water capture and minimizing water losses through extensive and prolific root systems, stomatal regulation, reduction in canopy size and duration, leaf rolling, increasing

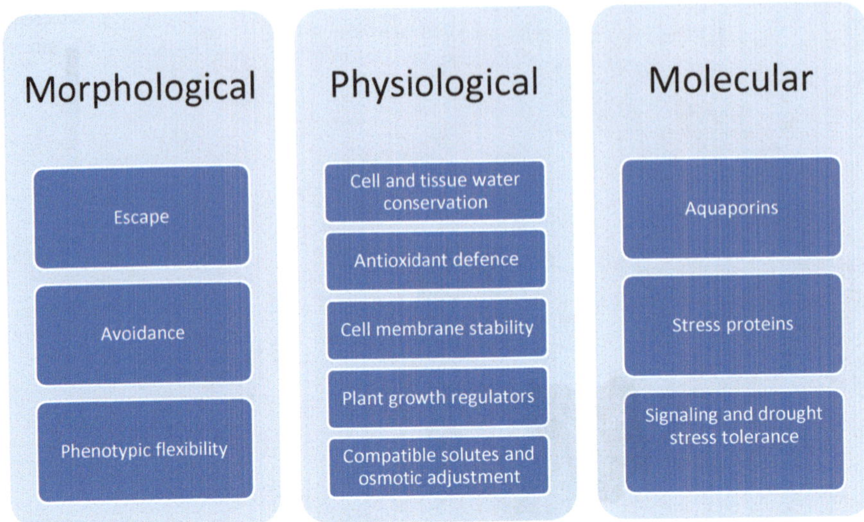

Fig. 3.2 Possible mechanisms through which Bambara groundnut avoid, escape or tolerate drought

wax accumulation on the leaf surface and heliotropism among other mechanisms. Consequently, photosynthesis reduces through reduced radiation capture and a reduced accumulation of antioxidants (Mayes et al., 2019).

Phenotypic Flexibility The critical components of plant adaptation to drought are the shoots and roots (Farooq et al., 2009). Consequently, the root growth, density, proliferation and size are vital responses of plants to drought stress. The distribution and structure and not quantity of roots determine the most efficient strategy for extracting water during the crop-growing season.

3.2.2 *Physiological Mechanisms of Drought Tolerance*

Cell and Tissue Water Conservation Osmotic adjustment allows the cell to decrease osmotic potential and as a consequence, increases the gradient for water influx and maintenance of turgor. Achieving improved tissue water status through osmotic adjustment and cell wall elasticity changes, thereby maintaining physiological activity for extended drought periods is possible (Farooq et al., 2009).

Antioxidant Defense Oxidative damage in the plant tissue is alleviated by a concerted action of both enzymatic and non-enzymatic antioxidant systems such as β-carotenes, ascorbic acid, α-tocopherol, reduced glutathione and enzymes including superoxide dismutase, peroxidase, ascorbate peroxidase, catalase, polyphenol oxidase and glutathione reductase (Hasegawa et al., 2000; Prochazkova et al., 2001).

Cell Membrane Stability Biological membranes are the first target of many abiotic stresses. The maintenance of integrity and stability of membranes under water stress is a significant component of drought tolerance in plants (Bajji et al., 2002).

Plant Growth Regulators *Compatible solutes and osmotic adjustment*: Overproduction of different compatible organic solutes such as low molecular weight, highly soluble compounds that are usually nontoxic even at high cytosolic concentrations protect plants from stress through various means such as contribution towards osmotic adjustment, detoxification of reactive oxygen species, stabilization of membranes and native structures of enzymes and proteins (Farooq et al., 2009).

3.2.3 Molecular Mechanisms of Drought Tolerance

Signaling and Drought Stress Tolerance The ability of plants to sustain a level of physiological activity under terminal drought stress through the regulation of thousands of genes and their networks. Consequently, improving their osmotic adjustment and increase the cell wall elasticity to maintain tissue turgidity. Furthermore, a series of metabolic pathways to reduce or repair the resulting stress damage occur to reduce the accumulation of harmful by-products (Mayes et al., 2019).

BGN has been reported to exhibit the morphological, physiological and molecular response to water stress through complex mechanisms (Table 3.1) including (1) drought escape by completing plant life cycle before severe water deficit e.g. earlier flowering in annuals species before the onset of severe drought; (2) drought avoidance through enhancing the capacity of getting water, e.g. developing root systems or conserving it such as reduction of stomata and leaf area/canopy cover; (3) drought tolerance mainly through improving osmotic adjustment ability and increasing cell wall elasticity to maintain tissue turgidity; (4) drought resistance by altering metabolic path for life survives under severe stress (e.g. increased antioxidant metabolism); (5) drought abandon by removing a part of the individual, e.g. shedding elder leaves under water stress; (6) drought-prone biochemical-physiological traits for plant evolution under long-term drought condition through genetic mutation and genetic modification (Farooq et al., 2009; Xu et al., 2010; Sean et al., 2019). BGN, like most legumes, has the potential to contribute to climate change mitigation by reducing greenhouse gas emissions, use of fossil fuels, accelerate carbon sequestration rates and provide a valuable source of raw materials for biorefineries (Jensen et al., 2011; Sambo, 2014). They can biologically fix 33–46 Tg N each year (Jensen et al., 2011). They are adaptive and resilient; have low water demand, and tolerate drought, rising temperature and reduced precipitation (Tsoata et al., 2015, 2017a, b; Sambo, 2014; Nordin & Singh, 2015). BGN (*Vigna subterranea* (L.) Verdc.) is an African indigenous crop used for human and animal nutrition and medicinal plant. Due to its ecology, distribution, nutritional and medicinal properties, it could be a potential future crop to be exploited for food security, climate change mitigation

Table 3.1 Bambara groundnut mechanisms for drought tolerance

Mechanism	Consequence	Author
1. Adapted to contrasting environments, from cold temperatures at night and high temperatures during the day in an arid environment in Botswana to a far milder and more humid environment in Indonesia	Promotion in areas currently drought-prone and an essential future crop in regions where climate change projection shows an increased frequency and intensity in droughts.	Mayes et al. (2019)
2. Yield and water productivity of BGN will increase by approximately 37.5% and 33.0%, respectively in response to projected climate change in South Africa.	Confirm the resilience of the crop under climate change.	Mabhaudhi et al. (2018a)
3. BGN can maintain leaf turgor pressure through a combination of osmotic adjustment, reduced leaf area, and active stomatal regulation of water loss.	BGN depends on a range of complementary physiological and morphological attributes for drought resistance.	Collinson et al. (1997) and Mabhaudhi and Modi (2013)
4. Minimizing water losses through stomatal closure, lower plant height, leaf number and leaf area index.	BGN showed a degree of phenological plasticity in response to drought	Mabhaudhi and Modi (2013) and Collinson et al. (1997)
5. Reduced canopy size, early flowering and maturity and maintaining high water use efficiency under stress. Brown and red landraces more resilience than light brown.	Effect of seed colour on possible drought tolerance.	Mabhaudhi et al. (2013)
6. Biomass accumulation and partitioning		Mabhaudhi and Modi (2013) and Mabhaudhi et al. (2013)
7. Leaf temperature-transpiration and epicuticular wax.		Collinson et al. (1997)

and poverty alleviation. It is well adapted to various agro-ecological conditions and climate change (Temegne et al., 2018).

Dark-coloured seeds such as the brown and red landraces showed better adaption to drought stress compared to the light brown landrace due to the presence of tannins, which are polyphenols that act as antioxidants under stress conditions (Mabhaudhi & Modi, 2013). BGN can improve food and nutrition security, increase agricultural diversification and minimize environmental degradation. However, there are some challenges in maximizing the use of BGN for households and industrial applications. These include (1) the Hard-to-cook phenomenon and (2) Photoperiod sensitivity of pod filling.

Many pulse legumes are 'hard-to-cook- HTC', representing the increased effort required to prepare seeds for meals with consequent more significant fuel cost (Mayes et al., 2019). The HTC develops during seed storage, particularly under humid and hot conditions, and several theories explain components of this storage trait (Juliet et al., 2017). Some genetic constraints reported for BGN include (1) grain yield, (2) fertility, pod set and photoperiod sensitivity issues, (3) the relatively long physiological maturity period, (4) seed characteristics such as testa colour, cookability, antinutritional factors, seed size, and associated effects on grain yield, crop architecture (a bunch, semi-bunch or spreading), phenology (determinacy or indeterminacy), growth habit and podding (below ground or at the soil surface) and (6) susceptibility to anoxic/hypoxic stress factors (sensitivity to flooding) (Siise et al., 2014). When grown under long photoperiod, BGN produces more leaves with pod formation decrease than those grown under short photoperiod (Berchie et al., 2013).

3.3 Quantitative Trait Loci (QTL) and Genetic Linkage Map Involved in Bambara Groundnut Agronomic and Drought-Related Traits

BGN landraces remain the primary source of planting material used by farmers. Co-dominant marker analysis revealed that BGN is highly inbred, with seed from a single plant essentially representing an inbred line, although with high genetic variation levels (Molosiwa et al., 2015). Different molecular marker systems have been developed and applied to BGN landraces to assess breeding systems, diversity and population origins. The use of linkage maps and marker-assisted selection (MAS) is commonly applied in molecular marker to speed breeding programmes. Comprehensive genetic maps have been reported for chickpea [*Cicer arietinum* (L.)], pigeon pea (*Cajanus cajan*) and peanut [*Arachis hypogaea* (L.)]. The first genetic linkage map of BGN between domesticated landraces and QTL analysis over two generations was reported by Ahmad et al. (2016). The map comprises 29 SSR and 209 DArT array markers grouped into 21 linkage groups (LGs) with a total of 608.3 cM. A total of 36 significant QTLs were detected for 19 traits. Two stable QTLs were mapped for internode length and growth habit. The identified QTLs could be considerable for MAS in BGN breeding programmes. Chai et al. (2017) identified the QTL involved in agronomic and drought-related traits of BGN using gene expression markers (GEM). QTL discovered related to stomatal conductance, carbon isotope discrimination analysis and stomatal density of the drought assessment traits.

3.4 Bambara Groundnut Crop for Food and Nutrition Security in the Face of Climate Change

Food security exists 'when all people, at all times, have physical and economic access to sufficient, safe and nutritious food to meet their dietary needs and food preferences for an active and healthy life' (FAO, 2006). Four aspects of food security include (1) food availability at the local or national level, (2) food access (consumption) at household and individual level, (3) food utilization resulting in a good nutritional status which is the ultimate goal and (4) stability of availability, accessibility and proper utilization condition over time. Food and nutrition security, therefore "exists when all people at all times have physical, social and economic access to food of sufficient quantity and quality in terms of variety, diversity, nutrient content and safety to meet their dietary needs and food preferences for an active and healthy life, coupled with a sanitary environment, adequate health, education and care" (FAO, 2017). Therefore, food and nutrition security suggest that food security elements operate at different interfaces to provide access to food of sufficient quantity and quality in terms of variety, diversity, nutrient content, and safety (Fig. 3.3). At the interface of food availability and utilization is crop and product diversity. Crop diversification is responsible for a viable market system, creates the opportunity to earn more, and a decisive step toward nutrition security. BGN, among plant-based traditional foods/dishes has a high potential for global diet diversification and food and nutrition security in developing economies. Such diets align with the Lancet Commissions' review on sustainable food systems (Willett et al., 2019). A growing interest in the so-called underutilized crops' potential to address food, nutritional, and income security challenges is motivated by the growing concerns

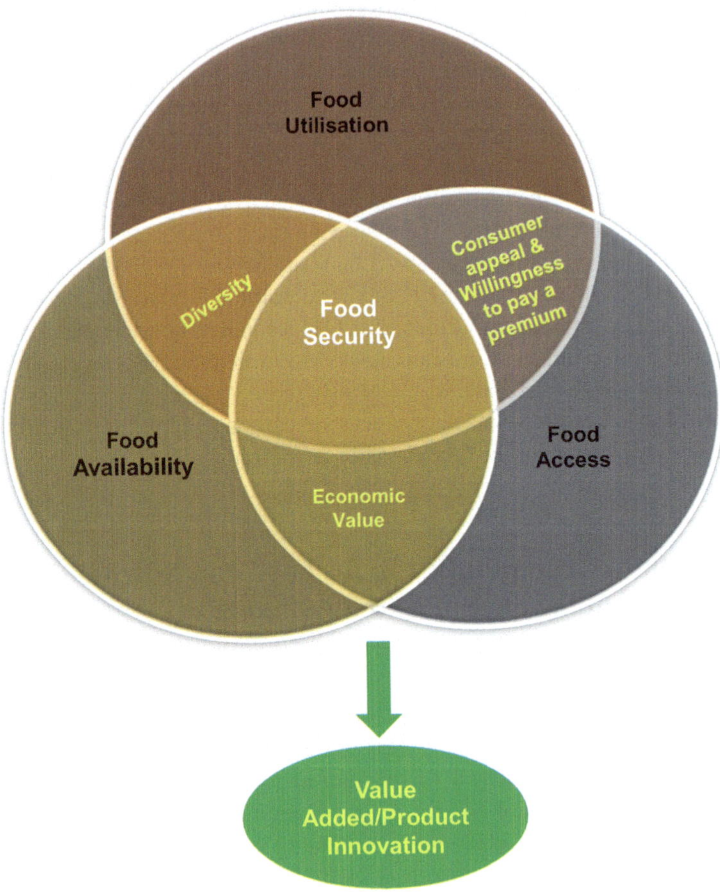

Fig. 3.3 Food and nutrition interface (Jideani & Jideani, 2016)

about the likely implications of climate change, the long-term sustainability of conventional agricultural approaches and biodiversity loss (Adzawla et al., 2016). Agronomic and nutritional benefits like drought tolerance, conservation of biodiversity, micro-nutrient content and socio-economic benefits (food and income security) are some of the advantages of underutilized crops. Adzawla et al. (2016) investigated the socio-economic factors characterizing the production of BGN in Ghana and the impact of its production on farmers welfare using a multi-stage sampling technique and observed (i) the probability of adopting BGN was greater for unmarried farmers; farmers in larger households; farmers with little or no formal education; and farmers who had no access to credit; (ii) the production of BGN led to increased household welfare (Fig. 3.4), as measured by the level of household per capita expenditure/consumption and (iii) that while further research and support for BGN production could contribute to addressing high poverty levels in the region (Fig. 3.5), many of the basic assumptions underlying current advocacy of underutilized crops need rigorous empirical verification.

However, product diversification is realized through value addition to agricultural commodities through product innovation. At the interface of food access and utilization is the need for products with consumer appeal and their willingness to pay a premium price. This cannot be achieved except the agricultural commodities through product innovation are turned into new nutritious food products with consumer appeal. At the interface of food availability and access is the economic value of the food. Hence, an agricultural commodity like BGN cannot be economically viable without food product innovation.

Food product innovation transforms agricultural raw materials into safe, convenient, good-tasting and nutritious products for consumers in a profitable and sustainable approach. Hence, crop diversification and product innovation are the two essential pillars of nutrition security and profit maximization. Value addition or product innovation is when a high price is realized through processing, packing, upgrading the quality, or other such methods for the same volume of a primary product. Value-added agriculture refers to the manufacturing process that increases

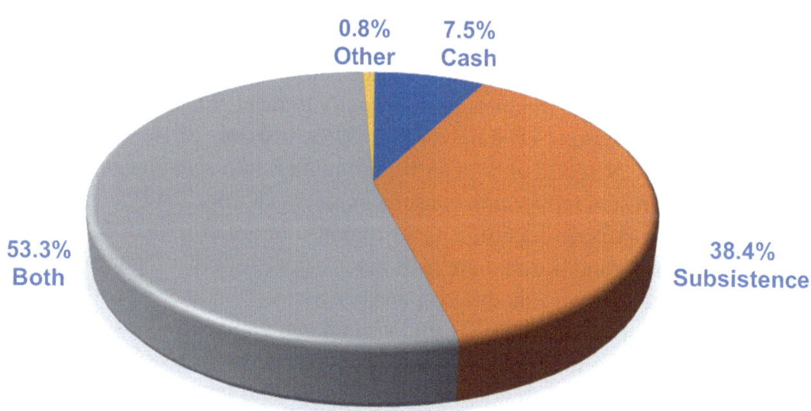

Fig. 3.4 Reasons for cultivating Bambara groundnut (Adzawla et al., 2016)

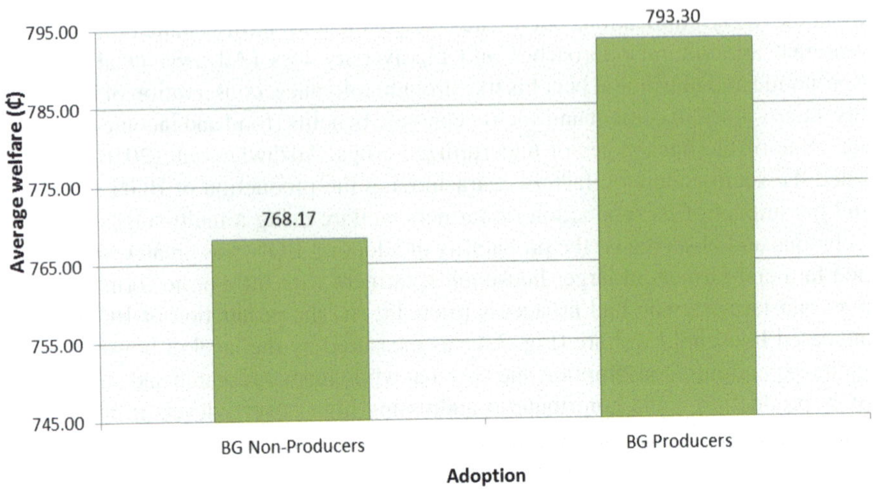

Fig. 3.5 Adoption and average welfare (GH₵) from the investigation of Adzawla et al. (2016) showing the estimated positive effect of adoption on welfare. *BG* Bambara groundnut

the value of primary agricultural commodities, and it is an effective development strategy for sustainable and healthy food systems.

In the era of climate change the food and beverage sector will be impacted in diverse ways (1) reliance on a stable supply of raw materials sensitive to the climate and vulnerable to changes in the natural environment which multiplies the level of unpredictability in availability, quality and prices; (2) heightened demand on the farmers' ability to track, understand and respond to changes in weather that they are not prepared to manage and (3) pressure on and opportunity for the sector to meet the nutritional needs of a growing population projected to reach 9 billion by 2050 (Wong & Schuchard, 2012). The hard-to-cook phenomenon in BGN provides an opportunity to diversify its use into protein concentrates, high protein beverages, special dietetic foods, canning, milling, popping, puffing and snack foods could lift it into many new markets. Product innovation with BGN will require a combination of traditional values such as origin and ethnicity with modern science such as processing and technology, novelty and change and comfort (Fig. 3.6). The raw materials, cooking methods and manufacturing processes for BGN food products must diversify to meet global consumers' needs and different ethnic groups (Gabriel et al., 2013; Jideani, 2016; Kwon, 2017). Furthermore, the BGN food culture can be formulated into scientifically empirical stories of great interest in influencing consumer choices (Kwon, 2017).

BGN, like most African legumes, is a drought-tolerant crop and thrives under harsh environmental conditions where soils are too poor for other leguminous crops to succeed, thus making it a valuable crop in such areas (Duodu & Apea-Bah, 2017; Ho et al., 2017; Atoyebi et al., 2017). BGN plant can capture and fix atmospheric nitrogen contributing to increasing yields of subsequent crops (Mohamadou et al., 2018) and is resistant to harsh conditions including high temperatures (Azam-Ali et al., 2001; Odongo et al., 2015). However, high temperature negatively affects the

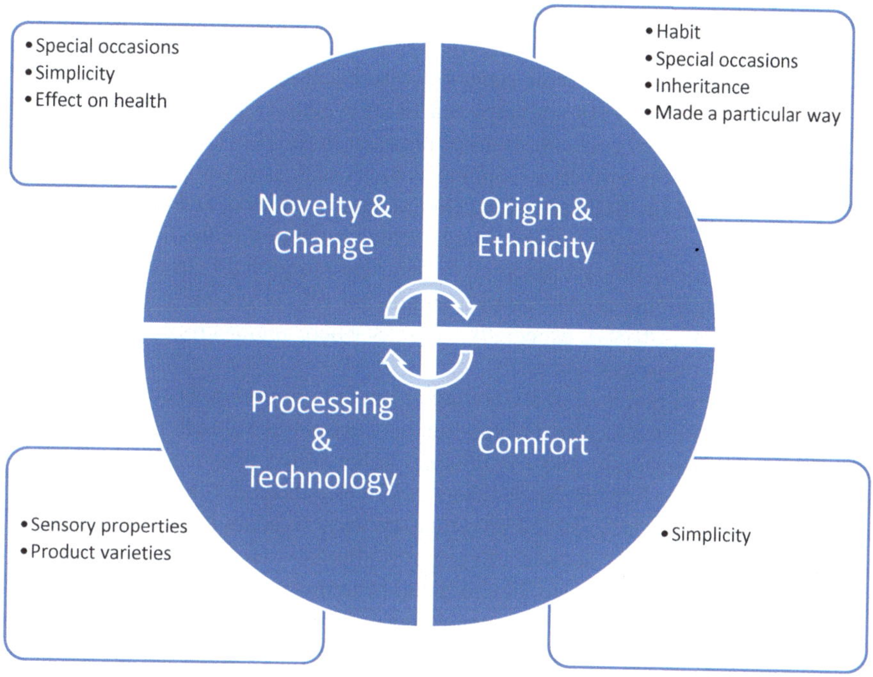

Fig. 3.6 Strategy for value addition to BGN, innovation concept interfaced with tradition

yield of plant crops (Agnolucci et al., 2020). Many current food systems are unsustainable because they cause significant resource depletion and unacceptable environmental impacts (IPCC, 2014; Agnolucci et al., 2020). The food system's technological control could help the transition from fossil to sustainable, considering the growing global population (Holden et al., 2018). The sustainable development goals focus on some critical global issues outlined in Goal 2 ('end hunger, achieve food security and improved nutrition and promote sustainable agriculture'); Goal 12 ('ensure sustainable consumption and production patterns') and Goal 13 ('take urgent action to combat climate change and its impacts'). All are intimately related to the need to transition global food systems from fossil to sustainable food systems (Holden et al., 2018; United Nations, 2015).

3.5 Transforming Bambara Groundnut into a Crop of the Future

The production of mature seeds, nitrogen fixation activity and nutritional profile of BGN support further research investment and its broader inclusion in agricultural systems. Production of mature seeds could reduce the demand and dependence on major crops such as soybean to produce value-added products like tofu, and plant-based milk (Murevanhema & Jideani, 2015). In low-input systems, there is potential to grow BGN on a relatively wide range of soils, because it tends to nodulate with different Rhizobia and its efficient nitrogen-fixation activity. Furthermore, the ability of BGN to store high concentrations of nitrogen in its biomass makes it possible to incorporate the crop into other cropping systems such as rotation intercropping.

The mature seeds of BGN are an excellent source of protein and other nutrients, making it a potential legume crop for large-scale production in the humid and sub-humid tropics with significant potential to contribute to global food and nutritional security. The nutritional profile plays a significant role in people's diet improvement (Adeyi et al., 2019a), especially in areas where major crops require high-inputs, limiting consumers' access.

The mature seed potential of BGN range from (1) improving its cookability; (2) reducing antinutritional factors; (3) gaining more economic value through introduction into food mainstream and (4) produce value-added products (Adeyi et al., 2019b; Diedericks & Jideani, 2015; Jideani, 2016; Maphosa & Jideani, 2016). Food technologists have already started investigating ways to adapt processing methods already used for crops like soybean, to generate alternative food products with various nutritional profiles and taste, like tofu and bean milk (Chap. 11). The outcomes will increase BGN seed products' appeal and potentially introduce them to larger global markets (Jideani, 2016).

Improvement programmes for BGN should keep in mind the prevalent practices used in traditional production regions, especially BGN cultivation methods and ways of consumption. Research efforts should prioritize the (1) development of high-yielding cultivars of high palatability and nutritional value (Maphosa & Jideani, 2016) and (2) improvement of yield, nutritional value, and marketability of mature seeds. It is vital to track viable material shared or retrieved from genebanks, with a labelling system consistent across research groups, and utilize a standard set of descriptors and procedures to phenotype the germplasm. Molecular tools genotyping by sequencing, RNA-sequencing and genome-wide association studies can then be used to accelerate the efforts to confront critical traits. The generation of winged bean mapping populations, linkage maps, and genome sequencing would boost information translation and the dissection of features of interest in BGN.

The BGN value chains are currently poorly developed creating opportunities to diversify the current food systems to creating new employment, improving access to nutritious foods, increasing market and distribution opportunities, and promoting self-directed pathways out of poverty (Mabhaudhi et al., 2019). Opportunities for gender empowerment through women involvement in the BGN food system exist because of women's significant role in production and conservation (Mabhaudhi et al., 2018b, 2019). Promoting gender equality and women empowerment is inextricably linked to the strengthening of sustainable food systems to fight hunger and malnutrition and improve rural populations' livelihoods. Including BGN as part of transformative adaptation allows for adapting to the climatic, ecological, and natural limits in which resource-poor farmers, in particular, reside. Orphan crops like

BGN can, therefore, offer opportunities for 'fitting to' or 'fitting with' the socio-ecological environment while trying to sustainably maintain the natural products or processes needed for sustainable food systems (Mabhaudhi et al., 2019).

There has been increasing research on the global view on climate change and climate-smart crops (CSCs). Several factors, singly or in combination, militate against crops such as incidence of insect pests in a warm climate (Deutsch et al., 2018), crop yield simulation with rising CO_2 concentration (Long et al., 2006) and extreme climatic conditions (Ho et al., 2017). BGN is a hardy plant being highly resistant to drought and produces comparable yields when grown under extreme climatic conditions (Shanono & Muhammad, 2015; Ho et al., 2017).

Understanding the impact of temperature changes and precipitation on crop yields is vital in developing policy and management options to feed the world (United Nations, 2015; Agnolucci & De Lipsis, 2019; Nilsson et al., 2016). Most studies on the effect of climate change are limited to a few staple crops. Agnolucci et al. (2020) implemented global statistical models to examine the influence of weather and management practices on the yields of 18 crops (accounting for 70% of crop production by area and 65% by calorific intake), focusing on the impact of temperature, they observed (i) considerable heterogeneity in the responses of yields across crops and countries, (ii) irrigation alleviated negative implications from temperature increases, (iii) countries where increasing temperature causes the most negative impacts are typically the most food insecure, with the lowest calorific food supply and average crop yield and (iv) international action must be coordinated to raise yields in these countries through improvement and modernization of agricultural practices to counteract future adverse impacts of climate change.

3.6 Conclusion

Genetic diversity of BGN related to climate change and food innovation are options in making BGN economically viable. Drought tolerance is a plant's ability to grow, flower and displays economic yield under a suboptimal water supply. BGN responds and adapts to and survives under drought stress through morphological, biochemical and physiological mechanisms. BGN landraces remain the primary source of planting material used by farmers. Co-dominant marker analysis revealed that BGN is highly inbred, with seed from a single plant essentially representing an inbred line, although with high genetic variation levels. Qualitative trait loci related to BGN adaptation to climate change include stomatal conductance, carbon isotope discrimination and stomatal density. Food, nutrition and economic security contribution of BGN are possible through food product innovation. The hard-to-cook phenomenon in BGN provides an opportunity to diversify its use into protein concentrates, high protein beverages, special dietetic foods, canning, milling, popping, puffing and snack foods could lift this orphan crop into many new markets.

References

Adeyi, O., Ikhu-Omoregbe, D. I. O., & Jideani, V. A. (2019a). Physical stability characteristics of sunflower oil-in-water emulsion containing sodium chloride, stabilized by gelatinized Bambara groundnut flour. *Cogent Engineering, 6*, 1–13.

Adeyi, O., Ikhu-Omoregbe, D. I. O., & Jideani, V. A. (2019b). Effect of citric acid on physical stability of sunflower oil-in-water emulsion stabilized by gelatinized Bambara groundnut flour. *International Journal of Civil Engineering and Technology, 10*, 2260–2273.

Adzawla, W., Donkoh, S. A., Nyarko, G., O'Reilly, P. J., Olayide, O. E., Mayes, S., Feldman, A., & Azman, H. R. (2016). Adoption of Bambara groundnut production and its effects on farmers' welfare in Northern Ghana. *African Journal of Agricultural Research, 11*(7), 583–594. https://doi.org/10.5897/AJAR2015.10568

Agnolucci, P., & De Lipsis, V. (2019). Long-run trend in agricultural yield and climatic factors in Europe. *Climatic Change*. https://doi.org/10.1007/s10584-019-02622-3

Agnolucci, P., Rapti, C., Alexander, P., De Lipais, V., Holland, R. A., Eigenbrod, F., & Ekins, P. (2020). Impacts of rising temperatures and farm management practices on global yields of 18 crops. *Nature Food, 1*, 562–571. https://doi.org/10.1038/s43016-020-00148-x

Ahmad, N. S., Redjeki, E. S., Ho, W. K., Aliyu, S., Mayes, K., Massawe, F., Kilian, A., & Mayes, S. (2016). Construction of a genetic linkage map and QTL analysis in Bambara groundnut (*Vigna subterranea* (L) Verdc.). *Genome, 59*, 459–472.

Aliyu, S., Massawe, F., & Mayes, S. (2016). Genetic diversity and population structure of Bambara groundnut (*Vigna subterranea* (L.) Verdc.): Synopsis of the past two decades of analysis and implications for crop improvement programmes. *Genetic Resources and Crop Evolution, 63*, 925–943.

Atoyebi, O. J., Osilesi, O., Adebawo, O., & Abberton, M. (2017). Evaluation of nutrient parameters of selected African accessions of Bambara groundnut (*Vigna subterranea* (L.) Verdc.). *American Journal of Food and Nutrition (SCIEP Series), 5*(3), 83–89. https://doi.org/10.12691/ajfn-5-3-1

Azam-Ali, S. N., Sesay, A., Karikari, S. K., Massawe, F. J., Aguilar-Manjarrez, J., Brennan, M., & Hampson, K. J. (2001). Assessing the potential of an underutilized crop – A case study using Bambara groundnut. *Experimental Agriculture, 37*, 433–472.

Bajji, M., Kinet, JM. & Lutts, S. (2002). The use of the electrolyte leakage method for assessing cell membrane stability as a water stress tolerance test in durum wheat. *Plant Growth Regulation 36*, 61–70. https://doi.org/10.1023/A:1014732714549

Berchie, J. N., Amelie, G., McClymont, S., Raizada, M., Adu-Dapaah, H., & Sarkoie-Addo, J. (2013). Performance of 13 Bambara groundnut (*Vigna substerranea* (L.) Verdc.) landraces under 12 H and 14 H photoperiod. *Journal of Agronomy, 12*, 20–28.

Chai, H. H., Ho, W. K., Graham, N., May, S., Massawe, F., Mayes, S., Hui Hui, C., Wai Kuan, H., Neil, G., Sean, T. M., Festo, M., & Sean, M. (2017). A cross-species gene expression marker-based genetic map and QTL analysis in Bambara groundnut. *Genes, 8*, 84.

Collinson, S. T., Clawson, E. J., Azam-Ali, S. N., & Black, C. R. (1997). Effects of soil moisture deficits on the water relations of Bambara groundnut (*Vigna subterranea* L. Verdc). *Journal of Experimental Botany, 48*, 877–884.

Deutsch, C. A., et al. (2018). Increase in crop losses to insect pests in a warming climate. *Science, 361*, 916–919.

Diedericks, C. F., & Jideani, V. A. (2015). Physicochemical and functional properties of insoluble dietary fiber isolated from Bambara groundnut (*Vigna subterranea* [L.] Verdc.). *Journal of Food Science, 80*, C1933–C1944.

Duodu, K. G., & Apea-Bah, F. B. (2017). Chapter 9: African legumes: Nutritional and health-promoting attributes, pp. 223–269. In J. Taylor & J. Awika (Eds.), *Gluten-free ancient grains: Cereals, pseudocereals, and legumes: Sustainable, nutritious, and health-promoting foods for the 21st century* (1st ed., pp. 356). Woodhead Publishing. eBook ISBN: 9780081008911.

https://www.elsevier.com/search-results?query=Bambara%20groundnut&labels=books&page=1
FAO. (2006). *Food security*. FAO Policy Brief.
FAO. (2017). *Africa regional overview of food security and nutrition 2016. The challenges of building resilience to shocks and stresses*. Food and Agriculture Organization.
FAO. (2018). FAOSTAT [Internet document]. Food and Agriculture Organisation of the United Nations, Rome, Italy. http://www.fao.org/faostat/en/#home. Accessed 28 Aug 2020.
Farooq, M., Wahid, A., Kobayashi, N., Fujita, D., & Basra, S. M. A. (2009). Review article plant drought stress : Effects, mechanisms and management. *Agronomy for Sustainable Development, 29*, 185–212.
Gabriel, E. G., Jideani, V. A., & Ikhu-omoregbe, D. I. O. (2013). Investigation of the emulsifying properties of Bambara groundnut flour and starch. *International Journal of Food Science and Engineering, 7*, 539–547.
Halimi, A. R., Barkla, B. J., Mayes, S., King, G. J., Razlin Azman, H., Bronwyn, J. B., Sean, M., & Graham, J. K. (2019). The potential of the underutilized pulse Bambara groundnut (*Vigna subterranea* (L.) Verdc.) for nutritional food security. *Journal of Food Composition and Analysis, 77*, 47–59.
Hasegawa P.M., Bressan, R.A., Zhu J-K., and Bohnert, H. J. (2000) P C M R H S. Annual *Review of Plant Physiology and Plant Molecular Biology 51*(1):463-499
Ho, W. K., Chai, H. H., Kendabie, P., Ahmad, N. S., Jani, J., Massawe, F., Kilian, A., & Mayes, S. (2017). Integrating genetic maps in Bambara groundnut [*Vigna subterranea* (L) Verdc.] and their syntenic relationships among closely related legumes. *BMC Genomics, 18*, 192. https://doi.org/10.1186/s12864-016-3393-8
Holden, N. M., White, E. P., Lange, M. C., & Oldfield, T. L. (2018). Review of the sustainability of food systems and transition using the Internet of Food. *npj Science of Food, 2*, 18. https://doi.org/10.1038/s41538-018-0027-3
IPCC. (2014). Intergovernmental Panel on Climate Change [Core Writing Team, Pachauri, R. K. & Meyer, L]. (2014) Climate Change 2014 Synthesis Report Summary for Policymakers. https://www.ipcc.ch/pdf/assessment-report/ar5/syr/AR5_SYR_FINAL_SPM.pdf. Accessed 01 Oct 2018.
Jensen, E. S., Peoples, M. S., Boddey, R. M., Gresshoff, P., Hauggaard-Nielsen, H., Alves, B. J. R., & Morrison, M. J. (2011). Legumes for mitigation of climate change and the provision of feedstock for biofuels and biorefineries. A review. *Agronomy for Sustainable Development, 32*(2), 329–364.
Jideani, V. A. (2016). Utilizing Bambara groundnut in value-added products. *Food Technology, 70*, 48–52.
Jideani, V. A., & Jideani, A. I. O. (2016). Versatility of Bambara groundnut (*Vigna subteranea*) for food and nutrition security. In *Pan-African Grain Legume & World Cowpea conference*.
Jørgensen, S. T., Aubanton, M., Harmonic, C., Dieryck, C., Jacobsen, S. P., Simonsen, H. L., Ntundu, W. H., Stadler, F., Basu, S., & Christiansen, J. L. (2009). Identification of photoperiod neutral lines of Bambara groundnut (*Vigna subterranea*) from Tanzania. *IOP Conference Series: Earth and Environmental Science, 6*, 372023.
Juliet, M., Vincenzo, F., Cathrine, C., Anita, R. L., Mubaiwa, J., Fogliano, V., Chidewe, C., & Linnemann, A. R. (2017). Hard-to-cook phenomenon in Bambara groundnut (*Vigna subterranea* (L.) Verdc.) processing: Options to improve its role in providing food security. *Food Reviews International, 33*, 167–194.
Khan, M. H., Md. Rafii, M. Y., Ramlee, S. I., Jusoh, M., & Mamun, A. l. (2020). Genetic variability, heritability, and clustering pattern exploration of Bambara groundnut (*Vigna subterranea* L. Verdc) accessions for the perfection of yield and yield-related traits. *BioMed Research International*, Article ID 2195797, 31 pages. https://doi.org/10.1155/2020/2195797
Kumar, J., & Abbo, S. (2001). Genetics of flowering time in chickpea and its bearing on productivity in semiarid environments. In *Advances in agronomy* (pp. 107–138). Elsevier. ISBN 978-0-12-374361-9.

Kwon, D. Y. (2017). Ethnic foods and globalization. *Journal of Ethnic Foods, 4*, 1–2.

Long, S. P., Ainsworth, E. A., Leakey, A. D. B., Nosberger, J., & Ort, D. R. (2006). Food for thought: Lower-than-expected crop yield simulation with rising CO2 concentration. *Science, 312*, 1918–1921.

Mabhaudhi, T., & Modi, A. T. (2013). Growth, phenological and yield responses of a Bambara groundnut (*Vigna subterranea* (L .) Verdc .) landrace to imposed water stress under field conditions. *South African Journal of Plant and Soil, 30*, 69–79.

Mabhaudhi, T., Modi, A. T., & Beletse, Y. G. (2013). Growth, phenological and yield responses of a Bambara groundnut (*Vigna subterranea* L . Verdc) landrace to imposed water stress: II. Rain shelter conditions. *Water SA, 39*(2), 191–198.

Mabhaudhi, T., Chibarabada, T. P., Chimonyo, V. G. P., & Modi, A. T. (2018a). Modelling climate change impact: A case of Bambara groundnut (*Vigna subterranea*). *Physics and Chemistry of the Earth, Parts A/B/C, 105*, 25–31. https://doi.org/10.1016/j.pce.2018.01.003

Mabhaudhi, T., Chibarabada, T. P., Chimonyo, V. G. P., Murugani, V. G., Pereira, L. M., Sobratee, N., Govender, L., Slotow, R., & Modi, A. T. (2018b). Mainstreaming underutilized indigenous and traditional crops into food systems: A South African perspective. *Sustainability (Switzerland), 11*, 172.

Mabhaudhi, T., Chimonyo, V. G. P., Hlahla, S., Massawe, F., Mayes, S., Nhamo, L., & Modi, A. T. (2019). Prospects of orphan crops in climate change. *Planta, 250*(3), 695–708.

Maphosa, Y., & Jideani, V. A. (2016). Physicochemical characteristics of Bambara groundnut dietary fibers extracted using wet milling. *South African Journal of Science, 112*, 72–79.

Mayes, S., Ho, W. K., Chai, H. H., Gao, X., Kundy, A. C., Mateva, K. I., Zahrulakmal, M., Hahiree, M. K. I. M., Kendabie, P., Licea, L. C. S., Massawe, F., Mabhaudhi, T., Modi, A. T., Berchie, J. N., Amoah, S., Faloye, B., Abberton, M., Olaniyi, O., & Azam-Ali, S. N. (2019). Bambara groundnut: An exemplar underutilized legume for resilience under climate change. *Planta, 250*, 803–820.

Mohamadou M, Ngakou A and Tchuenguem F-N F (2018) Influence de l'interaction mycorhizes-insectes floricoles sur les rendements de Vigna subterranea variété rouge (Fabaceae) à Dang (Ngaoundéré, Cameroun) *International Journal of Biological and Chemical Sciences, 12*(4), 1897-1913. DOI: https://dx.doi.org/10.4314/ijbcs.v12i4.30 Available online at http://www.ifgdg.org

Molosiwa, O. O., Aliyu, S., Stadler, F., Mayes, K., Masawe, F., Kilian, A., & Mayes, S. (2015). SSR marker development, genetic diversity and population structure analysis of Bambara groundnut [*Vigna subterranea* (L.) Verdc.] landraces. *Genetic Resources and Crop Evolution, 62*, 1225–1243.

Murevanhema, Y. Y., & Jideani, V. A. (2015). Production and characterization of milk produced from Bambara groundnut (*Vigna subterranea*) varieties. *Journal of Food Processing and Preservation, 39*(6), 1485–1498.

Nilsson, M., Griggs, D., & Visbeck, M. (2016). Policy: Map the interactions between sustainable development goals. *Nature, 534*, 320–322.

Nordin, N., & Singh, A. (2015). Seedling emergence of Bambara groundnut (*Vigna subterranea* (L.) Verdc.) landraces at various temperature conditions. *Procedia Environmental Sciences, 29*, 201–202.

Odongo, F. O., Oyoo, M. E., Wasike, V., Owuoche, J. O., Karanja, L., & Korir, P. (2015). Genetic diversity of Bambara groundnut (*Vigna subterranean* (L) Verde.) landraces in Kenya using microsatellite markers. *African Journal of Biotechnology, 14*(4), 283–291.

Prochazkova D, Sairam R K Srivastava G.C. and Singh D.V (2001) Oxidative stress and antioxidant activity as the basis of senescence in maize leaves. *Plant Science 161*(4): 765-771 https://doi.org/10.1016/S0168-9452(01)00462-9

Sambo, B. E. (2014). Endangered, neglected, indigenous resilient crops: A potential against climate change impact for sustainable crop productivity and food security. *IOSR JAVS, 7*(2), 34–41.

Sean, M., Wai Kuan, H., Hui Hui, C., Xiuqing, G., Aloyce Callist, K., Kumbirai, I. M., Muhammad, Z., Mohd Khairul Izwan Mohd, H, Presidor, K., Luis Carlos Cejudo, L., Festo,

M., Tafadzwanashe, M., Albert Thembinkosi, M., Joseph Nketiah, B., Stephen, A., Ben, F., Michael T., A., Oyatomi, O., Sayed, A. -A., Mayes, S., Ho, W. K., Chai, H. H., Gao, X., Kundy, A. C., Mateva, K. I., Zahrulakmal, M., Hahiree, M. K. I. M., Kendabie, P., Licea, L. C. S., Massawe, F., Mabhaudhi, T., Modi, A. T., Berchie, J. N., Amoah, S., Faloye, B., Abberton, M., Olaniyi, O, & Azam-Ali, S. N. (2019). Bambara groundnut: An exemplar underutilized legume for resilience under climate change. *Planta, 250*, 803–820.

Shanono, I. M., & Muhammad, Y. Y. (2015). Effect of cooking on proximate and anti-nutritional content of Bambara groundnut. *Journal of Advances in Food Science & Technology, 2*(4), 176–180. Retrieved from http://www.ikppress.org/index.php/JAFSAT/article/view/4222

Siise, A., Festo, M., Sean, M., Aliyu, S., Massawe, F., & Mayes, S. (2014). Beyond landraces developing improved germplasm resources for underutilized species – A case for Bambara groundnut. *Biotechnology and Genetic Engineering Reviews, 30*, 127–141.

Suhairi, T. A. S. T. M., Jahanshiri, E., Nizar, N. M. M., Ebrahim, J., & Nijin, N. (2018). Multicriteria land suitability assessment for growing underutilized crop, Bambara groundnut in Peninsular Malaysia. *IOP Conference Series: Earth and Environmental Science, 169*, 012044. https://doi.org/10.1088/1755-1315/169/1/012044

Temegne, N., Gouertoumbo, W. F., Wakem, G. A., Nkou, F. T. D., Youmbi, E., & Ntsomboh-Ntsefong, G. (2018). Origin and ecology of Bambara groundnut (*Vigna subterranea* (L.) Verdc): A review. *Journal of Ecology & Natural Resources, 2*, 1–10.

Tsoata, E., Njock, S. R., Youmbi, E., & Nwaga, D. (2015). Early effects of water stress on some biochemical and mineral parameters of mycorrhizal *Vigna subterranean* (L.) Verdc. (*Fabaceae*) cultivated in Cameroon. *International Journal of Agronomy and Agricultural Research, 7*(2), 21–35.

Tsoata, E., Temegne, N. C., & Youmbi, E. (2017a). Analysis of early biochemical criterion to screen four Fabaceae plants for their tolerance to drought stress. *International Journal of Current Research, 9*(1), 44568–44575.

Tsoata, E., Temegne, N. C., & Youmbi, E. (2017b). Analysis of early growth criterion to screen four Fabaceae plants for their tolerance to drought stress. *Research Journal of Life Sciences, Bioinformatics, Pharmaceutical and Chemical Sciences, 2*(5), 94–109. https://doi.org/10.26479/2017.0205.09

Uko, E. B. (2017). Food security potentials of Bambara groundnut (*Vigna subterranea* (L.) Verdc.). *International Journal of Development and Sustainability, 6*(12), 1919–1930. ISSN: 2186-8662. www.isdsnet.com/ijds

United Nations. (2015). *Transforming our world: The 2030 agenda for sustainable development*. United Nations.

Willett, W., Rockström, J., Loken, B., Springmann, M., Lang, T., Vermeulen, S., Garnett, T., Tilman, D., DeClerck, F., Wood, A., Jonell, M., Clark, M., Gordon, L. J., Fanzo, J., Hawkes, C., Zurayk, R., Rivera, J. A., De Vries, W., Majele Sibanda, L., Afshin, A., Chaudhary, A., Herrero, M., Agustina, R., Branca, F., Lartey, A., Fan, S., Crona, B., Fox, E., Bignet, V., Troell, M., Lindahl, T., Singh, S., Cornell, S. E., Srinath Reddy, K., Narain, S., Nishtar, S., & Murray, C. J. L. (2019). Food in the Anthropocene: The EAT-Lancet Commission on healthy diets from sustainable food systems. *Lancet, 393*(10170), 447–492. https://doi.org/10.1016/S0140-6736(18)31788-4

Wong, J., & Schuchard, R. (2012). *Adapting to climate change: A guide for the food, beverage, and agriculture companies* (BSR industry series). BSR.

Xu, Z., Zhou, G., & Shimizu, H. (2010). Plant responses to drought and rewatering. *Plant Signaling & Behavior, 5*, 649–654.

Chapter 4
Ethnonutritional and Ethnomedicinal Uses of Bambara Groundnut

4.1 Introduction

Bambara groundnut [BGN] is consumed almost solely as food, mostly by poor people living in rural areas with minimal resources. The rapid urban population growth and the associated rise of street food sales lead to expanded use in the metropolitan area (Greenhalgh, 2000). The seed is a complete food eaten in different forms, either immature or fully ripe. The fresh and semi-ripe seeds are more palatable than the dried mature ones. Information on processing techniques either passed on from older family members or shared from women cooperatives include boiling, soaking, milling, roasting, fermentation and several combinations of these methods (Greenhalgh, 2000; Mubaiwa et al., 2018a). Various processing techniques for BGN include boiling, roasting, and frying (Ndidi et al., 2014). Enhancement in quality and safety of BGN with increasing boiling or roasting time reduces the antinutritional properties below the permissible limits, especially with moist heat compared with dry heat. The significant reduction of antinutrients during processing suggests that the nutritional components of BGN will be more bioavailable (Ndidi et al., 2014).

4.2 Traditional Bambara Groundnut-Based Food Products by Boiling and Soaking

In sub-Saharan Africa, the seeds of Bambara groundnut (BGN) are pounded and mixed with onions, tomatoes and oil to relish (Misheck et al., 2013). In northern Ghana, the fresh immature beans are boiled, grilled, or roasted and eaten as a snack with salt. The dry beans are also boiled, mashed and made into cakes or balls, which are then fried and used to prepare stews. In southern Ghana the beans are usually soaked overnight, boiled until soft to produce a kind of porridge/blancmange; pepper and salt may be added during the boiling to make bean stew (*aboboi*). The bean

V. A. Jideani, A. I. O. Jideani, *Bambara groundnut: Utilization and Future Prospects*, https://doi.org/10.1007/978-3-030-76077-9_4

stew *aboboi*, is served with roasted, grated cassava (gari) or a mashed, fried, ripe plantain (Ntundu, 1995).

The following culinary uses of the crop have been reported in South Africa. 'Sekome' (Sesutho), 'tihove' (Shangaan) or 'tshidzimba' (Venda) is prepared by adding BGN beans and peanuts, or just one of the two, to maize or millet-meal and boiling the mixture until it forms a stiff dough, salted and pounded into a ball, and will often keep fresh for several days. Bambara groundnut is boiled and then stirred to make a thin porridge known as 'tshipupu' (in Venda). Like maize, they may also be added to 'lupida', a porridge made from peanuts. Immature 'Njugo' beans boiled until soft and shelled. When quite dry and hard, they are generally shelled, and then boiled to make a stiff porridge. Bambara groundnut can be cooked with maize and pounded into a thick, sticky dough known as 'dithaku' (in Sesutho) (Ntundu, 1995).

In Zimbabwe, immature seeds are consumed fresh or grilled, boiled, either shelled or unshelled and eaten as a meal or mixed with immature peanuts or green maize (Greenhalgh, 2000). Ripe or dry seeds are pounded to flour and boiled to a stiff porridge or soaked and then boiled. The porridge keeps well and is traditionally used on journeys. The dry seeds are roasted, broken into pieces, boiled, mashed and eaten as a relish with maize-meal porridge ("sadza") or with peanut butter (rupiza) (Greenhalgh, 2000; Mubaiwa et al., 2018a). Rupiza, made from boiled BGN grits and mixed with peanut butter before serving. Bambara groundnut rupiza takes 45 min to 1 h to boil and this cooking time is comparable to the boiling time of freshly harvested BGN pods. Hence, using BGN grits is a 75% reduction in cooking time than the long cooking time of whole BGN seeds with alkaline salts (Mubaiwa et al., 2018a). The fresh seed can be boiled to prepare Mukove or Nyimo nyoro. The seeds can be soaked overnight before boiling to reduce the cooking time. Boiling until soft can take up to 5 h, but can be reduced to about 3 h with soaking. An estimated 2 kg BGN will feed about ten people. Fresh BGN is famous as a snack when the crop is in season (Greenhalgh, 2000). BGN is boiled in water for approximately 2–4 h to make a snack or substitute product for bread called mutakura (Mubaiwa et al., 2018a). Soaking before drying, roasting and milling are employed to remove the seed coats in the preparation of pre-soaked mutakura. Optional ingredients added during boiling include cowpea and peanuts and or maize to make an assorted mutakura. Mutakura may also be mixed with peanut butter or cooking oil and other ingredients like tomatoes, onion and paprika to make a relish eaten with stiff maize porridge. Furthermore, Jambalaya is a product of mutakura mixed with Irish potatoes or sweet potatoes. Porridge from mashed peanut-based mutakura is recommended as a complementary and weaning food. Dried BGN-based soup is prepared from solar-dried mutakura flour mixed with fresh, grated tomatoes, onions, and spices to make a dough that is further dried in a solar tent dryer to the desired moisture content. Subsequently, the dried mutakura mixture is roasted (until golden brown) and ground with a mortar and pestle, and finally sieved to make a soup. There is a reported improvement of flavour and nutritional attributes of dishes after adding the soup. Despite all these diverse dishes, the long cooking time involved in boiling BGN remains a significant setback to its utilization (Mubaiwa et al., 2018a).

Table 4.1 Dikgobe formulation in Botswana based on Bambara groundnut

Name	Maize grain	Legumes (beans or peas)
Dikgobe tsa lechotlho(Fig. 4.1)	Lechotlho	Letlhodi, cowpea (Fig. 4.2c) or Bambara groundnut
Dikgobe tsa lebutho	Lebutho	Letlhodi, cowpea, or Bambara groundnut (Fig. 4.2d).
Dikgobe tsa mmidi	Dried whole grain maize	Letlhodi, cowpea, or Bambara groundnut
Dikgobe tsa satampa (samp) (Fig. 4.1)	Samp	Sugar beans, butter beans, cowpeas

Source: Bultosa et al. (2020)

In Botswana, BGN is predominantly grown for human consumption. Consumers often prefer the immature seeds boiled in the pod, salted, and consumed, either alone or mixed with maize seeds. When the seeds are ripe and dry, they are pounded into flour and used to make various cakes, or integrated with cereals porridge. Livestock, especially goats, are very fond of the haul, which they are allowed to graze on at the end of the season after harvesting the pods. The seeds of mature black landrace are used in traditional medicine (Ntundu, 1995). Bultosa et al. (2020) as presented in Table 4.1 reported that (i) Dikgobe *tsa lechotlho* causes constipation in some individuals, and (ii) Dikgobe *tsa satampa* cause heartburn and may also result in constipation and bloating in some individuals. Most of these dishes (Fig. 4.1) are available throughout the year, but the young generation seems to have limited knowledge and preparation skills about them.

The Luhya tribe in Kenya consumes BGN by boiling the fresh pods in salted water, reported as the tastiest BGN snack, and is rated even more highly than the peanut. The dry seeds may be boiled with maize and served as a snack, especially with tea or boiled, then mixed with boiled sweet potatoes and mashed (a popular children's dish) or boiled, mashed and consumed plain by this tribe. It may also be roasted. Unshelled pods are boiled, fried and served with potatoes, bananas or 'ugali'. The Luo in Kenya uses BGN in the same way as kidney beans, to make 'nyoyo' (a mixture of beans and maize boiled together), or plain boiled, served with tea, 'nyuka' (porridge), or on its own (Ntundu, 1995).

In semi-arid parts of Tanzania, BGN is considered insurance or a security crop and is consumed in various ways from freshly cooked pods to the cooked dry grains. It is an adjunct to main dishes like cooked plantains and cereals. The vegetative parts of the plant, and sometimes the haulms, are used to feed livestock. The crop is also useful in crop rotations, as it contributes nitrogen to the soils, which benefits subsequent crops (Ntundu, 1995).

In Angola and Mozambique, boiled salted BGN seeds are served as appetizers in restaurants (Greenhalgh, 2000).

In western Niger, BGN is consumed as boiled fresh pods (loubatou), roasted seeds (dawalé), seeds processed into small grits (béroua), gabda, touwo (stiff porridge from flour), kalapaté goudjia and maringuidé (Ibrahim et al., 2018). Loubatou and dawalé are the most important and preferred BGN end products. Loubatou is

Fig. 4.1 Bambara groundnut (BGN) and pulse ingredient for processing of Dikgobe (**a**) Lechotlho (maize mealies-fresh maize grains at around physiological maturity is cooked with its sheath while on its cob and then dried), ingredient for processing of Dikgobe; (**b**) Samp (Setampa-decorticated and cracked maize grains), ingredient for processing of Dikgobe; (**c**) Black-eye cowpeas, pulse ingredient for processing of Dikgobe; (**d**) Bambara groundnut, pulse ingredient for processing of Dikgobe; € Dikgobe tsa lechotlho (black-eye cowpea and lechotlho-cooked); (**f**) Dikgobe tsa setampa (decorticated maize and Bambara ground nut-cooked) (Bultosa et al., 2020)

eaten as a meal by village farmers for the night before staple food crops' maturity. Matankari village farmers use BGN flour to make spaghetti. The long cooking time of BGN is a constraint to utilizing BGN (Ibrahim et al., 2018; Mubaiwa et al., 2018a). Soaking overnight and use of bicarbonate were the two approaches to reduce the cooking time. The leaves are rich in phosphorus and are used for livestock feed (Ibrahim et al., 2018; Temegne et al., 2018).

In Malawi, BGN is a side dish with maize thick porridge (nsima). Farmers cultivate BGN for diet diversity, reliable yield, medicinal and ritualistic purposes. BGN is consumed during the ceremonies for the birth of a woman's first child and initiation rituals (Forsythe et al., 2015).

4.3 Bambara Groundnut-Based Food Products by Milling, Roasting and Steaming

The mature dried seeds pounded into flour are mixed with oil or butter to form a porridge or pulverized to make a soup. In Zambia, the flour is used for making bread while some parts of the plant for animal feed (Greenhalgh, 2000).

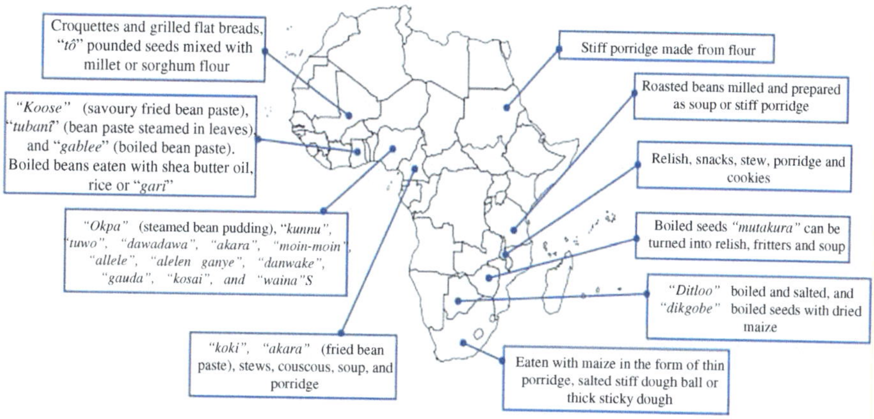

Fig. 4.2 Traditional local delicacies in different African countries prepared from Bambara groundnut (Tan et al., 2020)

In Ghana, the range of uses includes consumption as food and next year's crop, traditional performances-funeral rites, medicinal purposes, gifts for weddings or sale (Greenhalgh, 2000; Oppong-Sekyere, 2013). A traditional product "Tubani" made from BGN flour paste, wrapped in leaves and steamed, is consumed at funerals in northern Ghana (Greenhalgh, 2000; Oppong-Sekyere, 2013). Products made from dried BGN flour paste for various fried or steamed products, such as 'akara' or 'kose' [BGN paste fried in hot oil] and 'okpa' [BGN batter wrapped in banana leaves and steamed] are common in Nigeria (Greenhalgh, 2000; Oppong-Sekyere, 2013).

Among the Kambe and Giriama peoples of the Coast Province in Kenya, BGN is typically cooked or used when vegetables are in short supply. When the seeds are dry, they are pounded in a mortar to remove the seed coat (seeds do not break), winnowed and boiled until they are soft. They are then mashed using a serving spoon from the coconut shell, or a wooden spoon and added coconut juice. The mixture is then boiled until cooked, stirred with a wooden stirrer until smooth and served with rice or 'ugali' (a stiff maize meal porridge). When the seeds are still green, the seed coat is peeled off by hand, and the seeds are prepared in the same way as the dry ones (Ntundu, 1995).

In Cameroon the seeds are also ground into flour, sometimes after roasting, to make a porridge. The flour is added to cereals for the production of "baby lion" porridge. Bambara groundnut seeds are eaten raw, boiled, roasted; used for making cakes (koki, dough, pounded/crushed, etc.), stews, couscous, soups, porridge and as a snack. They are also added to corn flour to enrich traditional foods (Temegne et al., 2018).

The intensity of dry roasting differentiates between the production of coffee and flour. Dark roasting produces coffee. Bambara groundnut-based coffee is a substitute for tea in Zimbabwe (Mubaiwa et al., 2018a, b).

Dry seeds can either be pounded or ground and the resultant meal made into some stew (or sauce) by the Luhya tribe in Kenya. BGN is mainly cooked for consumption at ceremonies such as weddings or dignitaries. The leaves are for animal fodder (Ntundu, 1995). The stew/sauce is then added to the traditionally prepared leafy vegetables, particularly 'sikhubi'/'elikhubi' (cowpea) and cooked, served with 'ugali' or potatoes. The dry seeds are salted while roasting. Dried, ground using a grinding stone or pounded in a mortar, then cooked in the same way as green grams, in a sauce traditionally known as 'ogira' is practiced by the Luo in Kenya and served with other foods, such as potatoes. It can be roasted like the peanut, although not common because the seeds remain hard after roasting.

In northern Nigeria, most of the BGN produced is consumed locally. No industrial use of the crop has been reported. The haulm is used for livestock feed, while the fresh pods are boiled and eaten as a snack. Seeds are often crushed into flour, to prepare the following dishes: 'alele', 'alelen ganye', 'danwake', 'gauda', 'kosai', 'kunu', 'tuwo' and 'waina'. The fresh immature seeds are sometimes eaten raw (Brough et al., 1993; Ndidi et al., 2014; Linnemann, 1988; Linnemann & Azam-Ali, 1993; Oguntunde, 1985). "Okpa," a BGN flour-palm oil-water batter wrapped and steamed in banana leaves, is a favourite eastern Nigerian dish. Fresh BGN can be eaten by boiling the seeds (Hillocks et al., 2012). Dried seeds of BGN can be milled into flour, which can make flat cakes, biscuits, and porridge alone or mixed with other cereals. Bambara groundnuts are pounded and mixed with onions, tomatoes and oil for relish in Zimbabwe and Nigeria.

Despite its status as a marginal crop, BGN is esteemed by Togolese consumers and sells fast at local markets and those in urban centres, usually at a higher price than beans. The local producers consume the local crop produced (80%), and markets are generally undersupplied. Consumers prefer the ivory-cream coloured grains, of 6–8 mm diameter; small grains and black-coloured varieties are not particularly popular. Bambara groundnut is a complete food, which is more palatable when fresh, semi-dry, or not fully mature, as it becomes hardy when dry. It is consumed boiled or fried, and fried or boiled puddings may be prepared with a dough/paste made from BGN flour, which may also be used as a base for many other dishes (Ntundu, 1995).

People living in most villages of developing economies still have cultural knowledge and consume traditional foods and beverages. There is a concern of knowledge and information scarcity or a lack of the culture of traditional BGN dishes and beverages. The new generation's ignorance about traditional foods and beverages and the vast importation and preference for foreign foods and beverages are challenges that infringe on the utilization of BGN. Some BGN dishes are not produced like before, limiting their knowledge and culture for the new generation. Figure 4.2 summarises the traditional local delicacies prepared from BGN in sub-Saharan Africa as Tan et al. (2020) reported.

4.4 Network for the Reported Ethnofood Uses of Bambara Groundnut

The connectedness of the nodes indicates three communities (Fig. 4.3). The utilization of the boiled or dry and fresh seeds for a snack as practiced in Kenya, Zimbabwe and Southern Africa account for 46.2% of the network. The dry seeds of BGN when milled and utilized for preparing meals and porridge as practiced in Cameroon, Nigeria and Zambia account for 41.0% of the network. Soaking before boiling, boiling and mashing as utilized in stews as practiced in Ghana accounts for 12.8% of the network. The larger the node size, the more influential the node. Hence, the utilization of BGN is more in the dry form or milled into flour for meals and porridges.

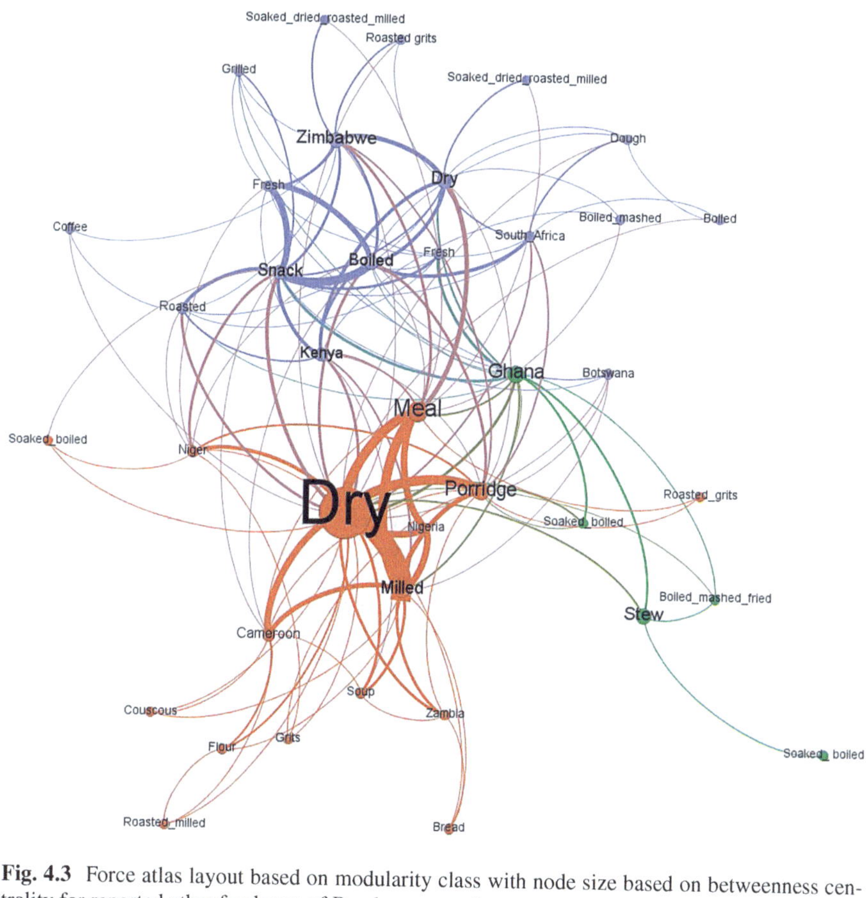

Fig. 4.3 Force atlas layout based on modularity class with node size based on betweenness centrality for reported ethnofood uses of Bambara groundnut

4.5 Ethnomedicinal Importance of Bambara Groundnut

Consumers now have increased awareness of the health benefits and potential disease-preventing and health-promoting compounds in foods. Hence, there is a corresponding market demand and outlets for health promotion and disease prevention of natural products, mostly traditional foods. Apart from BGN benefits, the roots, leaves and seeds are used in traditional medicine (Ademiluyi et al., 2015; Basu et al., 2007; Jideani & Diedericks, 2014; Maphosa & Jideani, 2016).

Bouchenak and Lamri-Senhadji (2013) reported that legumes play a role in cardiometabolic risk prevention. The world is experiencing an increasing cardiac arrhythmias burden, especially in Africa (Bonny et al., 2019). As Mirko Montuori (UN Food and Agriculture Organization) states in his Commentary, '…our health depends on plants while the health of plants depends on us.'

In Ghana's upper east region, the white landraces are mixed with guinea fowl meat and *dawadawa* (fermented locust bean seed), ground for the control of diarrhoea. The black BGN seeds are used in traditional medicine in several countries, including Namibia, Nigeria and Zimbabwe (Greenhalgh, 2000). The black landraces are ground and mixed with water and drank to treat a sick child. They are ground and applied to a child's skin to cure rashes or chewed to control swollen jaw diseases (Oppong-Sekyere, 2013).

The seeds and leaves of BGN are used in traditional medicine. In Senegal, leaf preparations are used to treat abscesses and infected wounds. The juice extracted from the leaf is applied to the eyes to treat epilepsy and the roots are sometimes used as an aphrodisiac. Crushed seeds, mixed with water, are administered to treat cataracts. The Ibo of Nigeria uses the plant to treat venereal diseases (Jideani, 2016; Oppong-Sekyere, 2013).

BGN is used to fight amoebic dysentery, sore throat, headaches, stomach pain, joint pain and bone decalcification in Cameroon. It helps in digestion through its laxative properties. It stimulates milk production in breastfeeding women and the healing of the mother after childbirth (Temegne et al., 2018).

In Malawi, a cup of BGN is given to newly married women to encourage reproduction. BGN is rubbed on the scrotum of infant boys to both ensure fertility and a successful BGN harvest. The link between men's testicles and the BGN is due to their similarity in appearance. On the contrary, BGN is thought to cause infertility or prevent promiscuity in adult men according to spiritual beliefs and, therefore, threaten adult male fertility. Consequently, there is the dual life/death meaning of the groundnut. Hence, young men are discouraged from eating BGN during the coming of age ceremonies because it is a symbolic reference to eating their testicles to protect their fertility. BGN is also associated with fertility and mating among animals, particularly pigeons, birds that mate for life. In these practices, two black BGN are used and mixed with other secret ingredients for the pigeons to drink or placed under the pigeon's house. BGN is also a key ingredient in love potions, which women use to limit men's promiscuity and their ability to reproduce with multiple women. The potions are believed to entice men to fall in love with a specific person and are commonly used by women on a man to prevent infidelity. Young men who visit young women's homes during courtship are discouraged from eating BGN to prevent them from becoming 'stuck' to that particular person. Traditional healers in Malawi administer various herbal remedies and treatments using BGN to prevent and treat various illnesses, diseases, and "spiritual attacks". The most common health and spiritual issues that traditional healers treat with BGN are sexually transmitted diseases, infertility, epilepsy, and spiritual attacks. The treatment involves mixing several secret herbs and plants with the BGN, and occasionally using animal parts, for the patient to consume, or to place in small straight line cuts made in the skin by the healer. Other uses include preventing children from having nightmares and contracting measles, treating abdominal pain, and as an abortifacient. Potions and ornaments with BGN are also used as a talisman to protect physical possessions, such as the home or assets such as livestock especially when threatened by witchcraft due to someone's jealousy (Forsythe et al., 2015).

In Kenya, the Luo tribe believes that drinking the water from the boiled maize and BGN mixture treats diarrhoea. Pounded BGN leaves with those of *Lantana trifolia* L. ('nyabend winyo', 'nyamrithi') mixed with water as a wash for prevention against livestock ticks. This solution is also used as a pesticide on vegetables. The leaves are combined with those of 'nyajagra' (Mexican marigold) and *L. trifolia*, pounded, and water added. This mixture can also be used as an insecticide. When applying the solution to vegetables, care is taken to apply it to the ground, and not pour it on the leaves to avoid scalding the leaves. When dry, the leaves are pounded with traditional salt ('mbala', harvested at Sindo and Homalime), and fed to cattle infected with 'tuoolao' (a type of mouth disease). The leaves treat and heal the animals' wounds (Ntundu, 1995).

In South Africa, chewing and swallowing raw BGN is alleged to check nausea and vomiting often to treat morning sickness in pregnant women (Ntundu, 1995).

4.6 Network for the Reported Ethnomedicinal Uses of Bambara Groundnut

The connectedness of the nodes for the ethnomedicinal benefits of BGN is indicated in Fig. 4.4. The network consists of 42 nodes and 59 edges. The edge thickness suggests the frequency of usage. Bambara groundnut (BGN) seeds are at the centre of the ethnomedicinal uses of BGN. Malawi, Senegal, Kenya and Cameroon utilize Bambara groundnut seeds for its therapeutic benefits more than others. Epilepsy for which Malawi and Senegal utilize BGN seeds and diarrhoea for which Kenya, Cameroon and Ghana utilize BGN seeds are the significant ailments BGN seeds resolved. Malawi and Nigeria reported using BGN seeds to heal venereal disease. Others include leaf preparation and the use of black seeds.

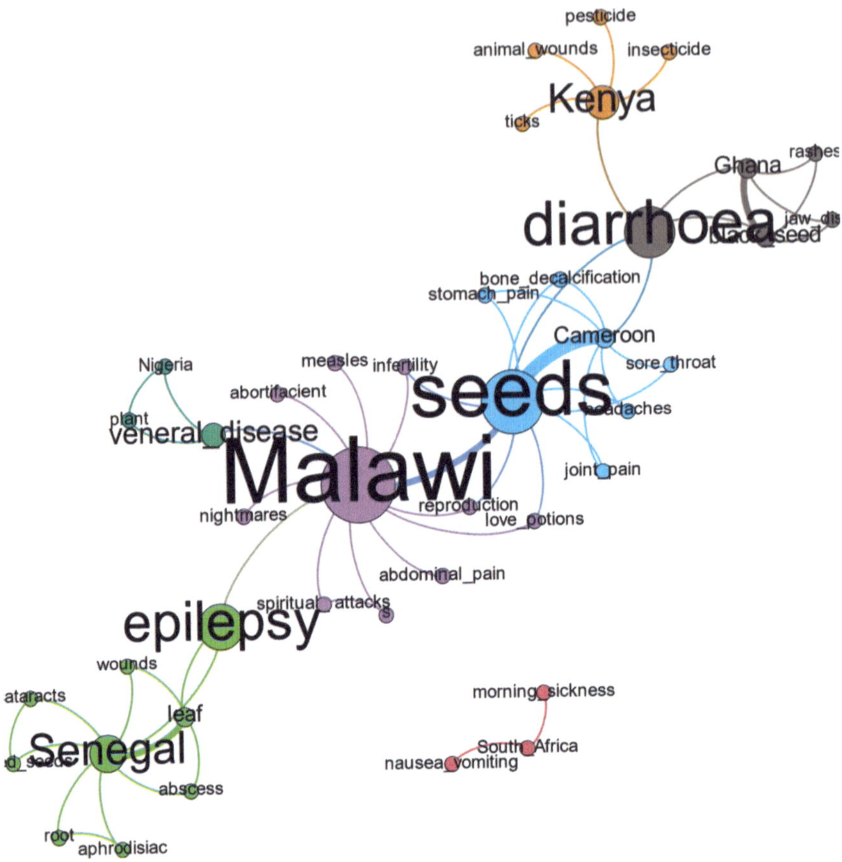

Fig. 4.4 Force atlas layout based on modularity class with node size based on betweenness centrality for reported ethnomedicinal uses of Bambara groundnut

4.7 Taboos Associated with Bambara Groundnut

Bambara groundnut has a symbolic significance of both life and death in Malawi. In northern Malawi production of BGN is restricted to older women who have lost a child. It is believed that BGN invites death into a household, children and men being particularly vulnerable. Consequently, women who have already experienced a child's death are immune from death believed to be associated with BGN. In some communities, men and youth can perform some agronomical activities, such as making ridges for BGN but they must not touch the seed (Forsythe et al., 2015).

In South Africa, when a chieftain dies, or in predicted drought, planting of the crop was prohibited for periods varying from one to five seasons. BGN might not be planted before the maize crop had germinated, and was pulled up if the rains were late. Males were forbidden to walk through 'njugo' fields, to prevent low yield.

Furthermore, land belonging to males was only planted with BGN every other year, for fear that the male landowners might otherwise die, or some evil comes to them or their families. Land belonging to women, however, was allowed to be planted annually with the crop.

Young people were not permitted to eat the fruit of unmounded BGN, as it was believed that permanent or temporary sterility would result if this rule was not observed. Planting of the crop in virgin soil was also forbidden. It is not known to what extent these traditional beliefs persist, but the Venda still considers that BGN should not be planted before the maize crop. Many of these beliefs may have a practical explanation. For example, the taboo against young people crossing BGN fields probably benefited the small, intensively cultivated plots. On the other hand, it seems likely that the taboo against eating unmounded BGN fruit served to ensure that women grew the crop properly (Ntundu, 1995).

References

Ademiluyi, A. O., Oboh, G., Boligon, A. A., & Athayde, M. L. L. (2015). Dietary supplementation with fermented legumes modulate hyperglycemia and acetylcholinesterase activities in Streptozotocin-induced diabetes. *Pathophysiology, 22*, 195–201.

Basu, S. M., Mayes, S., Davey, M., Roberts, J. A., Azam-Ali, S. N., Mithen, R., & Pasquet, R. S. (2007). Inheritance of 'domestication' traits in Bambara groundnut (*Vigna subterranea* (L.) Verdc.). *Euphytica, 157*, 59–68. https://doi.org/10.1007/s10681-007-9396-4

Bonny, A., Ngantcha, M., Scholtz, W., Chin, A., Nel, G., Anzouan-Kacou, J.-B., Karaye, K. M., Damasceno, A., & Crawford, T. C. (2019). Cardiac arrhythmias in Africa: Epidemiology, management challenges, and perspectives. *Journal of the American College of Cardiology, 73*(1), 100–109. https://doi.org/10.1016/j.jacc.2018.09.084

Bouchenak, M., & Lamri-Senhadji, M. (2013) Nutritional Quality of Legumes, Their Role in Cardiometabolic Risk Prevention. *Journal of Medicinal Food, 16*, 185–198. http://dx.doi.org/10.1089/jmf.2011.0238

Brough, S. H., Taylor, A. J., & Azam-Ali, S. N. (1993). The potential of Bambara groundnuts (*Vigna subterranea*) in vegetable milk production and basic protein functionality systems. *Food Chemistry, 47*, 277–283.

Bultosa, G., Molapisi, M., Tselaesele, N., Kobue-Lekalake, R., Haki, G. D., Makhabu, S., Sekwati-Monang, B., Seifu, E., & Nthoiwa, G. P. (2020). Plant-based traditional foods and beverages of Ramotswa village, Botswana. *Journal of Ethnic Foods, 7*, 1. https://doi.org/10.1186/s42779-019-0041-3

Forsythe, L., Nyamanda, M., Mwangwela, A. M., & Bennett, B. (2015). Beliefs, taboos and minor crop value chains: The case of Bambara groundnut in Malawi. *Food, Culture and Society, 18*, 501–517.

Greenhalgh, P. (2000). *The market potential for Bambara groundnut*. FRI/NRI/DFID Project on Marketing and processing of Bambara (West Africa). https://assets.publishing.service.gov.uk/media/57a08d57e5274a31e00017d8/R7581006.pdf, Visited 9-02-2021.

Hillocks, R. J., Bennett, C., & Mponda, O. M. (2012). Bambara nut: A review of utilisation, market potential and crop improvement. *African Crop Science Journal, 20*(1), 1–16.

Ibrahim, A. R., Dansi, A., Salifou, M., Ousmane, A., Alzouma, A., & Alou, W. (2018). Farmers' practices, utilization, conservation and marketing of Bambara groundnut (*Vigna subterranea* (L.) Verdc.) in Dosso Region, Western Niger. *Genetic Resources and Crop Evolution, 65*, 1907–1914.

Jideani, V. A. (2016). Utilizing Bambara groundnut in value-added products. *Food Technology, 70*, 48–52.

Jideani, V. A., & Diedericks, C. F. (2014). Nutritional, therapeutic, and prophylactic properties of *Vigna subterranea* and *Moringa oleifera*. In O. Oguntibeju (Ed.), *Antioxidant-antidiabetic agents and human health* (pp. 187–207). https://doi.org/10.5772/57338

Linnemann, A. R. (1988). Cultivation of Bambara groundnut in Northern Nigeria. *Tropical Crops and Communications, 15*, 1–14.

Linnemann, A. R., & Azam-Ali, S. (1993). Bambara groundnut (*Vigna subterranea* (L.) Verdc.). In J. T. Williams (Ed.), *Underutilized crops series 2. Vegetables and pulses* (pp. 13–58). Chapman & Hall.

Maphosa, Y., & Jideani, V. A. (2016). Physicochemical characteristics of Bambara groundnut dietary fibers extracted using wet milling. *South African Journal of Science, 112*, 72–79.

Misheck, C., Chagonda, I., & Makuvaro, V. (2013). Utilisation of common grain crops in Zimbabwe. *African Journal of Food Science, 7*(9), 253–257. https://doi.org/10.5897/AJFS2013.1011

Mubaiwa, J., Fogliano, V., Chidewe, C., Bakker, E. J., & Linnemann, A. R. (2018a). Utilization of Bambara groundnut (*Vigna subterranea* (L.) Verdc.) for sustainable food and nutrition security in semi-arid regions of Zimbabwe. *PLoS One, 13*, e0204817.

Mubaiwa, J., Fogliano, V., Chidewe, C., & Linnemann, A. R. (2018b). Bambara groundnut (*Vigna subterranea* (L.) Verdc.) flour: A functional ingredient to favour the use of an unexploited sustainable protein source. *PLoS One, 13*, 1–20.

Ndidi, U. S., Ndidi, C. U., Aimola, I. A., Bassa, O. Y., Mankilik, M., & Adamu, Z. (2014). Effects of processing (boiling and roasting) on the nutritional and antinutritional properties of Bambara groundnuts (*Vigna subterranea* [L.] Verdc.) from Southern Kaduna, Nigeria. *Journal of Food Processing*. https://doi.org/10.1155/2014/472129

Ntundu, W. H. (1995). Country report- Tanzania. In J. Heller, F. Begemann, & J. Mushonga (Eds.), *Conservation and improvement of Bambara groundnut (Vigna subtesrranea (L.) Verdc.)* (pp. 101–118). International Plant Genetic Resources Institute.

Oguntunde, A. O. (1985). Development of new food products from readily available raw materials. In *Proceedings of the Nigerian Institute of Food Science and Technology Training Workshop, Ibadan, Nigeria.*

Oppong-Sekyere, D. (2013). Assessing the level of cultivation and utilization of Bambara groundnut (*Vigna subterranea* (L.) Verdc.) in the sumbrungu community of Bolgatanga, Upper East Region, Ghana MM Akpalu 1, IA Atubilla. *International Journal of Plant, Animal and Environmental Sciences, 3*, 68–75.

Tan, X. L., Azam-Ali, S., Goh, E. V., Mustafa, M., Chai, H. H., Ho, W. K., Mayes, S., Mabhaudhi, T., Azam-Ali, S., & Massawe, F. (2020). Bambara groundnut: An underutilized leguminous crop for global food security and nutrition. *Frontiers in Nutrition, 7*, 601496. https://doi.org/10.3389/fnut.2020.601496

Temegne, N., Gouertoumbo, W. F., Wakem, G. A., Nkou, F. T. D., Youmbi, E., & Ntsomboh-Ntsefong, G. (2018). Origin and ecology of Bambara groundnut (*Vigna subterranea* (L.) Verdc): A review. *Journal of Ecology & Natural Resources, 2*, 1–10.

Chapter 5
Physical and Engineering Properties of Bambara Groundnut Seed

5.1 Introduction

Bambara groundnut [BGN] seeds like most legumes are essential in global human nutrition. Processes such as soaking, drying and decortication are necessary processing steps in converting BGN into value-added food products. The engineering characteristics of BGN seeds are the vital determining factor for the behavior of the seeds during rehydration and equipment design for handling, harvesting, processing and storage. Cleaning, grading, and separation equipment are designed based on grains engineering properties, which affects the conveying characteristics of solids by air or water and cooling (Baryeh, 2001). Some engineering properties such as physical (geometric, gravimetric), mechanical (angle of repose, coefficient of friction, internal friction), aerodynamic and hydrodynamic are directly related to equipment design (Krishnappa et al., 2017). The physical properties (geometry- dimensions, size, geometric mean diameter) and dry beans' mechanical behavior is vital for designing the grading, harvesting and grinding machines. Physical (gravimetric- bulk density, porosity, and pressure ratio) and mechanical (angle of repose, internal friction, coefficient of friction) properties affect the structural load (vertical and horizontal acting by-product on silo wall). They are essential in designing drying and aeration, storage systems (silos and warehouses) and transporting structures. The friction coefficient (static and dynamic) of the grains against various surfaces is necessary for designing conveying, grain flow (mass and hopper) and storage structures. The angle of internal friction is vital in developing hopper openings and wall pressures in storage structures (Kibar et al., 2014; Mpotokwane et al., 2008). Water absorption kinetics of grains are influenced by their intrinsic (physical and chemical) and extrinsic (temperature, soaking solutions, etc.) characteristics (Baryeh, 2001; Kaptso et al., 2008). The physical and engineering properties of BGN seed have received some scientific attention (Baryeh, 2001; Jideani & Mpotokwana, 2009; Ntundu, 1995; Ntundu et al., 2006; Mandizvo & Odindo, 2019; Mpotokwane et al., 2008). This chapter is a collection of the physical and mechanical properties of Bambara groundnut.

V. A. Jideani, A. I. O. Jideani, *Bambara groundnut: Utilization and Future Prospects*, https://doi.org/10.1007/978-3-030-76077-9_5

5.2 Bambara Groundnut Agro-ecology Plant Population and Trait

Bambara groundnut is a prostrate, herbaceous annual legume with a compact tap-root system bearing abundant geotropic lateral roots. Traits responsible for agro-morphology diversity among BGN landraces are phenological (days to emergence, days to first flowering, days to maturity), vegetative (number of leaves per plant, petiole length, terminal leaflet length, terminal leaflet width, internode length, number of stems per plant, plant height, canopy spread, growth habit) and yield types [number of pods per plant, pod length, seed width, shelling percentage, 100-seed weight] (Amadou et al., 2015; Bonny et al., 2019; Ntundu et al., 2006; Ouedraogo et al., 2008; Zenabou et al., 2014).

BGN vegetative growth habit is characterized by bunchy, semi-bunch and spreading growth habits (Abu & Buah, 2011; Koné et al., 2011; Ntundu et al., 2006) (Fig. 5.1). Bunchy growth habit is characterized by short stems and internodes, resulting in plants with tightly clustered leaves. The branches make an acute angle with the erect main shoot. Spreading or open growth habit types have stems with long internodes resulting in plants with a much larger plant foliar crown diameter. The branches are spreading while the main shoot may be erect or bent. The semi-bunchy growth habit is between the bunchy and spreading growth types. Both bunch and semi-bunch growth habits are more popular among farmers (Ntundu et al., 2006). The typical characteristics of these plants are short stems and internodes that produce plants with tightly clustered leaves.

During harvest, most pods remain attached to the stem crown after the plant is pulled up. The spreading growth habit is not very common because the stretching stems usually long internodes with a much larger diameter to the plant foliar crown, resulting in difficulties during harvest and increased yield losses. Many of the mature pods may remain in the ground. Spreading landraces have a more prolonged vegetative phase, take longer to mature, and low in yield due to the savannah zone's low soil moisture levels. Late flowering also has a detrimental effect on the seed yield of the BGN in the savannah zone. Local names of landraces are based on their maturity period, growth habit and seed coat color or their use. The fast-cooking, early varieties of BGN with large, cream-coloured seeds are preferred by local farmers (Abu & Buah, 2011). Landraces with high seed yield are early flowering and maturity. However, landraces with spreading growth habits could be used in intercropping as they would form a more rapid ground cover and help suppress weed growth. The bunch and semi-bunch types would be preferred in monocultures where optimum yield would be achieved at higher plant populations. The agro-morphological diversity of BGN in the Ivory Coast is structured according to geographical origin (Bonny et al., 2019).

Agronomic characteristics of 22 landraces from the Upper West Region of Ghana differ in morphology, physiology and agronomic traits such as days to maturity, 100-seed weight, shelling, canopy spread and grain. The essential character influencing grain yield and shelling is the 100-seed weight. The northern BGN

Fig. 5.1 Some qualitative variables of Bambara groundnut. (Khan et al., 2020)

population is characterized by dry climate conditions, more diversified and typical of the bunch type with lower phenological, vegetative and yield traits. The western, central and eastern populations are less diversified, located in wet climate conditions and the spreading type with higher phenological, vegetative and yield traits. Thus, rainfall and temperature are the most important climatic factors that differentiate BGN landraces.

Bambara groundnut pods develop underground and may reach up to 3.7 cm depending on the number of seeds they contain (Fig. 5.2). Most varieties have single-seeded pods, but three-seeded pods are frequent in Congo. Mature pods are indehiscent, often wrinkled, ranging from a yellowish to a reddish dark brown color

Fig. 5.2 Bambara groundnut plant and underground pods

or purple (Fig. 5.3a). Among 213 landraces of BGN, 102 pod types were identified with pointed and round pod shape the highest frequency (47%) (Fig. 5.3b). Four pod color types (Fig. 5.4) were observed. About 76% were yellowish, and only 4.7% were reddish-brown. In terms of pod texture majority (72%) had little grooved texture and <1% had a much-folded surface (Fig. 5.5). Absence and presence of an "eye" were about 59% for no eye and 41% for an eye present (Mohammed, 2014).

Ntundu (2006) reported the genetic diversity of BGN landraces from Tanzania based on morphological characters. Accessions differed significantly in phenological (days to 50% flowering), vegetative (peduncle length, number of leaves per plant, terminal leaflet width, terminal leaflet length, petiole length, plant spread, plant length), and yield (pod width, seed length, seed width, number of pods per plant, shelling percentage) traits. A considerable diversity for morphological and agronomic characteristics useful for germplasm management and utilization into crop improvement. Similar characters were reported for agro-morphological variability in 12 BGN accessions in Cameroon (Zenabou et al., 2014). Ouedraogo et al. (2008) reported characteristics of BGN accessions from Burkina Faso. Vegetative traits (canopy spread, number of pods per plant, number of seeds per pod, seed length, seed width and 100-seed weight) are positively correlated to yield per plant. A negative correlation exists between phenological trait (days to 50% flowering)

(a)

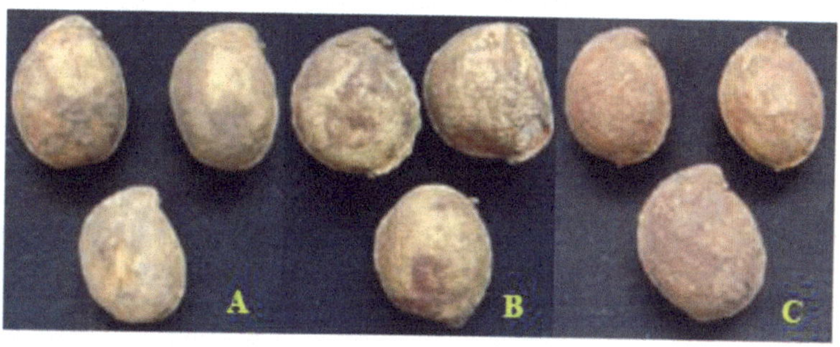

(b)

Fig. 5.3 Mature pods of Bambara groundnut [BGN] (**a**) Yellowish pod (top left), Dark purple pod (top right); Light purple pod (bottom left) and Green pod (bottom right); (**b**) BGN pod shapes (A) point and round; (B) point and nook and (C) point and point (Mohammed, 2014)

and yield per plant; indicating that the plants with more time at the vegetative stage have less yield in semi-arid conditions.

Agro-morphological variability in 14 BGN morphotypes cultivated in Niger differ in vegetative (plant height and plant diameter) and yield (pod weight per plant, seed weight per plant and number of pods per plant) traits. Grain yield per plant is positively related to the yield (pod weight per plant, number of pods per plant) and vegetative (number of leaves per plant, plant height and plant diameter) traits. The cream and red-coloured landraces have higher pod weight, seeds weight, the number of pods per plant and can be selected for high yield per plant compared to other landraces (Amadou et al., 2015).

BGN seed coat modulates the gas exchange with the environment and water absorption during the germination process and the supply of nutrients to the embryo during seed development (Mandizvo & Odindo, 2019). Seeds of landraces with a low hue degree have thick seed coats (Mandizvo & Odindo, 2019).

Fig. 5.4 Bambara groundnut dry pod colour: yellowish (top left); purple (top right); brown (bottom left) and reddish-brown (bottom right) (Mohammed, 2014)

Fig. 5.5 Bambara groundnut dry pod surface texture (**a**) smooth; (**b**) little grooves; (**c**) much grooves and (**d**) much folded (Mohammed, 2014)

5.3 Physical and Mechanical Properties of Bambara Groundnut Seed Landraces

The seed colour of BGN varies, from white to creamy, yellow, brown, purple, red or black (Fig. 5.6). Various testa patterns exist, including mottled, blotched or striped, and the predominantly uniformly coloured seeds (Ntundu, 1995). The author evaluated and characterized 1384 BGN accessions for 38 characters and identified nine

Fig. 5.6 Divers colours of Bambara groundnut seeds

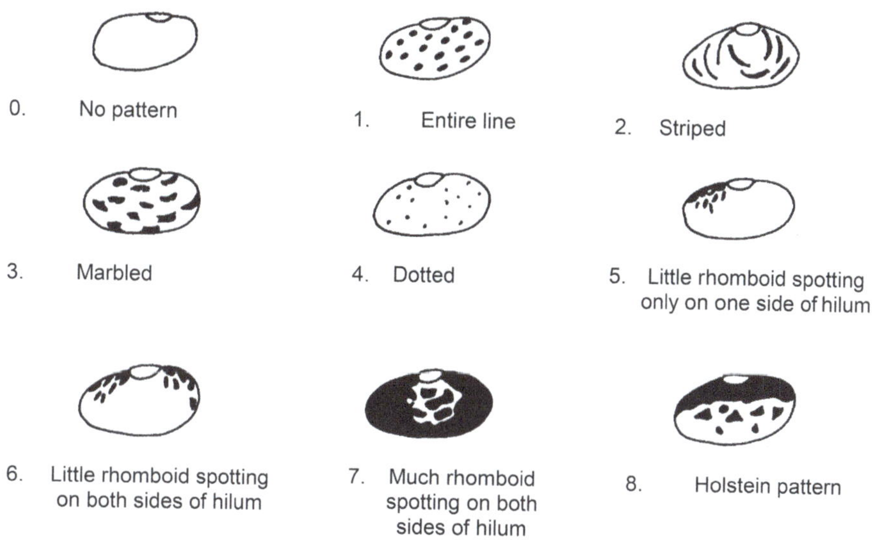

Fig. 5.7 Bambara groundnut seed testa patterns (Ntundu, 1995)

testa patterns as detailed in Fig. 5.7. The majority of the seeds have no testa pattern, followed by the marbled, dotted and rhomboid spotting on both sides of the hilum. Eight eye patterns were identified as outlined in Fig. 5.8. The majority of the accessions had no eye pattern followed by eye as a thin circle around hilum; eye as two thin lines on both sides of the hilum; eye as two thick lines on both sides of the hilum; eye with an almost triangular shape, large shapeless eye becoming frayed and eye like a butterfly.

The majority of the accessions did not have pigmentation around the seed eye, whereas 47% had. About 55% of the accessions had a creamy seed coat, while 86% had a dark brown-eye. Inverse document frequency of the terms used by Ntundu et al. (2006), Mpotokwane et al. (2008), Touré et al. (2012), Zenabou et al. (2014)

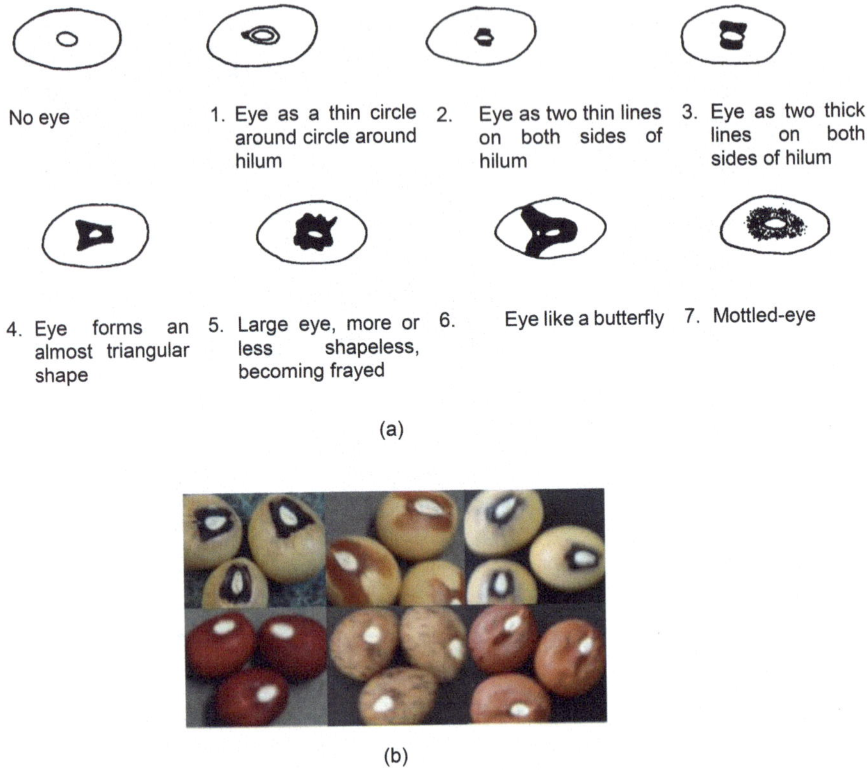

Fig. 5.8 Bambara groundnut (BGN) eye patterns (**a**) BGN seeds eye patterns (Ntundu, 1995); (**b**) Presence and absence of eye on the seeds, top row, eye present and bottom, eye absent (Mohammed, 2014)

and Bonny et al. (2019) to describe BGN testa and eye pattern are detailed in Fig. 5.9. The majority of the seeds have cream testa. The cream testa may have black rhomboid spots or mottled with eye or no eye colour. Other see colours are black, red and brown.

Mpotokwane et al. (2008) reported the physical properties of BGN seeds landraces from Botswana with mean moisture content (dry basis) of 7.17–8.99%. The landraces were Zimbabwe Red (National Tested Seed Red- NTSR), Botswana Red (BotR), Diphiri cream (DipC1 & DipC2) and speckled (AS17). The NTSR seeds are oval with a mean length of 11.8 mm and width of 9.8 mm with a white hilum but no peculiar eye characteristic (Fig. 5.10).

The seeds are uniform in colour with very little segregation into different colours, although different red colour shades appear. Botswana red [BotR] is light reddish-brown and is similar to NTSR, but the seeds are shorter in width with no peculiar eye characteristics. Diphiri cream (DipC) has the characteristics of uniform cream colour with eye as a thin circle around the hilum appearing as black-eye (Dip1) and brown-eye (DipC2). The speckled (AS17) is cream in colour with little rhomboid

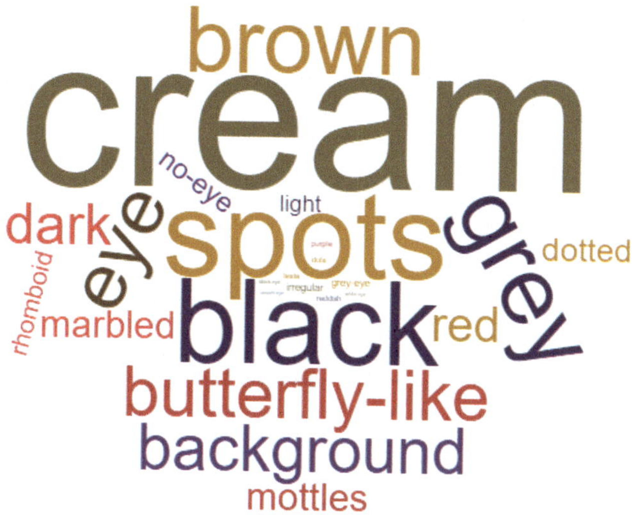

Fig. 5.9 Inverse document frequency of terms used to describe Bambara groundnut testa and eye patterns

Fig. 5.10 Grains of Vigna subterranean Rouge (Mohamadou et al., 2018)

spotting on both sides of the hilum and black-eye as a thin circle around the hilum (Ntundu, 1995). The linear dimension of the seeds varied in length (11.01–11.81 mm), width (9.41–9.96 mm) and thickness (9.05–9.48 mm). Consequently, BGN seeds can be graded as large (>10.5 mm), medium (9.50–10.49 mm) and small (<9.50 mm).

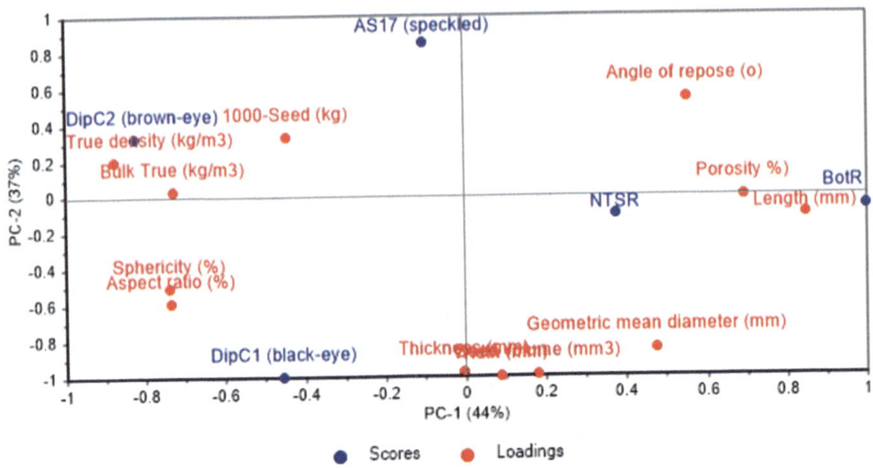

Fig. 5.11 Principal component scores and loadings for the physical properties of Bambara groundnut landraces from Botswana

Based on the data published by Mpotokwane et al. (2008), the principal component scores and loading bi-plot are detailed in Fig. 5.11. The NTSR and BotR are positively correlated to the seed frictional (angle of repose), gravimetric (porosity) and geometric (length and geometric mean diameter) properties. DipC1 (black-eye) is low in these properties but high in geometric [sphericity (88.3%) and aspect ratio (88.1%)] properties. DipC1 is more spherical and occupies more volume and area when compared to the other landraces. DipC2 (brown-eye) and AS17 are high in gravimetric (1000-seed mass and true density) properties. Bambara groundnut seeds are highly spherical (oval) with a high aspect ratio and would roll rather than slide. Small-sized seeds are higher in sphericity and aspect ratio than large seeds. Large seeds are higher in geometric mean diameters, volume and surface area than small seeds. The speckled seeds (AS17) were significantly heavier than the others followed by the DipC1 and DipC2. The seeds of BGN are heavier than water and therefore would not float in water. Not only does the seed not float, but the hulls also are thick and do not float, a challenge in the separation of the hulls from the cotyledons during processing.

5.4 Bambara Groundnut Seed Moisture Content and Water Absorption

The moisture content (dry basis) of Bambara groundnut reported by Mpotokwane et al. (2008) is in the range of 7.17–8.99%. Dry legumes require rehydration by soaking in water or salt solution before processing to gelatinize the starch either through conditioning below the gelatinization temperature and then cooking above the gelatinization temperature or through direct cooking above the gelatinization temperature (Jideani & Mpotokwana, 2009). Palatability, digestibility, and reduce cooking time are other reasons for soaking the seeds (Seyhan-Gürtaş et al., 2001). Hydration kinetics of five BGN seed landraces [NTSR (Red), BotR (Reddish brown), DipC1 (Black-eye), DipC2 (Brown-eye) and AS17 (cream with rhomboid spotting on both sides of the hilum)] from Bostwana, soaked in water at 25–100 °C for 0.5–24 h was reported by Jideani and Mpotokwana (2009). The water absorption rate increased, and water absorption capacity decreased with an increase in temperature. Changes during the hydration of the seeds are associated with exothermic and energetically favourable transformation. The red, reddish-brown, black-eye and brown-eye experienced an immense change and hydration more influenced by temperature. Simultaneously, the cream with rhomboid spotting on both sides of the hilum is more thermally stable, and hydration changes are less affected by temperature. An activation energy (E_a) of 37.83, 39.60, 33.02 and 33.33 kJ/mol was estimated for the red, reddish-brown, black-eye and brown-eye seeds, respectively, a lower value of 16.46 kJ/mol was estimated for AS17. Indicating that the rate of water absorption is faster for AS17 compared to the other varieties. The optimum soaking time for BGN seeds for all the varieties is 6 h at a temperature ≤75 °C.

Baryeh (2001) reported the physical properties of BGN seeds as affected by grain moisture content from 5% to 35% wet basis. The seed's linear dimensions (grain surface area, seed volume and coefficient of friction) increased non-linearly with the moisture content increase. The length, width, thickness and geometric diameter of the seeds increased with moisture up to 25% moisture content, beyond which there was no appreciable dimensional change. The 1000-seed weight increased linearly with an increase in moisture content. The bulk density and sphericity decrease non-linearly with the increase in moisture content, while the true density decreases linearly with moisture content. The porosity and repose angle increase with the increase in moisture content to a maximum and then decreased with further moisture content. The true density is higher than bulk density at all the grain moisture contents. The principal component scores and loading based on the Baryeh (2001) data on the effect of moisture content on the physical properties of BGN are in Fig. 5.12. Seed porosity, density and 1000-seed weight explain the variation in the moisture influence on BGN seeds.

Bambara groundnut seeds with 10–20% moisture content correlate to high porosity and seed density. Increased moisture content beyond this level does not appreciably affect the porosity and seed density. However, increased moisture content (28–35%) increases the 1000-seed weight. A low moisture content (5%) does not affect the physical properties of BGN.

The dark-coloured seed coats possess semi-permeability, greater epidermal thickness, higher concentrations of phenolic compounds and lignin giving them superior characteristics compared to the light seed coat (Kaptso et al., 2008; Mandizvo & Odindo, 2019). The semi-permeability to water confers more excellent resistance to deterioration in the field after maturity and slower hydration during the germination process. Consequently, the dark-coloured landraces result in less mechanical damage than traditionally favoured, light-coloured permeable seed coats (Mandizvo & Odindo, 2019). However, the authors reported that seed coat thickness and colour alone do not account for hydration pattern in BGN and the seed viability may not imply good seed vigour.

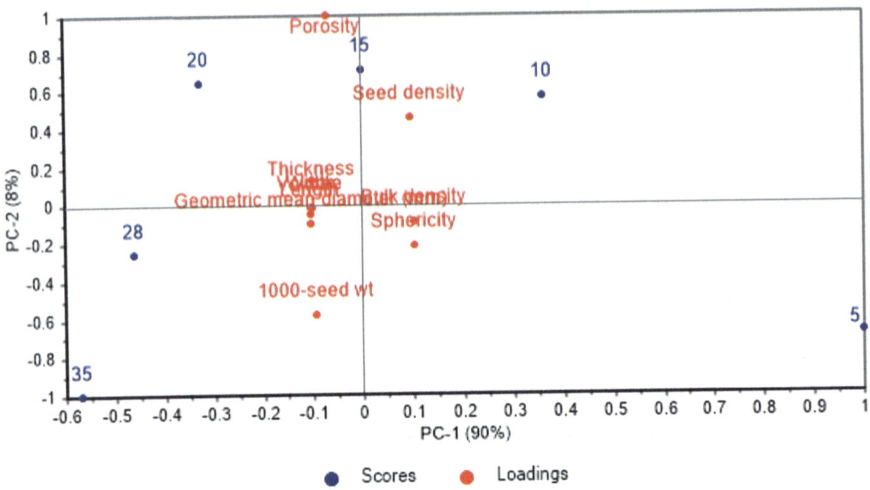

Fig. 5.12 Principal component scores and loadings for the physical properties of as a function of seed moisture content (5–35%)

The outer layer of dark coloured seed coats may consist of a gelatin-like substance that rapidly takes up free humid water. In contrast, light-coloured seed coats may have waxy material deposits that prevent water absorption (Mandizvo & Odindo, 2019). Small cuticular cracks on the seed coat's surface are responsible for the seed coat permeability to water. The dark coloured seeds are rich in phenolics and antioxidants than the light coloured seeds. Dark coloured seeds imbibe water rapidly compared to the light-colored seeds due to a high matrix potential [due to the tendency of proteins (phenolics) to attract water] (Mandizvo & Odindo, 2019).

Nyau (2019) studied the swelling capacity, swelling index, hydration capacity and hydration index of the two market classes from Zambia and reported that lower values of these properties suggest longer cooking time. The red and brown BGN being comparable in cooking times. However, the brown BGN is significantly higher in 100-seed mass than the red and will be a better choice in yield and market weight. This trend is in agreement with the 1000-seed weight reported by Mpotokwane et al. (2008).

5.5 Effect of γ-irradiated on Engineering Properties of Bambara Groundnut

The physical technique of food preservation, γ-Irradiation is reported to have the potential to protect food grains from insect infestation and microbial contamination during storage but may also have significant effects on the physicochemical and functional properties of macromolecules as well as improvement in nutritive value due to inactivation of antinutritional factors (Falade & Adebiyi, 2015; Falade &

Nwajei, 2015). Falade and Adebiyi (2015) reported the effect of γ-irradiation dose (2, 4, 6, 8 and 10 kGy) on cooking, functional and pasting properties of four BGN cultivars (TvSu 315, TvSu 571, TvSu 760 and white). The sphericity, volume, surface area and aspect ratio of BGN seed increased significantly with increased γ-irradiation dose and cooking time. The volume and surface area of seeds of TvSu 315 and white cultivars increased with increasing cooking time, while sphericity and aspect ratio showed a mixed effect. At lower doses, increased γ-irradiation resulted in decreased cooking time, elongation ratio and water uptake ratio of the seeds in all cultivars. However, at higher doses, cooking time increased, while the increase in γ-irradiation showed little effect on seed elongation and water uptake ratio in all cultivars. Generally, the water absorption capacity of flours increased significantly with increased γ-irradiation dose in all cultivars, while oil absorption capacity increased until the 4 kGy and then declined. Foaming capacity, foam stability and swelling capacity of flours did not show a dose-dependent increase in all cultivars. In contrast, Falade and Nwajei (2015) reported that the irradiated flours were higher than the non-irradiated. Peak, trough, final and setback viscosities of flours decreased significantly with increased γ-irradiation dose, while breakdown viscosity and pasting temperature did not show a dose-dependent increase. However, the irradiated flours had lower viscosities compared to the non-irradiated.

References

Abu, H., & Buah, S. (2011). Characterization of Bambara groundnut landraces and their evaluation by farmers in the Upper West Region of Ghana. *Journal of Developments in Sustainable Agriculture, 6*, 64–74.

Amadou, H. I., Doumma, A., Katsileros, A., Zoubeirou, A. M., & Nourou, S. S. M. (2015). Agromorphological variability in fourteen Bambara groundnut (*Vigna subterranea* (L.) Verdc.) morphotypes cultivated in Niger. *Scholars Academic Journal of Biosciences (SAJB), 3*, 774–781.

Baryeh, E. A. (2001). Physical properties of Bambara groundnuts. *Journal of Food Engineering, 47*, 321–326.

Bonny, B. S., Adjoumani, K., Seka, D., Koffi, K. G., Kouonon, L. C., Koffi, K. K., & Zoro Bi, I. A. (2019). Agromorphological divergence among four agro-ecological populations of Bambara groundnut (*Vigna subterranea* (L.) Verdc.) in Côte d'Ivoire. *Annals of Agricultural Science, 64*, 103–111.

Falade, K. O., & Adebiyi, A. O. (2015). Effect of γ-Irradiation on cooking, functional and pasting properties of Bambara groundnut (*Vigna subterranea* [L.] Verdc.) cultivars. *Journal of Food Process Engineering, 38*, 452–466.

Falade, K. O., & Nwajei, C. P. (2015). Physical, proximate, functional and pasting properties of four non- and γ-irradiated Bambara groundnut (*Vigna subterranea*) cultivars. *International Journal of Food Science and Technology, 50*, 640–651.

Jideani, V. A., & Mpotokwana, S. M. (2009). Modeling of water absorption of Botswana Bambara varieties using Peleg's equation. *Journal of Food Engineering, 92*, 182–188.

Kaptso, K. G., Njintang, Y. N., Komnek, A. E., Hounhouigan, J., Scher, J., & Mbofung, C. M. F. (2008). Physical properties and rehydration kinetics of two varieties of cowpea (*Vigna unguiculata*) and Bambara groundnuts (*Voandzeia subterranea*) seeds. *Journal of Food Engineering, 86*, 91–99.

Khan, M. H., Md. Rafii, M. Y., Ramlee, S. I., Jusoh, M., & Mamun, A. I. (2020). Genetic variability, heritability, and clustering pattern exploration of Bambara groundnut (*Vigna subterranea* L. Verdc) accessions for the perfection of yield and yield-related traits. *BioMed Research International*, Article ID 2195797, 31 pages. https://doi.org/10.1155/2020/2195797

Kibar, H., Öztürk, T., & Temizel, K. E. (2014). Propriedades efetivas de engenharia nas estruturas de armazenagem pós-colheita dos grãos secos de feijão. *Acta Scientiarum Agronomy, 36*, 147–158.

Koné, M., Paice, A. G., & Touré, Y. (2011). Bambara groundnut [*Vigna subterranea* (L.) Verdc: (Fabaceae)] usage in human health. In *Nuts and seeds in health and disease prevention* (pp. 189–196). https://doi.org/10.1016/B978-0-12-375688-6.10022-2

Krishnappa, M., Maddi, A., Matamari, V., Kulkarni, P., Desai, B., & Vasudevan, S. N. (2017). Engineering properties of selected groundnut (*Arachis hypogea* L.) varieties (*Arachis hypogea* L.) varieties. *International Journal of Agricultural Science and Research, 7*(4), 203–216.

Mandizvo, T., & Odindo, A. O. (2019). Seed coat structural and imbibitional characteristics of dark and light coloured Bambara groundnut (*Vigna subterranea* L.) landraces. *Heliyon, 5*, 1–21.

Mohamadou, M., Ngakou, A., & Tchuenguem F-N, F. (2018). Influence de l'interaction mycorhizes-insectes floricoles sur les rendements de *Vigna subterranea* variété rouge (*Fabaceae)* à Dang (Ngaoundéré, Cameroun). *International Journal of Biological and Chemical Sciences, 12*(4), 1897–1913. https://doi.org/10.4314/ijbcs.v12i4.30. Available online at http://www.ifgdg.org

Mohammed, M. S. (2014). *Pre-breeding of Bambara groundnut (Vigna subterranean [L.] Verdc.)*. Doctor of Philosophy (Ph.D.) in Plant Breeding thesis. University of KwaZulu-Natal, Pietermaritzburg Campus, South Africa. Accessed on 11 January 2021 at https://researchspace.ukzn.ac.za/bitstream/handle/10413/10989/Mohammed_Sagir_Mohammed_2014.pdf?sequence=1&isAllowed=y

Mpotokwane, S. M., Gaditlhatlhelwe, E., Sebaka, A., & Jideani, V. A. (2008). Physical properties of Bambara groundnuts from Botswana. *Journal of Food Engineering, 89*, 93–98.

Ntundu, W. H. (1995). Country report- Tanzania. In J. Heller, F. Begemann, & J. Mushonga (Eds.), *Conservation and improvement of Bambara groundnut (Vigna subtesrranea (L.) Verdc.)* (pp. 101–118). International Plant Genetic Resources Institute.

Ntundu, W. H., Shillah, S. A., Marandu, W. Y. F., & Christiansen, J. L. (2006). Morphological diversity of Bambara groundnut [*Vigna subterranea* (L.) Verdc.] landraces in Tanzania. *Genetic Resources and Crop Evolution, 53*, 367–378.

Nyau, V., Mwelwa-Zgambo, L., Chirwa-Moonga, T., Nthani, D., Prakash, S., Rodrigues, J., & Farrant, J. (2019). Some nutritional and physical properties of different Zambian market classes of Bambara groundnut (*Vigna subterranea*). *Journal of Food Research, 9*, 34.

Ouedraogo, M., Ouedraogo, J., Tignere, J., Bilma, D., Dabire, C., & Konate, G. (2008). Characterization and evaluation of accessions of Bambara groundnut (*Vigna subterranea* (L.) Verdcourt) from Burkina Faso. *Sciences & Nature, 5*, 191–197.

Seyhan-Gürtaş, F., Ak, M. M., & Evranuz, E. Ö. (2001). Water diffusion coefficients of selected legumes grown in Turkey as affected by temperature and variety. *Turkish Journal of Agriculture and Forestry, 25*, 297–304.

Touré, Y., Koné, M., Kouakou Tanoh, H., & Koné, D. (2012). Agromorphological and phenological variabilities of 10 Bambara groundnut [*Vigna subterranea* (L.) Verdc. (Fabaceae)] landraces cultivated in Ivory Coast. *Tropicultura, 30*, 216–221.

Zenabou, N., Martin, B. J., Ernest, F. P., Bassiaka, O., Claude, S., & Didier, D. (2014). Agromorphological variability in twelve Bambara groundnut (*Vigna subterranea* (L.) Verdc.) accessions in Cameroon. *Sciences, Technologies et Developpement, 16*, 38–45.

Chapter 6
Nutritional Composition of Bambara Groundnut

6.1 Introduction

Nutritionally, Bambara groundnut (BGN) is superior to other legumes as well as other food crops (Table 6.1) and is a preferred food source by many individuals (Mohamadou et al., 2018; Teye et al., 2020). Koné et al. (2011) stated that BGN produces exceptional nutritional quality. Thus, it plays a vital role in building and maintaining a solid foundation for good health with the seeds having the right nutrient balance and are a rare example of complete plant food. BGN accessions are significantly different from one another in physicochemical composition, namely protein, fat, carbohydrate, mineral, sugar, amino acids, etc. (Atoyebi et al., 2017; Poulter, 1981; Purseglove, 1992). The variation is most likely to have an impact on the functional properties of the flour. The variation in available data on the chemical composition (Table 6.1) and nutritional properties of both commonly used (commercial) and newly developed varieties is due to the variety of genetic origin, climatic conditions, soil, pesticides, and fertilizers employed in the production of the crop (Adeleke et al., 2018) as well as the method of analysis used in quantification.

Some selected BGN accessions revealed nutrition potentials in some lines (Atoyebi et al., 2015, 2017; Amarteifio & Moholo, 1998). BGN seeds contain 16–21% proteins, 4.5–6.5% fats, and 50–60% carbohydrates (Purseglove, 1992; Brough & Azam-Ali, 1992; Brough et al., 1993; Mashau et al., 2020). Atoyebi et al. (2017) investigated selected accessions of BGN from the International Institute of Tropical Agriculture, Nigeria for moisture, carbohydrate, protein, crude fiber, total sugar, fat, starch, ash, minerals, free and essential amino acids and observed (i) high, medium and below medium nutrient accessions, and (ii) protein values with the range 15.88 ± 0.05–24.91 ± 0.01 g/100 g; carbohydrate content from 42.77 ± 0.01 to 62.76 ± 0.23 g / 100 g (Table 6.2).

Ojimelukwe (1999) reported the proximate composition of four varieties of Bambara groundnut (BGN) with white, elliptical hilum and yellow, round, wrinkled

V. A. Jideani, A. I. O. Jideani, *Bambara groundnut: Utilization and Future Prospects*, https://doi.org/10.1007/978-3-030-76077-9_6

Table 6.1 Composition of Bambara groundnut (BGN) in comparison with other staple crops

Crop	Protein (%)	Energy Kcal	Calcium mg/100 g fw	Iron mg/100 g fw	Vit. A (mg)	Thiamine (mg)	Riboflavin (mg)	Niacin (mg)	Vit. C (mg)
BGN flour[a]	15–23	413–417	30.2 ± 1.6	8.8 ± 0.6					
Sorghum flour	9.5	353[a]	28	10.0	20	28	0.9	3.4	0
Cassava (fresh)	1.2	128	68	1.0	45	0.04	0.05	0.6	31
Yam (fresh)	1.9	111	52	0.8	25	0.11	0.02	0.3	6
Plantain (raw)	1.2	128	8	1.3	170	0.08	0.04	0.6	20

Adapted from FAO and ICRISAT (1996). The World Sorghum and Millet Economies: Facts, Trends and Outlook. http://www.fao.org/docrep/W1808E/w1808e00.htm on 11/10/2012
[a]From other sources

pods (1) red, smooth near round seeds (2) black smooth round seeds (3) cream smooth round seeds and (4) russet-brown seeds some of which are round while others are cylindrical. BGN seeds are characterized by 18–21% protein, 6–8% fat, 8–10% moisture, 1.2–2.6% ash and 54–60% carbohydrates (Fig. 6.1). The varieties were similar in proximate composition except in the carbohydrate content for which the cream coloured cultivar (60%) was significantly higher than that of the black coloured seed cultivar (54%). The tannin contents of the red and black coloured seed cultivars were substantially higher than that of the cream and brown seed cultivars.

6.2 Bambara Groundnut Carbohydrate

The BGN flour carbohydrate content of 63.07% (Ndidi et al., 2014) is significantly low compared to maize, wheat, millet, and cassava flours (Ikegwu et al., 2020). The low carbohydrate content is expected for BGN being a legume (Table 6.2). Accessions with high carbohydrate content make BGN consumption a ready supply of glucose to the body systems, making it efficient in the protection against ketone and keto-acidosis conditions (Afolabi, 2012; Atoyebi et al., 2017; Ashogbon, 2014; Bernhoft, 2010; Duthie & Crozier, 2000; Huang et al., 2015). Further details of the BGN starch (Sirivongpaisal, 2008) are presented in Chap. 8 – Bambara groundnut starch.

Table 6.2 Nutritional composition of Bambara groundnut (BGN), flour, modified starch, formulated diet

Properties (%)	Temegne et al. (2018) Seed	Igbabul et al. (2013) Seed	Atoyebi et al. (2017, 2018) Seed	Adeleke et al. (2018) Seed	Ndidi et al. (2014) Seed	Ikegwu et al. (2020) Flour	Afolabi et al. (2017) Flour	Kelechi-Ebisike et al. (2020) Diet		Afolabi et al. (2017) Native starch
								A	B	
Moisture	–	10.01%	–	7-19 -11-87	6.54 ± 0.28	5.04 ± 0.13	9.15 ± 0.01	2.45 ± 0.15	1.14 ± 0.12	14.11 ± 0.25
Ash	–	3.49%	2.80–4.50	4.18–5.15	3.05 ± 0.08	4.04 ± 0.12	2.92 ± 0.096	2.58 ± 0.59	2.43 ± 0.31	0.21 ± 0.15b
Protein	19%	16.44%	15.88–24.91	19.50–20.73	19.07 ± 0.45	23.54 ± 0.13	16.88 ± 0.01	15.84 ± 0.59	16.72 ± 0.67	1.77 ± 0.00b
Fat	6.5%	5.75%	4.35–7.68	6.01 – 8-73	6.05 ± 0.23	5.50 ± 0.08	6.98 ± 0.04	7.05 ± 0.08	8.41 ± 0.21	2.59 ± 0.01b
Carbohydrate	63%	62.87%	42.77–62.76	52.73 60.48	63.07 ± 1.04	57.80 ± 0.06	55.66	71.67 ± 0.28	68.80 ± 0.48	79.11
Crude fiber	–	1.45%	2.50–5.50	2.11–3.64	4.67 ± 0.13	4.10 ± 0.01	6.41 ± 0.01	2.95 ± 0.2	3.47 ± 0.15	2.21 ± 0.01b
Energy value (kcal/100 g)	–	–	–	412.0–608.0	163.68 ± 2.71	–	–	413.46 ± 1.84	417.81 ± 1.69	–
Total sugar	–	–	2.56–7.01	–	–	–	–	–	–	–
Amylose	–	–	–	–	–	–	6.10 ± 0.04	–	–	27.70 ± 0.05a
Amylopectin	–	–	–	–	–	–	93.9	–	–	72.3

In most cases, values are expressed as mean ($n = 3$) ± standard deviation. The significant differences in the proximate composition could be attributed to the species/variety, soil type, environmental condition as well as biotechnology retooling (Ikegwu et al., 2020). Diet A = Millet:Sweet Potatoes:Groundnut:BGN (35:25:30:10% w/w), Diet B – Millet:Tiger nut:Groundnut:BGN (35:25:30:10% w/w)

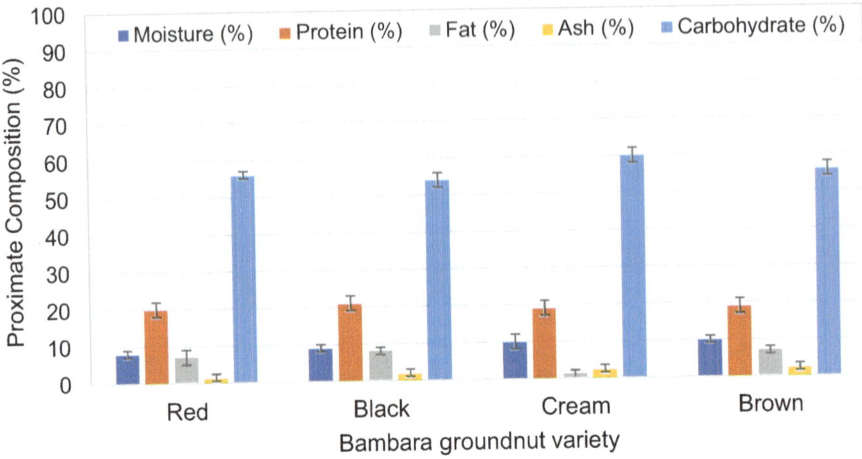

Fig. 6.1 Proximate composition of Bambara groundnut varieties (Adapted from Ojimelukwe, 1999)

6.3 Bambara Groundnut Ash Content and Quality

Bambara groundnut (BGN) flour is high in ash ranging from 3 to 5% (Table 6.2) compared to cereal grains. The trend could be associated with high mineral constituents in the pericarp and aleurone layers of BGN. High ash content is disadvantageous in food processing because of ash in food product's colour stability. In pasta production, high ash contents of flour are discouraged as it reduces food appearance stability over time. The ash content contributes significantly to the quality of flour products (Ikegwu et al., 2020; Chen & Yang, 2010). Hence, flour mills supply low ash flour to noodle manufacturers for colour stable, bright, and white noodles during storage. The high ash content of BGN flour may cause changes in colour with time for products like noodles; hence, composite flour formulation with BGN flour may have poor color stability in storage. Ikegwu et al. (2020) explain flour ash content association to less brightness and more visible bran specks in flour products. Still, with a bleaching agent (colourant) in noodle production in some countries, the inclusion of BGN may be necessary due to high protein and crude fat contents.

Phosphorus is the primary mineral in BGN, followed by Mg and Ca; their levels (335.8 ± 5.9, 136.0 ± 2.0 and 30.2 ± 1.6 mg/100 g fw, respectively) are generally consistent with those of other varieties grown in Southern Africa when expressed on a dry basis. The Ci12 landrace of BGN also contains trace elements like Fe, Cu and Zn (8.8 ± 0.6, 0.5 ± 0.0 and 1.9 ± 0.1 mg/100 g fw, respectively) (Yao et al., 2015).

6.4 Bambara Groundnut Fat Content and Quality

The fat content of BGN ranges from 4.4 to 8.4% (Table 6.2). Fatty acid components of BGN fat include the saturated acids [palmitic (16:0) and stearic (18:0)] and polyunsaturated components [linoleic (18:2) and linolenic (18:3)]. Linoleic acid (18:2) is the most abundant accounting for 44% of the fatty acid, palmitic acid (16:2) is second with an average of about 30% and linolenic acid (18:3) is in the third about 21%. Stearic acid (18:0) is at a low level of about 5% (Minka & Bruneteau, 2000). This fatty acid composition is like that of lipids isolated from pigeon pea, chickpea, garden pea, cowpea and black-eyed pea by other researchers (Minka & Bruneteau, 2000). However, (Yao et al., 2015) reported the presence of oleic acid (18:1 n-9) [23%] in addition to palmitic (21%) and linoleic acid (36%). Adeleke et al. (2018) observed that the monounsaturated fatty acid (oleic) was the most abundant fatty acid in BGN constituting more than half of the total fatty acids. Linoleic and oleic acids were present in generous quantities in BGN. Other fatty acids include linoleic palmitic, and linolenic (Enwere & Hung, 1996; Obizoba, 1990; Okonkwo & Opara, 2011; Minka & Bruneteau, 2000).

6.5 Bambara Groundnut Fibre Content and Quality

The high fiber content [1.5–6.4%] (Table 6.2) of BGN is an added advantage in developing functional foods. Although dietary fiber helps regulate bowel movement, softens faeces, and increases faecal bulk, preventing constipation, when in excess, it may decrease appetite in infants and young children (Abeshu et al., 2016). Further details of BGN fiber is in Chap. 7 – Physicochemical and functional properties of BGN dietary fibers.

6.6 Antinutrients and Bioactives in Bambara Groundnut and Reduction from Processing

Antinutrients or antinutritional factors are compounds or substances that reduce nutrient intake, digestion, absorption and utilization and may produce other adverse effects (Aberoumand, 2012). Antinutrients found in some plant foods include oxalate, phytate, alkaloid, saponin and tannin. These compounds create an aversion to the presence of gluten and other allergenic ingredients in some plant foods. Generally, plants rich in polyphenols tend to have protective effects on human health. Trypsin inhibitor and phytic acid may play a role as an antioxidant in plants and benefit health when consumed by humans. However, being an antinutritional factor is not an intrinsic characteristic of a compound but depends upon the digestive process of the ingesting animal. Trypsin inhibitors, which are antinutritional factors for monogastric animals, do not negatively affect ruminants because they are degraded in the rumen (Aberoumand & Deokule, 2010).

Although BGN contains beneficial bioactive compounds(Mashau et al., 2020; Nyau et al., 2015, 2017), it also has some compounds that can be regarded as acute poisons and as contaminants having some risks to animals and humans (Atoyebi et al., 2018). However, it is the dose of these antinutrients that makes the poison as they affect the bioavailability of important food nutrients. For instance, a high intake of phytate will inhibit zinc absorption by chelating this and other micro-minerals in the gastrointestinal tract (Ross et al., 2012). Tannin forms complexes with proteins and hence limits the availability of proteins (Eka, 1985).

Despite the high nutritional value of BGN, antinutritional factors include tannins and trypsin inhibitors (Atoyebi et al., 2018; Bamishaiye et al., 2011). Atoyebi et al. (2018) quantified certain antinutrients in BGN. They observed that selected African accessions (i) had varying, but a high amount of cyanogenic glycosides in the form of hydrocyanic acid (0.34 mg/100 g) and trypsin inhibitors (18.97 mg/g), (ii) oxalic acid was not detectable (negligible), (iii) have observable values for cyanogenic glycosides, trypsin Inhibitors, oxalic, tannic and phytic acids, and (iv) apart from trypsin inhibitors, all the other antinutrients (cyanogenic glycosides, tannin, phytate and oxalate) do not show any potential risks. The authors recommended cream-coloured BGN for human nutrition and consumption due to a lower amount of these antinutrients.

Ndidi et al. (2014), working on the effect of processing (boiling and roasting) observed (i) a significant difference ($p < 0.05$) between the levels of crude lipid, crude fiber, gross energy, carbohydrate, and moisture content in the raw and processed BGN, (ii) no significant difference in protein content of processed BGN and the raw seeds, and (iii) magnesium, potassium, and phosphorus were the most abundant macrominerals in BGN with zinc being the most abundant. Some nutritional studies on BGN carried out including (i) the *in vitro* digestibility, amino acid profile and antioxidant activity of cooked BGN grain by Oyeyinka et al., 2019), and (ii) the blending of precooked sorghum with roasted BGN flours by Leopold et al. (2020) before fermentation yielding a good alternative instant flour having decreased total polyphenols, tannins and phytates contents and can be used for young children nutrition.

Studies have shown that different simple processing techniques such as soaking, cooking, germination, frying, blanching and fermentation can reduce or remove antinutritional factors (Barimalaa & Anoghalu, 1997; Ndidi et al., 2014; Odetokun, 2000; Martınez et al., 2012). Boiling, roasting and fermentation reduce the antinutritional factors in BGN (Barimalaa & Anoghalu, 1997, Ndidi et al., 2014,). The effect of processing (boiling and roasting) on the proximate, antinutrient and mineral composition of BGN seeds reported by Ndidi et al. (2014) showed (i) significant reduction of oxalate, tannins, phytate, trypsin inhibitor, and hydrogen cyanide contents (ii) that levels of crude protein, ash, moisture, and lipid were negatively affected by phytate, tannins, and oxalate and (iii) that processing significantly lowered the levels of antinutrients in BGN making it safer for consumption. The levels of oxalate, phytate, hydrogen cyanide and trypsin inhibitor activities (TIA) (which

Table 6.3 Antinutritional composition (mg/100 g) of different samples of Bambara groundnut (BGN) and products*

Sample	Oxalate	Tannin	Phytate	Trypsin	Trypsin inhibitors activities (TIA)	Hydrogen cyanide (HCN)
Fermented blend (Ijarotimi & Keshinro, 2013)						
Popcorn-African locust bean (FPA)	$1.925^{ab} \pm 0.023$	$0.026^{a} \pm 0.014$	$19.250^{ab} \pm 2.121$	$0.057^{ab} \pm 0.022$	–	–
Popcorn-BGN (FPB)	$0.018^{b} \pm 0.002$	$0.014^{a} \pm 0.001$	$18.01^{b} \pm 2.01$	$0.018^{b} \pm 0.006$	–	–
Popcorn-African locust-BGN (FPAB)	$0.015^{c} \pm 0.008$	$0.015^{a} \pm 0.005$	$15.212^{b} \pm 0.25$	$0.018^{b} \pm 0.035$	–	–
Maize–BGN complementary foods (Uvere et al., 2010)						
Untreated maize-BGN blends (Bu Mbu)	$0.08^{d} \pm 0.01$	$0.4^{e} \pm 0.03$	$0.22^{e} \pm 0.01$	–	–	–
Maize–wet dehulled BGN (MB MB)	$0.03^{b} \pm 0.00$	$0.3^{d} \pm 0.01$	$0.18^{d} \pm 0.01$	–	–	–
Maize–BGN (MBm MBm)	$0.03^{b} \pm 0.00$	$0.2^{b} \pm 0.02$	$0.14^{b} \pm 0.00$	–	–	–
Maize–BGN fermented by backslopping [(MB)f(MB)f]	$0.04^{c} \pm 0.01$	$0.23^{c} \pm 0.01$	$0.16^{c} \pm 0.02$	–	–	–
Maize–BGN fermented by backslopping [(MBm)f(MBm)f]	$0.02^{a} \pm 0.00$	$0.15^{a} \pm 0.01$	$0.10^{a} \pm 0.01$	–	–	–
Raw and processed BGN (Ndidi et al., 2014)						
Raw	0.62 ± 0.07^{c}	1861.0 ± 2^{b}	319.55 ± 1.0^{c}	–	7.11 ± 1.41^{c}	19.70 ± 1.41^{c}
Boiled	0.20 ± 1.00^{a}	533.0 ± 3^{a}	113.71 ± 0.8^{a}	–	1.05 ± 0.41^{a}	3.87 ± 0.64^{a}
Roasted	0.30 ± 0.04^{b}	533.0 ± 2^{a}	160.08 ± 1.0^{b}	–	1.89 ± 0.96^{b}	4.29 ± 0.49^{b}
Permissible limits	0.5	2000	250	–	0.7–3.0	5.0
BGN accessions (Atoyebi et al., 2017, 2018)						
Seed (g/100 g)	0.00 ± 0.00	2.00 ± 0.01	2.08 ± 0.02		14.39 ± 0.04	0.25 ± 0.02

*Results are the means of three replications. Values carrying different superscripts in the same column, within each authors' data, are significantly different ($P < 0.05$)

were above the permissible limits in the raw BGN) reduced significantly ($P < 0.05$) to acceptable levels by both boiling and roasting with boiling causing more significant reduction than roasting. Both processing methods, however, had the same significant ($P < 0.05$) reducing effect on the level of tannin (Table 6.3).

The ethnobotanical advantages of BGN need pharmacological studies for confirmation. There is a scanty of detailed and published biochemical studies on the toxicity and potential adverse effects associated with both food and medicinal usage of BGN. The content of antinutritional factors in BGN poses no significant problems. Like most other crops, traces of antinutrients can be removed or inactivated during food processing operations. Moreover, some antinutritional factors are phytochemical constituents with increasing evidence of their potential health benefits in diet-related non-communicable diseases (Duodu & Apea-Bah, 2017; Pale et al., 1997; Taylor & Awika, 2017).

Luzardo-Ocampo et al. (2018, 2017) reported gallic acid and ferulic acid present in common beans as important phenolics present in all steps after a gastrointestinal and colonic fermentation of products made of corn and common beans; and the same type of products has also shown anti-inflammatory potential, modulating inflammatory citokines. Apoptosis is a genetically programmed cell death mechanism that can be activated by various stimuli. The morphological features of apoptosis include chromatin condensation, cell shrinkage and nuclear fragmentation due to the accumulation or decrease of proteins related to this process. Different target proteins of the apoptotic process, cellular cycle and viability are explored to indicate the effectiveness of extracts. The mechanism of action of extracts will show differences in the modulation of proteins related to the cell proliferation and the apoptotic mechanism (Moreno-Jiménez et al., 2018). Moreno-Jiménez et al. (2018) concluded that common beans (*Phaseolus vulgaris* L.) processed by canning, contain phenolic compounds with the ability to inhibit the proliferation of colorectal cancer cells through the activation of signaling pathways, inducing apoptotic processes.

References

Aberoumand, A. (2012). Screening of phytochemical compounds and toxic proteinaceous protease inhibitor in some lesser-known food-based plants and their effects and potential applications in food. *International Journal of Food Science and Nutrition Engineering, 2*(3), 16–20. https://doi.org/10.5923/j.food.20120203.01. ISSN: 2166-5168 e-ISSN: 2166-5192.

Aberoumand, A., & Deokule, S. S. (2010). Elements evaluation of some edible vegetables and fruits of Iran and India. *Asian Journal of Agricultural Sciences, 2*(1), 35–37.

Abeshu, M. A., Lelisa, A., & Geleta, B. (2016). Complementary feeding: Review of recommendations, feeding practices, and adequacy of homemade complementary food preparations in developing countries – Lessons from Ethiopia. *Frontiers in Nutrition, 3*, 41. https://doi.org/10.3389/fnut.2016.00041.

Adeleke, O. R., Adiamo, O. Q., & Fawale, O. S. (2018). Nutritional, physicochemical, and functional properties of protein concentrate and isolate of newly-developed Bambara groundnut (*Vigna subterrenea* L.) cultivars. *Food Science & Nutrition, 6*, 229–242. https://doi.org/10.1002/fsn3.552.

Afolabi, T. A. (2012). Synthesis and physicochemical properties of carboxymethylated Bambara groundnut *(Voandzeia subterranea)* starch. *International Journal of Food Science and Technology, 47*, 445–451. https://doi.org/10.1111/j.1365-2621.2011.02860.x.

Afolabi, T. A., Opara, A. O., Kareem, S. O., & Oladoyinbo, F. O. (2017). In vitro digestibility of hydrothermally modified Bambara groundnut (*Vigna subterranean* L.) starch and flour. *Food Science & Nutrition, 6*, 36–46. https://doi.org/10.1002/fsn3.510.

Amarteifio, J. O., & Moholo, D. (1998). The chemical composition of four legumes consumed in Botswana. *Journal of Food Composition and Analysis, 11*(4), 329–333.

Ashogbon, A. O. (2014). Physicochemical properties of Bambara groundnut starch and cassava starch blends. *African Journal of Food Science, 8*(6), 322–329.

Atoyebi, O. J., Abberton, M., Adebawo, O.O., Osilesi, O., & Oyatomi, O. (2015). *Preliminary studies on variations in nutrient content in Bambara groundnut (Vigna subterranea) L. Verdc.* Presentation in Nov. at the annual R4D week, IITA, Ibadan, Nigeria.

Atoyebi, O. J., Osilesi, O., Adebawo, O., & Abberton, M. (2017). Evaluation of nutrient parameters of selected African accessions of Bambara groundnut (*Vigna subterranea* (L.) Verdc.). *American Journal of Food and Nutrition (SCIEP Series), 5*(3), 83–89. https://doi.org/10.12691/ajfn-5-3-1.

Atoyebi, J. O., Osilesi, O., Abberton, M., Adebawo, O., & Oyatomi, O. (2018). Quantification of selected antinutrients and bioactive compounds in African Bambara groundnut (*Vigna subterranea* (L.) Verdc.). *American Journal of Food and Nutrition, 6*(3), 88–95. https://doi.org/10.12691/ajfn-6-3-5.

Bamishaiye, O. M., Adegbola, J. A., & Bamishaiye, E. I. (2011). Bambara groundnut: An under-utilized nut in Africa. *Advances in Agricultural Biotechnology, 1*, 60–72.

Barimalaa, I. S., & Anoghalu, S. E. (1997). Effect of processing on certain antinutrients of Bambara groundnuts (*Vigna subterranea*) cotyledons. *Journal of the Science of Food and Agriculture, 73*, 186–188.

Bernhoft, A. (2010). *Bioactive compounds in plants: Benefits and risks for man and animals: Proceedings from a Symposium Held in Norwegian Academy of Science and Letters, Oslo, Norway, 13–14 November 2008*. Novus Forlag. ISBN: 8270995835, 9788270995837, 255 pages.

Brough, S. H., & Azam-Ali, S. N. (1992). The effect of soil moisture on the proximate composition of Bambara groundnuts (*Vigna subterranea* L. Verdc). *Journal of the Science of Food and Agriculture, 60*, 197–203.

Brough, S. H., Taylor, A. J., & Azam-Ali, S. N. (1993). The potential of Bambara groundnuts (*Vigna subterranea*) in vegetable milk production and basic protein functionality systems. *Food Chemistry, 47*, 277–283.

Chen, R., & Yang, D. (2010). Responding to compliments in Chinese: Has it changed? *Journal of Pragmatics, 42*, 1951–1963.

Duodu, K. G., & Apea-Bah, F. B. (2017). African legumes: Nutritional and health-promoting attributes (chapter 9). In J. Taylor & J. Awika (Eds.), *Gluten-free ancient grains: Cereals, pseudo-cereals, and legumes: Sustainable, nutritious, and health-promoting foods for the 21st century, 1st Edition, pp. 356* (pp. 223–269). Woodhead Publishing, eBook ISBN: 9780081008911 https://www.elsevier.com/search-results?query=Bambara%20groundnut&labels=books&page=1

Duthie, G., & Crozier, A. (2000). Plant-derived phenolic antioxidants. *Current Opinion in Clinical Nutrition and Metabolic Care, 3*(6), 447–451. https://doi.org/10.1097/00075197-200011000-00006.

Eka, O. U. (1985). Studies on level of oxalic acid and phytic acid in traditional food of northern Nigerians. *Journal of Biological and Applied Chemistry, 20*, 26–30.

Enwere, N. J., & Hung, Y. C. (1996). Some chemical and physical properties of Bambara groundnut (*Voandzeia subterranean* Thouars) seed and products. *International Journal of Food Science and Nutrition, 47*(6), 469–475. https://doi.org/10.3109/09637489609031875.

FAO/ICRISAT. (1996). *Basic Foodstuffs Service, FAO Commodities and Trade Division and the Socioeconomics and Policy Division, International Crops Research Institute for the Semi-Arid Tropics*. ICRISAT/FAO. https://www.feedipedia.org/node/4842

Huang, H., Chen, A. Y., Ye, X., Li, B., Rojanasakul, Y., Rankin, G. O., & Chen, Y. C. (2015). Myricetin inhibits proliferation of cisplatin-resistant cancer cells through a p53-dependent apoptotic pathway. *International Journal of Oncology, 47*(4), 1494–1502. https://doi.org/10.3892/ijo.2015.3133.

Igbabul, B., Adole, D., & Sule, S. (2013). Proximate composition, functional and sensory properties of Bambara nut (*Voandzeia subterranean*), cassava (*Manihot esculentus*) and soybean (Glycine max) flour blends for "Akpekpa" production. *Current Research in Nutrition and Food Science Journal, 1*(2), 147–155. https://doi.org/10.12944/CRNFSJ.1.2.06.

Ijarotimi, O. S., & Keshinro, O. O. (2013). Determination of nutrient composition and protein quality of potential complementary foods formulated from combination of fermented popcorn, African locust and Bambara groundnut seed flour. *Polish Journal of Food and Nutrition Science, 63*(3), 155–166. https://content.sciendo.com/view/journals/pjfns/63/3/article-p155.xml

Ikegwu, T. M., Nkama, I., & Okafor, I. G. (2020). Comparative studies of the proximate, microscopic and thermal properties of processed maize, wheat, millet, cassava and Bambara nut flours. *Acta Scientific Nutritional Health, 4*(8). ISSN:2582-1423.

Kelechi-Ebisike, V. O., Solomon, M. D., Jaryum, K. H., Ebisike, K. J., Mshelia, P., & Danjin, M. (2020). Energy/macronutrient density and sensory evaluation of formulated complementary foods for use in low-income populations in northern Nigeria. *Acta Scientific Nutritional Health, 4*(3), 01–06.

Koné, M., Paice, A. G., & Touré, Y. (2011). Bambara groundnut [*Vigna subterranea (*L.) Verdc. (Fabaceae)] usage in human health. Chapter 22. In V. Preedy & R. Watson (Eds.), *Nuts and seeds in health and disease prevention* (1st ed., pp. 189–196. Hardcover ISBN: 9780123756886, eBook ISBN: 9780123756893, p. 1226). Academic.

Leopold, S. S., Porcher, R., Gebhardt, M. C., Gioe, T. J., Manner, P. A., Rimnac, C. M., & Wongworawat, M. D. (2020). Editorial: Opposites attract at CORR®—Machine learning and qualitative research. *Clinical Orthopaedics and Related Research, 478*(10), 2193–2196. https://doi.org/10.1097/CORR.0000000000001466.

Luzardo-Ocampo, I., Campos-Vega, R., Gaytán-Martínez, M., Preciado-Ortíz, R., Mendoza, S., & Loarca-Piña, G. (2017). Bioaccessibility and antioxidant activity of free phenolic compounds and oligosaccharides from corn (*Zea mays* L.) and common bean (*Phaseolus vulgaris* L.) chips during in vitro gastrointestinal digestion and simulated colonic fermentation. *Food Research International, 100*, 304–311. https://doi.org/10.1016/j.foodres.2017.07.018.

Luzardo-Ocampo, I., Campos-Vega, R., Cuellar-Nuñez, M. I., Vázquez-Landaverde, P. A., Acosta-Gallegos, J. A., & Loarca-Piña, G. (2018). Fermented non-digestible fraction from combined nixtamalized corn (*Zea mays* L.)/cooked common bean (*Phaseolus vulgaris* L.) chips modulate anti-inflammatory markers on RAW 264.7 macrophages. *Food Chemistry, 259*, 7–17. https://doi.org/10.1016/j.foodchem.2018.03.096.

Martınez, C. J., Martınez, A. C., Ayala, A. L. M., Muzquiz, M., Pedrosa, M. M., & Davila-Ortiz, G. (2012). Changes in protein, non-nutritional factors, and antioxidant capacity during germination of *L. campestris* Seeds. *International Journal of Agronomy, 2012*. |Article ID 387407. https://doi.org/10.1155/2012/387407.

Mashau, M., Mabodze, T., Tshiakhatho, O., Silungwe, H., & Ramashia, S. (2020). Evaluation of the content of polyphenols, antioxidant activity and physicochemical properties of tortillas added with Bambara groundnut flour. *Molecules, 25*, 3035.

Minka, S. R., & Bruneteau, M. (2000). Partial chemical composition of Bambara pea. (*Vigna subterranea* L. Verde). *Food Chemistry, 68*, 273–276. https://doi.org/10.1016/S0308-8146(99)00186-7.

Mohamadou, M., Ngakou, A., & Tchuenguem, F.-N. F. (2018) Influence de l'interaction mycorhizes-insectes floricoles sur les rendements de *Vigna subterranea* variété rouge (Fabaceae) à Dang (Ngaoundéré, Cameroun) *International Journal of Biological and Chemical Sciences, 12*(4),1897–1913. https://doi.org/10.4314/ijbcs.v12i4.30. Available online at http://www.ifgdg.org

Moreno-Jiménez, M. R., López-Barraza, R., Cervantes-Cardoza, V., Perez-Ramirez, I. F., Reyna-Rojas, J. A., Gallegos-Infante, J. A., Estrella, I., Rojas-Contreras, J. A., Gonzalez-Laredo,

R. F., & Rocha-Guzman, N. E. (2018). Mechanisms associated to apoptosis of cancer cells by phenolic extracts from two canned common beans varieties (*Phaseolus vulgaris* L.). *Journal of Food Biochemistry, 2018*, e12680. https://doi.org/10.1111/jfbc.12680.

Ndidi, U. S., Ndidi, C. U., Aimola, I. A., Bassa, O. Y., Mankilik, M., & Adamu, Z. (2014). Effects of processing (boiling and roasting) on the nutritional and antinutritional properties of Bambara groundnuts (*Vigna subterranea* [L.] Verdc.) from southern Kaduna, Nigeria. *Journal of Food Processing, 2014. |Article ID 472129.* https://doi.org/10.1155/2014/472129.

Nyau, V., Prakash, S., Rodrigues, J., & Farrant, J. (2015). Antioxidant activities of Bambara groundnuts as assessed by FRAP and DPPH assays. *American Journal of Food and Nutrition, 3*(1), 7–11. https://doi.org/10.12691/ajfn-3-1-2.

Nyau, V., Prakash, S., Rodrigues, J., & Farrant, J. (2017). Profiling of phenolic compounds in sprouted common beans and Bambara groundnuts. *Journal of Food Research, 6*(6), 74–82. ISSN 1927-0887 E-ISSN 1927-0895. https://doi.org/10.5539/jfr.v6n6p.

Obizoba, I. C. (1990). Nutritive quality of blends of corn with germinated cowpeas (*Vigna unguiculanta*), pigeon pea (*Cajanus cajan*) and Bambara groundnut (Voandzei subterranean). *Cereal Chemistry, 67*(3), 230–232.

Odetokun, S. M. (2000). *Effect of fermentation on the physicochemical properties, anti-nutrients and in-vitro multi-enzyme digestibility of selected legumes*. Ph.D. dissertation, Federal University of Technology, Akure, Nigeria.

Ojimelukwe, P. C. (1999). Cooking characteristics of four cultivars of Bambara groundnuts seeds and starch isolate. *Journal of Food Biochemistry, 23*, 109–117.

Okonkwo, S. I., & Opara, M. F. (2011). The analysis of Bambara nut (*Voandzeia subterranea* (L) Thouars) for sustainability in Africa. *Research Journal of Applied Sciences, 5*(6), 394–396.

Oyeyinka, A. T., Pillay, K., & Siwela, M. (2019). In vitro digestibility, amino acid profile and antioxidant activity of cooked Bambara groundnut grain. *Food Bioscience, 31*, 100428.

Pale, E., Nacro, M., Vanhaelen, M., & Vanhaelen-Fastré, R. (1997). Anthocyanins from Bambara groundnut (*Vigna subterranea*). *Journal of Agricultural and Food Chemistry, 45*(9), 3359–3361. https://doi.org/10.1021/jf960897c.

Poulter, N. H. (1981). Properties of some protein fractions from Bambara groundnut. *Journal of Science and Food Agriculture, 32*, 44–50.

Purseglove, J. W. (1992). *Tropical crops (Dicotyledons)*. Longman House Burnt Mill, Harlow, Essex CM20 2JE England.

Ross, A. C., Caballero, B. H., Cousins, R. J., Tucker, K. L., & Ziegler, T. R. (2012). *Modern nutrition in health and disease: Eleventh edition*. Wolters Kluwer Health Adis (ESP), 1616 pages, ISBN (print) 9781605474618. https://jhu.pure.elsevier.com/en/publications/modern-nutrition-in-health-and-disease-eleventh-edition

Sirivongpaisal, P. (2008). Structure and functional properties of starch and flour Bambara groundnut. *Songklanakarin Journal of Science and Technology, 30*(1), 51–56.

Taylor, J., & Awika, J. (2017). *Gluten-free ancient grains: Cereals, pseudocereals, and legumes: Sustainable, nutritious, and health-promoting foods for the 21st century* (1st ed., p. 356). Woodhead Publishing. eBook ISBN: 9780081008911.

Temegne, N. C., Gouertoumbo, W. F., Wakam, G. A., Nkou, F. T. D., Youmbi, E., & Ntsomboh-Ntsefonng, G. (2018). Origin and ecology of Bambara groundnut (*Vigna subterranea* (L.) Verdc): A review. *Journal of Ecology & Natural Resources, 2*(4), 000140.

Teye, E., Deha, C. I., Dadzie, R., & MacArthur, R. L. (2020). Delivering the nutritional needs by food to food fortification of staples using underutilized plant species in Africa. *International Journal of Food Science, 2020*. https://doi.org/10.1155/2020/8826693.

Uvere, P. O., Onyekwere, E. U., & Ngoddy, P. O. (2010). Production of maize–Bambara groundnut complementary foods fortified pre-fermentation with processed foods rich in calcium, iron, zinc and provitamin A. *Journal of the Science of Food and Agriculture, 90*, 566–573. www.interscience.wiley.com. https://doi.org/10.1002/jsfa.3846.

Yao, D. N., Kouassi, K. N., Erba, D., Scazzina, F., Pellegrini, N., & Casiraghi, M. C. (2015). Nutritive evaluation of the Bambara groundnut Ci12 landrace [*Vigna subterranea* (L.) Verdc. (Fabaceae)] produced in Côte d'Ivoire. *International Journal of Molecular Sciences, 16*, 21428–21441.

Chapter 7
Physicochemical and Functional Properties of Bambara Groundnut Dietary Fibers

7.1 Introduction

As stated in Chap. 6, the high fiber content in BGN helps develop functional foods. The health benefit of dietary fiber includes reduced lifestyle diseases like obesity, diabetes, coronary heart disease, cancers, and haemorrhoids (Maphosa & Jideani, 2016). Dietary fiber from cowpeas, lentils and chickpeas is reported; however, not much is known about BGN dietary fibers. Hence, an understanding of BGN dietary fibers' physicochemical properties will highlight different food and human gastro-intestinal tract behaviour.

7.2 Physicochemical Properties of Bambara Groundnut Insoluble and Soluble Fibers

Diedericks and Jideani (2015) subjected four BGN varieties (brown, red, black-eye, brown-eye) to sequential enzymatic treatments followed by centrifugation to obtain the insoluble dietary fiber (IDF). BGN flour is characterized by the high total dietary fiber of 24.3% for red and 23.9% for black-eye both significantly higher than 21.0% in brown and 17.7% for brown-eye. These values confirm BGN as a good source of dietary fiber. The yield of insoluble fiber from the four varieties of BGN differed significantly and ranged from 37.4 for the brown-eye to 48.3% for the red. However, using the wet milling extraction method Maphosa and Jideani (2016) reported a lower yield of 12.0% for brown-eye to 15.6% for red.

All BGN-IDFs differed significantly in color due to variation in the seed coats' colour (Fig. 7.1). Polyphenols were predominant in red and brown BGN-IDFs (Diedericks & Jideani, 2015; Maphosa & Jideani, 2016). However, polyphenols were higher in the black-eye and brown-eye soluble dietary fibers (Maphosa & Jideani, 2016) extracted by wet milling. The major sugars in BGN-IDFs indicating

V. A. Jideani, A. I. O. Jideani, *Bambara groundnut: Utilization and Future Prospects*, https://doi.org/10.1007/978-3-030-76077-9_7

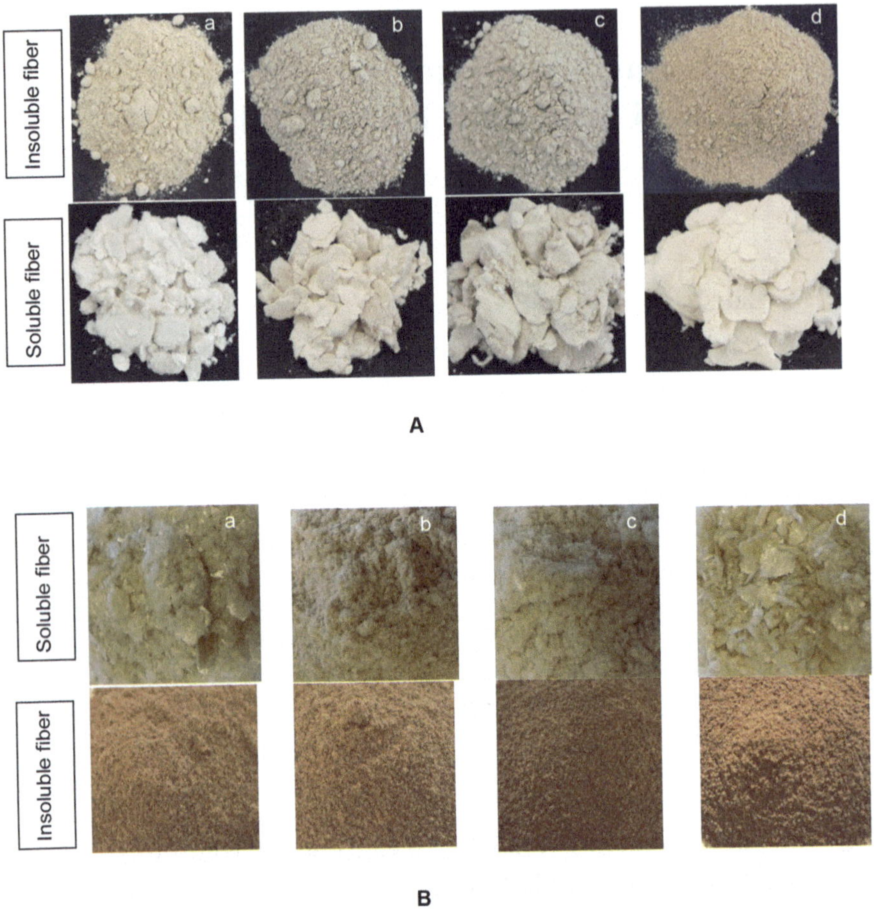

Fig. 7.1 Bambara groundnut dietary fibres: (**A**) Enzymatic extracted insoluble (BGNIF) and soluble (BGNSF) dietary fibers from four Bambara groundnut (BGN) varieties. **a:** black-eye, **b:** red, **c:** brown, **d:** brown-eye. (**B**) Wet milling extracted soluble (SDFs) and insoluble (IDFs) dietary fibers from four varieties of Bambara groundnut (BGN) (**a**) black-eye, (**b**) brown-eye, (**c**) brown, (**d**) red

a hemicellulosic polysaccharide include arabinose/galactose (31.0–37.1%), xylose (16.5–27.3%), and mannose (14.5–22.2%) through the enzymatic method of extraction (Table 7.1).

The range of these sugars was lower in the wet milling method arabinose/galactose (2.3–2.8%), xylose 11.3–13.8%) and mannose (4.7–6.6%) (Maphosa & Jideani, 2016). The high proportions of arabinose and galactose could be attributed to neutral arabinogalactan type polysaccharide linked to galacturonic residues (Diedericks & Jideani, 2015). The presence of xylose and arabinose is indicative of arabinoxylans as part of hemicellulose constituent. The glucose composition in BGN dietary fibers is low compared to other legume fibers where glucose constitutes the

Table 7.1 Neutral sugar composition of Bambara groundnut insoluble dietary fibers from the enzymatic and wet milling extraction methods

	Sugars (%)							
Variety	Arabinose/Galactose	Fructose	Fucose	Glucose	Mannose	Rhamnose	Xylose	Total sugar
Enzymatic method								
Black-eye	37.1 ± 9.3^{a}	2.4 ± 0.6^{a}	2.6 ± 1.0^{a}	2.9 ± 0.3^{ab}	14.5 ± 3.1^{a}	ND	16.5 ± 1.2^{a}	76.0
Brown-eye	31.0 ± 3.9^{a}	2.5 ± 0.5^{a}	2.0 ± 1.3^{a}	3.5 ± 0.3^{b}	17.4 ± 2.5^{ab}	ND	27.3 ± 2.4^{b}	83.7
Brown	33.1 ± 7.5^{a}	2.5 ± 0.4^{a}	1.9 ± 0.4^{a}	2.5 ± 0.4^{a}	22.2 ± 3.1^{b}	ND	22.9 ± 4.9^{b}	85.1
Red	30.4 ± 8.5^{a}	1.8 ± 0.1^{a}	2.9 ± 1.1^{a}	5.5 ± 0.8^{c}	20.4 ± 5.1^{ab}	ND	25.5 ± 2.8^{b}	86.5
Wet milling method								
Black-eye	2.8 ± 0.2^{a}	1.7 ± 0.1^{ac}	0.1 ± 0.0^{a}	0.6 ± 0.1^{a}	4.7 ± 0.4^{a}	2.6 ± 0.1^{a}	11.3 ± 1.2^{a}	23.8
Brown-eye	2.3 ± 0.1^{b}	1.7 ± 0.1^{a}	0.3 ± 0.1^{b}	0.7 ± 0.1^{ab}	6.6 ± 0.1^{ab}	1.0 ± 0.1^{b}	11.3 ± 1.9^{a}	23.9
Brown	2.5 ± 0.2^{ab}	1.5 ± 0.1^{b}	0.4 ± 0.1^{b}	0.9 ± 0.1^{c}	5.6 ± 0.4^{c}	1.2 ± 0.1^{c}	13.8 ± 2.3^{a}	25.9
Red	2.8 ± 0.2^{a}	1.9 ± 0.1^{c}	0.2 ± 0.1^{a}	0.8 ± 0.0^{b}	5.6 ± 0.6^{b}	1.1 ± 0.1^{c}	12.4 ± 0.8^{a}	24.8

Values are Mean ± SD of triplicate readings. Means within a column followed by the same letter (under the same heading) are not significantly [$p > 0.05$] different. *IDF* insoluble dietary fiber, *SDF* soluble dietary fiber, *ND* not detected

Adapted from Maphosa and Jideani (2016), Diedericks and Jideani (2015), Diedericks (2014)

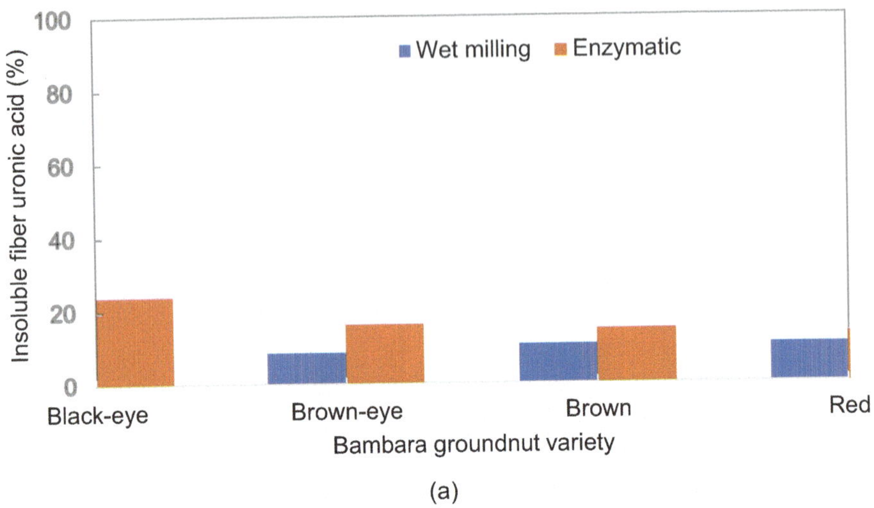

Fig. 7.2 Uronic acid content of Bambara groundnut dietary fiber based on extraction method (**a**) insoluble fiber and (**b**) soluble fiber

significant portion indicating cellulose as a predominant polysaccharide. BGN insoluble fiber's sugar composition shows hemicellulose as a major polysaccharide (Diedericks & Jideani, 2015; Maphosa & Jideani, 2016). BGN insoluble fiber's uronic acid range from 6.7% for the black-eye to 10.6% for the red (Maphosa & Jideani, 2016) [Fig. 7.2].

Black-eye and brown-eye dietary fibers had higher phenolic and total sugar content. Both the enzymatic and wet milling extraction methods indicate that BGN insoluble dietary fiber (BGN-IDF) is high in arabinose/galactose, mannose, xylose

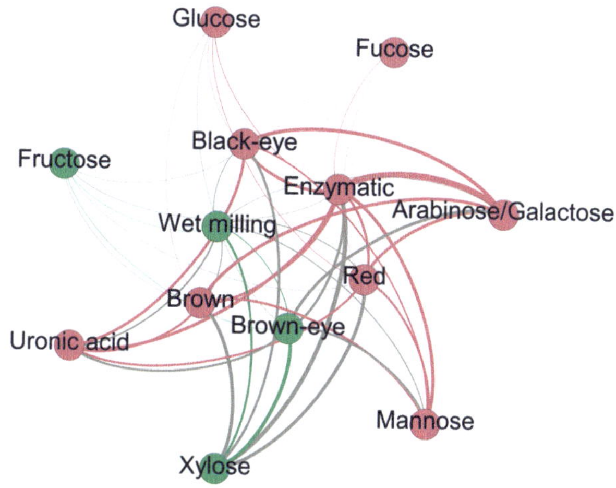

Fig. 7.3 BGN Insoluble fiber composition as affected by wet milling and enzymatic extraction methods. Yifan Hu layout on modularity class. The edges are weighted according to the sugar content

and uronic acid (Fig. 7.3). Two classes differentiate the two extraction methods based on the sugar contents of the insoluble dietary fibers. The wet milling extracted BGN-IDF is associated with brown-eye BGN with high xylose and low fructose. The enzymatic method relates to the black-eye, brown and red BGN dietary fibers with higher arabinose/galactose, mannose and uronic acid.

Table 7.2 details the sugar content of Bambara groundnut soluble fibers by the enzymatic and wet milling extraction methods. While Maphosa and Jideani (2016) reported soluble dietary fiber yield of 15.4% for red to 17.1% for brown-eye, Diedericks (2014) reported 12.7–13.9% with a ratio of soluble fiber to the insoluble fiber of 1:3. This high ratio of insoluble fiber is expected as the extraction starts with whole seed flour. The soluble dietary fiber of BGN is higher than reported for pea, broad pea and soybean (Maphosa & Jideani, 2016). Yao et al. (2015) reported that BGN dietary fiber of 10.3% for BGN flour could be related to the thick seed peels in the whole flour. The insoluble to soluble fiber ratio in legumes range from 5:1 to 5:19 (Mallillin et al., 2008) but Yao et al. (2015) reported 20:1.

The wet milling soluble dietary fiber relates to the red and its arabinose/galactose, uronic acid and fructose content (Fig. 7.4). In contrast, the enzymatic method relates to brown-eye, brown and black-eye soluble fibers higher in mannose and xylose. Uronic acids are sugar acids with carbonyl (C = O) and carboxylic acid (COOH) functional groups, their presence in high quantities in BGN dietary fibers is beneficial. Uronic acids form salts with some wastes in the human body, thereby facilitating their excretion. Furthermore, uronic acids react with less polar compounds such as steroids, bilirubin and drugs increasing their polarity, thereby aiding their removal through urine and bile, contributing to the body's detoxification process. Uronic acids are also responsible for forming mucopolysaccharides such as hyaluronic acid and heparin, significant mucus components, synovial fluid, cornea, cartilage and tendons. The presence of uronic acids in BGN dietary fibers also indicates the presence of pectin. Pectin is a thickening heteropolysaccharide and would be expected

Table 7.2 Neutral sugar composition of Bambara groundnut soluble dietary fibers from the enzymatic and wet milling extraction methods

	Sugars (%)							
Variety	Arabinose/Galactose	Fructose	Fucose	Glucose	Mannose	Rhamnose	Xylose	Total sugar
Enzymatic method								
Black-eye	7.9 ± 0.2^{a}	0.9 ± 0.5^{a}	4.8 ± 0.6^{a}	3.0 ± 0.5^{a}	24.9 ± 2.0^{a}	ND	54.5 ± 2.4^{a}	96.0
Brown-eye	10.8 ± 3.8^{a}	1.3 ± 0.4^{a}	1.7 ± 0.3^{c}	4.8 ± 0.4^{c}	37.5 ± 5.8^{b}	ND	38.5 ± 2.3^{b}	94.6
Brown	8.4 ± 2.0^{a}	1.3 ± 0.6^{a}	1.6 ± 0.3^{c}	7.4 ± 1.4^{b}	34.3 ± 4.4^{b}	ND	42.7 ± 6.8^{b}	95.7
Red	10.9 ± 0.1^{a}	3.6 ± 0.9^{b}	3.1 ± 1.0^{b}	2.8 ± 0.6^{a}	33.0 ± 3.1^{b}	ND	41.1 ± 4.6^{b}	94.5
Wet milling method								
Black-eye	19.6 ± 0.9^{a}	1.3 ± 0.1^{a}	0.3 ± 0.1^{a}	1.0 ± 0.1^{a}	6.1 ± 0.5^{a}	ND	16.6 ± 1.6^{a}	44.8
Brown-eye	15.2 ± 1.5^{b}	1.5 ± 0.1^{ab}	0.3 ± 0.1^{a}	0.8 ± 0.1^{ab}	6.5 ± 0.2^{a}	ND	15.6 ± 0.2^{ab}	39.8
Brown	9.4 ± 0.1^{c}	1.7 ± 0.1^{b}	0.1 ± 0.0^{b}	0.8 ± 0.1^{ab}	7.4 ± 0.6^{b}	ND	13.0 ± 1.8^{c}	32.5
Red	10.0 ± 0.6^{c}	1.6 ± 0.1^{b}	0.3 ± 0.0^{a}	0.6 ± 0.2^{b}	5.0 ± 0.3^{c}	ND	13.7 ± 0.8^{bc}	31.2

Values are Mean ± SD of triplicate readings. Means within a column followed by the same letter (under the same heading) are not significantly [$p > 0.05$] different. *IDF* insoluble dietary fiber, *SDF* soluble dietary fiber, *ND* not detected

Adapted from Maphosa and Jideani (2016), Diedericks and Jideani (2015), Diedericks (2014)

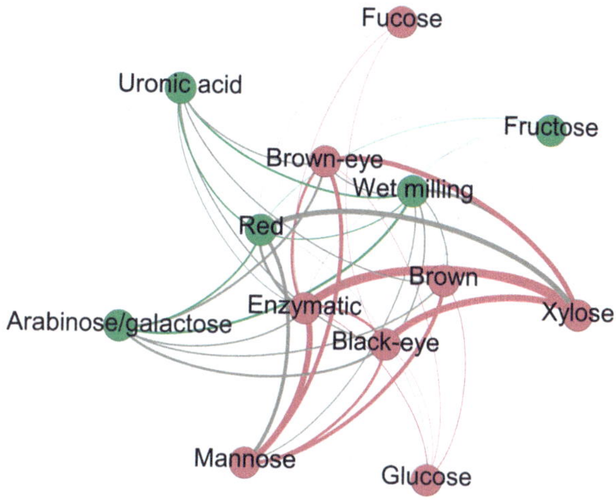

Fig. 7.4 BGN soluble fiber composition as affected by wet milling and enzymatic extraction methods. Yifan Hu layout on modularity class. The edges are weighted according to the content of each sugar for soluble dietary fiber

to increase food viscosity in the stomach, slowing down digestion and prolonging satiety. Furthermore, BGN dietary fibers could find use in food systems as thickeners by increasing the viscosity of food systems (Maphosa & Jideani, 2016).

7.3 Functional Properties of Insoluble and Soluble Bambara Groundnut Dietary Fibers

Maphosa and Jideani (2016) reported that the swelling capacities of brown-eye (6.5 g/mL) and black-eye (6.2 g/mL) fibers were significantly higher than those of red (6.0 g/mL) and brown (5.5 g/mL) fibers while the water holding capacities of black-eye and brown-eye fibers (2.84 g and 2.83 g water/g sample) were higher than those of brown and red fibers. Swelling capacity ranged from 6.37 to 7.72 mL/g with the brown-eye BGN-IDF (7.72 mL/g) significantly higher. All BGN-IDF varieties had a higher swelling capacity than commercial fibers, with black-eye BGN-IDF having a swelling ability comparable to apple fiber. Fat absorption for all BGN-IDFs was higher compared to commercial pea fibers. Water retention capacity ranged from 1.63 to 2.01 g water/g dry weight. Fat absorption for red BGN-IDF differed significantly. Black-eye and brown-eye fibers have superior physicochemical properties compared to the brown and red fibers as evidenced by their higher swelling capacities, water-holding capacities, oil binding capacities, phenolic, and total sugar content. Brown-eye and black-eye IDFs were lighter in colour, yellower, redder, more saturated and had higher hues than the red and brown IDFs. The physicochemical properties of BGN fibers make them valuable to the food industry as thickening agents, stabilizers, healthy ingredients, and cryoprotectants in frozen dairy products. BGN fibers can be considered suitable alternatives for commercial fibers such as pea, chickpea and lentil fibers.

The bulk densities of insoluble dietary fibers (IDFs) and soluble dietary fibers (SDFs) ranged between 0.57 g/mL (red) to 0.67 g/mL (brown-eye) and 0.46 g/mL (brown-eye) to 0.57 g/mL (black-eye), respectively. The oil binding capacities (OBCs) of SDFs ranged between 2.78 g oil/g sample (brown) and 4.03 g oil/g sample (brown-eye) while the OBC of all IDFs did not differ ranging between 1.52 g oil/g sample (brown) and 1.40 g oil/g sample (brown-eye and black-eye) (Maphosa & Jideani, 2016). BGN-IDF has potential in food systems as fat replacers, emulsion stabilizers, water binders, bulking agents, thickeners and nutritional additives.

The functional properties of the BGN-IDFs were studied in wheat bread (Diedericks & Jideani, 2015). The optimal formulation for incorporating the fibers in wheat bread involves 59.5% water, 4.3% yeast and 8.5% BGN-IDF. A good quality bread is characterized by lower gumminess and chewiness, and high-specific loaf volume. Higher BGN-IDF concentrations resulted in reduced gumminess and chewiness. Specific loaf volume was also slightly reduced with increased BGN-IDF, although comparable to loaves of bread with lower BGN-IDF content. The incorporation of BGN-IDF into white bread resulted in improved quality characteristics of white bread.

Soluble dietary fibers (SDFs) [30% (w/w)] from four varieties of Bambara groundnut (BGN), black-eye, brown-eye, brown and red were used to stabilize orange oil beverage emulsions containing 6% (w/w) orange oil (Maphosa et al., 2017). The volume-surface mean diameter ($d_{3,2}$) and equivalent volume-mean diameter ($d_{4,3}$) of the four emulsions ranged between 2.68–4.38μm and 17.09–18.62μm, respectively. Emulsions stabilized with black-eye-SDF, and brown-SDF possessed the least and highest $d_{3,2}$ and $d_{4,3}$, respectively. The emulsions were relatively stable to creaming and destabilized mainly by oil-droplet aggregation. The emulsions' backscattering flux ranged from 72.9% (brown-SDF stabilized emulsion) to 85.0% (black-eye-SDF stabilized emulsion). All four BGN SDFs greatly indicated their potential in stabilizing beverage emulsions. Thus, BGN-SDFs can be used to stabilize beverage emulsions. Black-eye and brown-eye SDFs form more stable emulsions than brown and red SDFs. Each emulsion behaves slightly differently due to compositional differences. The turbidity loss rate is largely influenced by the average droplet sizes of the emulsions.

The prebiotic properties of Bambara groundnut (BGN) soluble dietary fibers (SDFs) from black-eye, brown-eye, brown and red varieties were assessed using batch culture fermentation systems, monitoring short-chain fatty acids (SCFAs) as products of fermentation (Maphosa & Jideani, 2015). The growth rates of *Bifidobacterium bifidum, Bifidobacterium animalis* subsp. *animalis* and *Bifidobacterium breve* in the presence of BGN SDFs were assessed using optical densities (ODs) and microscopic counts and data modelled using the Gompertz kinetic equation. All four BGN SDFs supported the growth of the *Bifidobacterium* spp. which confirmed their prebiotic properties. The growth of *B. breve* was supported by black-eye SDF > brown-eye SDF > brown SDF > red SDF. The growth of *B. bifidum* was supported by brown-eye SDF > black-eye SDF > red SDF > brown SDF. The growth of *B. animalis* subsp. *animalis* was supported by black-eye

SDF > brown-eye SDF > brown SDF > red SDF. According to Gompertz kinetic parameters, the growth rate of *B. breve* ranged from 0.33 Δ log cfu/mL/h (brown SDF) to 0.46 Δ log cfu/mL/h (brown-eye SDF). *B. bifidum* had the highest and least growth rates in red SDF (1.75 Δ log cfu/mL/h) and black-eye SDF (0.46 Δ log cfu/mL/h). *B. animalis* subsp. *animalis* had the highest and least growth rates in the presence of red SDF (1.75 Δ log cfu/mL/h) and black-eye SDF (0.54 Δ log cfu/mL/h). The growth rate of *B. animalis* subspp. *animalis* as assessed by optical density experiments was in the range 0.076 OD units/h (red SDF) to 0.083 OD units/h (black-eye). The growth rate of *B. animalis* subspp. *animalis* in black-eye SDF and brown SDF were significantly similar ($p \geq 0.05$). *B. bifidum* showed the highest growth rate in the presence of brown-eye SDF (0.098 OD units/h) and the least growth rate in the presence of brown SDF (0.084 OD units/h). The growth rate of *B. bifidum* in black-eye SDF and brown-eye SDF was significantly similar ($p \geq 0.05$). The growth rate of *B. breve* was in the range 0.082 OD units/h (red SDF) to 0.090 OD units/h (black-eye SDF). The growth rate of *B. breve* in brown SDF and brown-eye SDF were similar. The fermentation of BGN SDFs by *Bifidobacterium* spp. produced short-chain fatty acids (SCFAs). *B. bifidum* showed a preference for black-eye SDF producing 2.9123 mMol propionic acid. Acetic acid was produced the most from the fermentation of brown-eye SDF (2.1322 mMol) by *B. animalis* subsp. *animalis* and the least from brown SDF fermentation by all three microorganisms. Fructooligosaccharides as control were utilised the most by all microorganisms. The control showed peaks of propionic and acetic acids below the limit of quantification (LOQ) in all microorganisms indicating that BGN SDFs were responsible for the SCFAs produced. Therefore, the bifidogenic nature of BGN SDFs suggests that their inclusion in food systems would impart prebiotic properties and hence increase the market value of that product.

7.4 Wet Milling and Enzymatic-Gravimetric Methods of Bambara Groundnut Dietary Fiber Extraction-A Comparison

The enzymatic-gravimetric method of Diedericks and Jideani (2015) differs from the wet milling method adopted by Maphosa and Jideani (2016). The former is an enzyme-based extraction process while wet milling is a water-based extraction method. The enzymatic-gravimetric method produced fibers of higher purity. The cost of a technique directly impacts the product's price, as such, BGN dietary fiber obtained from the enzymatic-gravimetric method would cost more than that from a wet-milling process. However, both methods result in BGN fibers with functional properties, making them valuable to the food industry as potential fortifiers, thickening agents, gelling agents, high-temperature stability, and cryoprotectants in frozen dairy products. BGN fibers are suitable alternatives for commercial fibers such as pea, chickpea and lentil fibers.

References

Diedericks, C. F. (2014). *Functional properties of Bambara groundnut (Vigna subterranea) non-starch polysaccharides in model and food systems*. Master of Technology in Food Technology thesis, Cape Peninsula University of Technology, South Africa.

Diedericks, C. F., & Jideani, V. A. (2015). Physicochemical and functional properties of insoluble dietary fiber isolated from Bambara groundnut (*Vigna subterranea* [L.] Verdc.). *Journal of Food Science, 80*, C1933–C1944.

Mallillin, A. C., Trinidad, T. P., Raterta, R., Dagbay, K., & Loyola, A. S. (2008). Dietary fiber and fermentability characteristics of root crops and legumes. *British Journal of Nutrition, 100*, 485–488.

Maphosa, Y., & Jideani, V. A. (2015). *Characterisation of Bambara groundnut (Vigna subterranean) non-starch polysaccharides from we milling method as prebiotics*. Master of Technology in Food Technology thesis, Cape Peninsula University of Technology, South Africa.

Maphosa, Y., & Jideani, V. A. (2016). Physicochemical characteristics of Bambara groundnut dietary fibers extracted using wet milling. *South African Journal of Science, 112*, 72–79.

Maphosa, Y., Jideani, V. A., & Adeyi, O. (2017). Effect of soluble dietary fibers from Bambara groundnut varieties on the stability of orange oil beverage emulsion. *African Journal of Science, Technology, Innovation and Development, 9*, 69–76.

Yao, D. N., Kouassi, K. N., Erba, D., Scazzina, F., Pellegrini, N., & Casiraghi, M. C. (2015). Nutritive evaluation of the Bambara groundnut Ci12 landrace [*Vigna subterranea* (L.) Verdc. (Fabaceae)] produced in Côte d'Ivoire. *International Journal of Molecular Sciences, 16*, 21428–21441.

Chapter 8
Bambara Groundnut and Starch

8.1 Physicochemical Characteristics of Bambara Groundnut Starch

Bambara groundnut (BGN) starch has been extensively studied (Sirivongpaisal, 2008; Afolabi, 2012; Otegbayo et al., 2013; Ashogbon, 2014; Falade & Adebiyi, 2015; Falade & Nwajei, 2015). Sirivongpaisal (2008) reported the structure and functional properties of starch and flour from BGN (*Voandzeia subterranea*). The BGN flour contained 58.38% of carbohydrate, 15.48% of protein, 7.90% of fat, 4.19% of ash and 2.54% fiber content. The low value may be from dehulling. The starch isolated by the alkaline method had a low content of protein (0.61%), fat (0.44%) ash (0.47%) and fiber (0.60%). The amylose content of BGN starch was 21.67%. Afolabi et al. (2018) reported the amylose content of BGN native starch as 27.7%. Yao et al. (2015) reported 50% of the nutrients in BGN as starch with 35% amylose for Ci12 landrace of Ivory Coast. Amylose is the primary determinant for a favourable glycemic response and resistant starch formation, with the consequent health benefits on glucose metabolism, energy intake and colonic health. The range of these components is presented in Table 6.2 in Chap. 6. A significant portion of BGN starch fraction is in the form of resistant starch.

The starch granule shape appeared to be oval and round, having an average diameter of 31.11 mm. The native BGN starch granule is oval with few mild ruptures having a diameter between 22 and 30μm in the range for pulses (Afolabi et al., 2018). The mild rupture is due to the isolation method of the starch. Annealing and heat-moisture treatment of BGN starch does not significantly affect the granule size, surface morphology and granule size distribution.

The starch showed good resistance to acid at a pH of 4.6–7.0. Compared with Bambara groundnut flour, the starch exhibited higher swelling power, breakdown and setback, but lower gelatinization temperature, pasting temperature, water and oil absorption capacity.

V. A. Jideani, A. I. O. Jideani, *Bambara groundnut: Utilization and Future Prospects*, https://doi.org/10.1007/978-3-030-76077-9_8

8.2 In Vitro Starch Digestibility of Cooked Bambara Groundnut Seeds and Cells

The hard-to-cook phenomenon in legumes results from extended storage at temperatures greater than 25 °C and 65% humidity leading to prolonged cooking times to attain a palatable texture (Juliet et al., 2017; Gwala et al., 2019). Gwala et al. (2019) explored the influence of the hard-to-cook (HTC) phenomenon on *in vitro* starch digestion of Bambara groundnut, taking into account three distinct HTC levels (seeds stored at 35 °C and 78% relative humidity for 0, 16 and 32 weeks). Stored BGN required prolonged cooking times, 5 h for 32-week stored seeds approximately 2.5 times longer than the control seeds. Increasing storage time led to a decrease in the rate constant of texture degradation, signifying the development of the HTC phenomenon. For cooking times of 60 and 120 min at 95 °C, longer stored seeds exhibited higher rate constants and starch digestion compared to the fresh sample. The higher rate of digestion was attributed to the high hardness that resulted in more significant cell rupture and faster access of amylase to starch. Adapting cooking times of BGN with distinct HTC levels to obtain equivalent hardness and microstructures resulted in comparable starch digestion kinetics. The implication of this results is that different starch digestibility could be obtained depending on the consumers' preference for hardness after processing BGN with different HTC. A consumer who prefers harder BGN may shorten the cooking time which upon mastication may rupture some cells releasing starch granules that might increase the rate of starch digestion. On the contrary, a consumer who prefers a softer texture may extend the cooking time, resulting in starch encapsulation within cell walls, consequently limiting starch digestion. One way of controlling starch digestion in pulses is to tailor processing steps to deliver the macronutrient in an encapsulated form to the gastrointestinal tract (Gwala et al., 2020). Limited small intestinal starch digestion is preferred in the context of controlling postprandial blood glucose levels and in increasing the amount of fermentable "resistant" starch in the large intestines. The authors demonstrated that hardness and interlinked cell failure (microstructure) determine starch digestion of BGN with distinct HTC levels.

Cooking pulses to achieve an edible texture (i.e., palatable hardness) is a process that preserves encapsulation of macronutrients to particular extents. Depending on the cooking intensity, partial or complete solubilisation of the pectin-rich middle lamella occurs with starch gelatinisation and protein denaturation. Insufficient pectin solubilisation, and upon subsequent mechanical disintegration, individual "intact" cells are obtained (Chigwedere et al., 2018). These particular undamaged individual cells from cooked pulses prevail after mastication (Pallares Pallares et al., 2019) and passage throughout the human digestive tract. Gwala et al. (2020) establishes the influence of processing intensity and sorting into hardness on the *in vitro* starch kinetics of individual cotyledon cells isolated from cooked BGN. There is a broad variation in hardness in cooked BGN seeds at specific cooking times, and hardness-based sorting could reduce the variation and increase uniformity. Starch digestion kinetics strongly depend on the hardness; consequently, hardness is a suitable food design variable to adjust starch digestion of BGN. Both sorted and non-sorted samples did not differ significantly in the cell wall barrier, reflected in the similar lag phases of starch digestion kinetics. Low hardness samples were digested faster than high hardness ones. The faster protein hydrolysis in the low hardness samples was accompanied by speedier starch digestion, indicating the possible role of the protein matrix barrier. Individual cells of comparable hardness obtained from the two different processing times had similar starch and protein digestion kinetics. The authors concluded that hardness is a suitable food design attribute to modulate starch and protein digestion kinetics of pulse cotyledon cells beyond cooking time.

Bambara groundnut flour has 33.0% resistant starch and 11.6% digestible starch. Heat-moisture treatment significantly increased the resistant starch content of the native starch. The utilization of BGN and the heat-moisture-treated starches in food and food products could alleviate obesity and diabetes due to its high resistant starch content (Afolabi et al., 2018).

8.3 Modification of Native Bambara Groundnut Starch

The development of a value-added product from starch depends on the knowledge of its structure and functional properties (Piyarat, 2008). Native starch can be modified by physical, enzymatic, and chemical modifications to produce functional starches with unique properties. Gulu et al. (2019) reported the functional properties of the BGN starch-catechin complex using solid dispersion techniques (microwave, co-evaporation and kneading) and cyclodextrins as initiators. The characteristics of BGN starch granules with cyclodextrin adhered to the surfaces indicating inclusion complex formation (Fig. 8.1). The swelling capacities of BGN starches ranged from 0.36 g/g (co-evaporation, β-cyclodextrin + catechin) to 4.84 g/g (microwave,

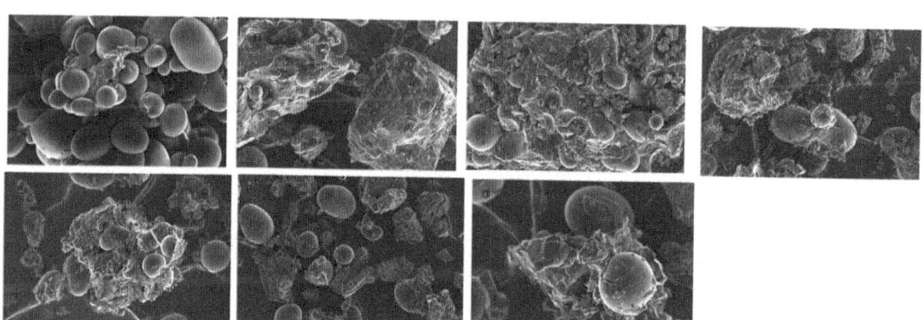

Fig. 8.1 Scanning electron micrographs of (**a**) native BGN starch, BGN starch-catechin complexes modified through; (**b**) microwave method using alpha cyclodextrin, (**c**) microwave method using beta cyclodextrin, (**d**) kneading method using alpha cyclodextrin, (**e**) kneading method using beta cyclodextrin, (**f**) co-evaporation method using alpha cyclodextrin, (**g**) co-evaporation method using beta cyclodextrin

β-cyclodextrin + catechin). The solid dispersion methods reduced the swelling capacity of native BGN starch. The solubility of BGN starches in water ranged from 0.61 g/100 g (native BGN starch) to 46.04 g/100 g (kneading, α-cyclodextrin). The water absorption of native and modified BGN starches ranged from 0.23 to 1.17 g/g, respectively. Starches complexed with β-cyclodextrin showed increased water absorption capacity compared to the starches complexed with α-cyclodextrin. The oil absorption capacities 1.02 g/g for native starch to 1.07 g/g for complexed starches, with no differences. The modified starches exhibited lower breakdown, setback, and final viscosities with a significant increase in gelatinisation temperature compared to the native BGN starch.

The effect of carboxymethylation on the functional and physicochemical properties of BGN starch has been reported (Afolabi, 2012). The starch yield on the dry weight of the seed basis was 40.4%, the degree of substitution was 0.78, while the reaction efficiency was 62.5%. The FTIR spectra confirmed the presence of the carboxymethylated group on the modified starch granule with new bands at 1590, 1455 and 1423 cm^{-1}. X-ray diffractogram revealed the loss of crystallinity as a result of carboxymethylation of the starch. Gelatinization temperature and enthalpy were significantly altered by the carboxymethylation with an increase in the paste viscosity from 278.58 to 300.83 RVU. The solubility and swelling power of the native and carboxymethylated starch increased with an increase in temperature, but with no significant effect on the swelling power of the starch in alkaline and acidic medium [pH 2–12] (Afolabi, 2012).

Oxidation and acetylation of BGN native starch reduced the ash, crude protein, crude fiber and crude fat. (Adebowale et al., 2002). Oxidised starch had the highest oil and water absorption capacity (3.04 and 2.04 g/g, respectively). The least gelation concentration for the starches ranged from 6 to 10% (w/v). Pasting temperature decreased with oxidation and acetylation with oxidised starch having the lowest pasting temperature. Both oxidation and acetylation lowered the peak viscosity, hot paste viscosity and cold paste viscosity. The authors concluded that chemical modification enhanced the stability of the starches with oxidised starch having the highest stability, followed by the acetylated starch.

Afolabi et al. (2018) investigated the in-vitro digestibility and functional properties of BGN flour, isolated native starch (BNS), hydrothermally (annealing at 50 °C for 48 h) modified (BAS) starch and heat-moisture treated (100 °C for 16 h at 25% moisture level) starch (BHS) and observed the following (i) amylose content of the flour and native starch was 6.10% and 27.70%, respectively, (ii) hydrothermal modification increased the gelatinization profile of the starch but reduces its pasting properties, (iii) swelling and solubility of the flour and starches increased with increase in temperature, (iv) X-ray diffraction revealed BNS and BAS exhibited "Type C" crystallinity pattern while BHS was "Type A", (v) BNS granule was oval, its diameter between 22 and 30μm, with no significant change in the morphology of BAS and BHS (Fig. 8.2), (vi) BG flour had 33% resistant starch and 11.63% digestible starch, and (vii) heat-moisture treatment enhanced the resistant starch content of the native starch significantly.

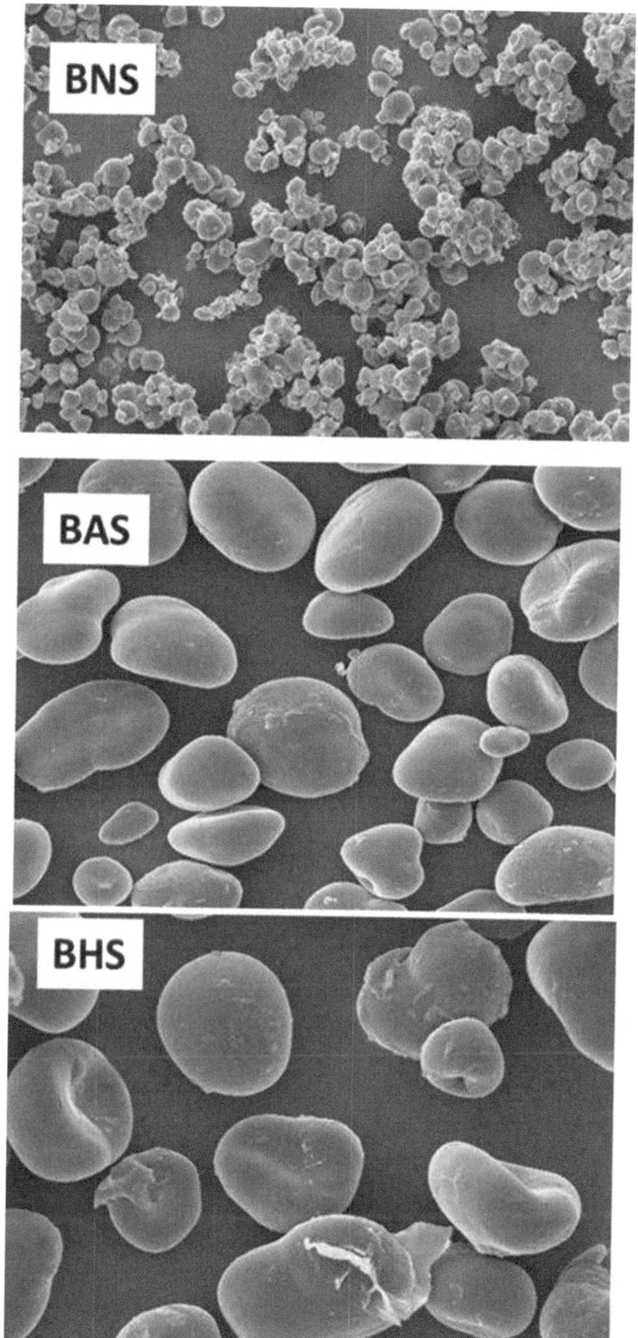

Fig. 8.2 The scanning electron micrograph of the native (BNS), annealed (BAS), and heat moisture treated (BHS) BGN starch at 500X magnification Afolabi et al. (2018). Purity of starch evaluated by determining their protein, fat, and ash content

8.4 Properties of Bambara Groundnut Starch Complexes

Oyeyinka et al. (2017a) studied the physicochemical and mechanical properties of biofilm prepared from BGN starch modified with varying concentrations of stearic acid (0%, 2.5%, 3.5%, 5%, 7%, and 10%). The BGN starch films modified with stearic acid (3.5%) showed a progressively rough surface compared to those with 2.5% stearic acid and the control. Fourier transform infrared spectroscopy spectra revealed a peak shift of approximately 31 cm^{-1}, suggesting the promotion of hydrogen bond formation between hydroxyl groups of starch and stearic acid. The addition of 2.5% stearic acid to BGN starch film reduced water vapor permeability by approximately 17%. BGN starch films modified with a higher concentration of stearic acid were more opaque and showed significantly high melting temperatures. However, the mechanical properties of starch films were most negatively affected by stearic acid. BGN starch film may be modified with 2.5% stearic acid for improved water vapor permeability and thermal stability with minimal tensile strength effect.

Amylose can form inclusion complexes with lipids for improved starch functionality. Oyeyinka et al. (2016) established the influence of high-pressure homogenization on the physicochemical properties of BGN starch complexed with lysophosphatidylcholine. Homogenization significantly increased the degree of complexation of BGN starch with lysophosphatidylcholine. BGN starch showed a higher complex index than corn and potato starch control samples. X-ray diffraction revealed the formation of V-amylose crystalline types with peaks at $2\Theta = 7.4$, 12.9 and 19.9°. Complexing BGN, corn and potato starches with lysophosphatidylcholine resulted in the formation of type I V-amylose complexes. These complexes had melting temperatures and enthalpies ranging from 91.5 to 98.6 °C and 1.4–5.5 J/g, respectively. Starch-lysophosphatidylcholine complexes displayed a low syneresis rate. Homogenized BGN starch complexes were non-gelling while unhomogenized types produced weak gels, with $G' > G''$ in the frequency range of 0.1–10 Hz. The non-gelling properties of homogenized BGN starch-lysophosphatidylcholine complex suggest that the modified starch could be used in the food industry to provide a smooth texture in frozen foods and desserts.

Oyeyinka et al. (2017) investigated stearic acid, linoleic acid, and lysophosphatidylcholine on complex formation and physicochemical properties of BGN starch in comparison with potato starch. The complexation index reached a maximum at 2% lipid concentration. BGN starch complexed better with stearic acid than with linoleic acid and lysophosphatidylcholine. A similar trend was observed for potato starch but to a lesser extent. All lipids significantly reduced the peak and setback viscosities of BGN starch but increased the final viscosity. Pasting of BGN and potato starches with lipids resulted in type-I V-amylose complexes, with melting temperatures ranging from ≅98 to 102 °C. X-ray diffraction of these complexes showed the crystalline V-amylose pattern with a significant peak at $2\theta = 19.9°$ and minor peaks at $2\theta = 7.4°$ and 12.9°. Modification of BGN starch with lipids resulted in reduced digestibility, suggesting their potential application in formulating foods to manage diabetes.

Oyeyinka et al. (2018) reported annealing on the complexing ability and functionality of BGN starch with palmitic acid. Corn starch was included as the reference. Annealing created cracks and pores on the surface of BGN and corn starches, respectively. Bambara groundnut starch had significantly higher amylose content, higher peak and final viscosities than corn starch. The peak viscosities of native BGN and corn starches were significantly reduced with palmitic acid. More significant peak viscosities reduction was observed when the annealed starches were complexed with palmitic acid, suggesting that more palmitic acid was complexed after annealing. XRD peaks and melting enthalpies confirmed this. Pasting of native BGN and corn starches with palmitic acid resulted in type I V-amylose complexes, while type II complexes were formed from annealed starches pasted with palmitic acid.

8.5 Structural and Crystalline Properties of Bambara Groundnut Starch

Oyeyinka et al. (2017b) studied the physicochemical properties of starches extracted from maroon, brown and cream BGN landraces. Starch yield (35%) from the maroon BGN was slightly higher compared to the brown and cream (≅29%). The amylose contents (31.5–34.6%) of the BGN starches were significantly different among the landraces; brown starch slightly higher (34.6%) than maroon (31.5%) and cream (32.9%) starches. The amylose contents (20–35%) varied significantly among genotypes (Oyeyinka et al., 2015). The peak gelatinization temperature (approx. 73 °C) of brown BGN starch was slightly low compared to maroon (approx. 78 °C) and cream (approx. 76 °C) BGN starch. Ojimelukwe (1999) reported gelatinization temperature of BGN starch isolates from 72 to 74 °C from red, black, cream and brown BGN. The BGN starches showed a substantially high proportion of resistant starch (71%) and a similar predicted glycemic index (40.1) among landraces. Bambara groundnut starch can potentially be used as a thickening agent in food products and ingredient development (Oyeyinka et al., 2017b).

Sirivongpaisal (2008) reported the structure and functional properties of starch and flour from BGN (*Voandzeia subterranea*). The BGN flour contained 58.4% of carbohydrate, 15.5% of protein, 7.9% of fat, 4.2% of ash and 2.5% of fiber content. The starch isolated by the alkaline method had a low content of protein (0.61%), fat (0.44%) ash (0.47%) and fiber (0.60%). The amylose content of BGN starch was 21.7%. However, Afolabi et al. (2018) reported slightly higher amylose content (27.7%) of BGN native starch and Oyeyinka et al. (2017b) reported a range of 31.5–34.6%. Thus, variation in landraces could be the reason for the different amylose contents. The starch granule shape appeared to be oval and round having an average diameter of 31.11 mm. X-ray diffraction pattern of the starch granules revealed an A-type with 43.7% of crystallinity. Oyeyinka et al. (2015) reported the physicochemical properties of starches extracted from five BGN genotypes. BGN starch granules were predominantly oval-shaped with a smooth surface and an average size of 26 ± 0.2μm. However, Afolabi et al. (2018) reported a type C crystallinity pattern for BGN native starch and annealed starch at 50 °C for 48 h while hydrothermally modified native starch exhibited type A crystallinity. X-ray diffraction revealed the C-type pattern for all starches with a relative crystallinity range of 29–35% (Oyeyinka et al., 2015).

The thermal transition temperature of the starch was 71.69 °C at onset (T_0) and gelatinisation enthalpy (ΔH) was 11.73 J/g. The water and oil absorption capacity of BGN starch were 1.67 and 1.01 ml/g, respectively. Bambara groundnut starch showed a two-stage swelling pattern indicating a fairly restricted swelling starch. The pasting properties of the starch showed pasting temperature, breakdown and setback of 77.7 °C, 170 BU, 220 BU, respectively. The starch showed good resistance to acid at pH 4.6–7.0. Compared with BGN flour, the starch exhibited higher swelling power, breakdown and setback, but lower gelatinisation temperature, pasting temperature, water and oil absorption capacity.

FTIR spectra of BGN starches showed variable peak intensities at 2931, 1655 and 860 cm^{-1}, which corresponds to C-H stretching, H_2O bending vibrations and C-O stretching, respectively. BGN genotype with the highest amylose content showed the lowest intensity at wavenumber 2931 cm^{-1} (Oyeyinka et al., 2015).

8.6 Swelling of Bambara Groundnut Seed and Starch

Except for similar oil absorption among the genotypes, swelling power, water absorption and paste clarity of starches were significantly different. Genotype with high amylose content showed restricted swelling, low paste clarity and remarkable ability to absorb water. All Bambara starches displayed shear-thinning behaviour ($n < 1$) (Oyeyinka et al., 2015). The swelling power of BGN flour and starch is reported to increase with the increase in temperature which is consistent with other legumes starches (Ojimelukwe, 1999; Afolabi et al., 2018). The swelling characteristics of starch isolates from the four BGN seed cultivars reported by Ojimelukwe (1999) are detailed in Fig. 8.3. The starch swelling power at 90 °C was significantly higher than the swelling power at 65 °C and 70 °C for all four seed cultivars. All the starch started swelling and gelling at 70 °C. The starch from the red variety exhibited higher resistance to swelling but with swelling power comparable to the swelling power of other starch isolates. The swelling power of all the starch isolates at 80 °C and 85 °C did not differ from their swelling power at 90 °C. The swelling usually involves a weakening of bonds that hold the starch molecules and depends on the affinity of starch for water. With the increase in temperature, there is a loss of birefringence and an increase in viscosity as the starch granules rupture (Ojimelukwe, 1999). A similar rise in swelling capacity with an increase in temperature was reported for native BGN, oxidised and acetylated starches with acetylated starch

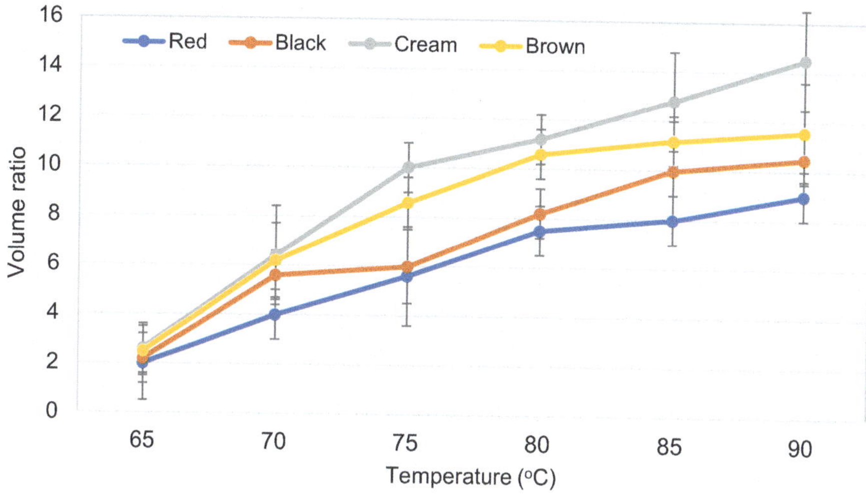

Fig. 8.3 Swelling power of isolates as affected by temperature. (Adapted from Ojimelukwe, 1999)

having the highest swelling power (Adebowale et al., 2002). There is increased swelling and solubility in the alkaline pH. Hydrothermal modification of BGN starch reduced the swelling power due to the increase in crystallinity and strengthening of intermolecular bonds due to the heat-moisture treatment leading to a restriction in the swelling of the starch granules. High amylose content in starch inhibits swelling during gelatinisation and also increase the viscosity. In the contrary starch with low total amylose content swell freely when heated. The reduction in the swelling power of annealed BGN starch could be due to the interaction between the degree of crystalline perfection and amylose-amylose or amylose-amylopectin iinteraction; which decreases the hydration of amorphous regions of starch, thereby decreasing granular swelling of starch (Zavareze & Guerra Dias, 2011). The swelling power of BGN starch is at its peak at pH 4 and 12. Hydrothermally treated starch swelling power is highest at pH 2–6, while the annealed starch at pH 8–12.

Final pasting temperatures of 8% starch solution from red, black, cream and brown BGN starch indicate loss of at least 98% of the total birefringence of the starch granules. The drop in viscosity was calculated as the difference between the peak viscosity and the final viscosity after holding for 15 min at 95 °C (shear thinning). The change in consistency is evaluated by the difference between the viscosity at 50 °C and the final viscosity at 95 °C. Starches from the black and brown seed exhibited the highest peak viscosity. The starch isolates from the black-colored seed cultivar had the poorest consistency. The ease of cooking is reflected in the shear-thinning behavior and is lowest for the starch isolates from the red-colored cultivars. In general, the starch from the black BGN is the least stable to mechanical treatment, while other seed cultivars have more stable starch pastes.

8.7 Conclusion

Different starch digestibility could be obtained from BGN depending on the consumers' preference for hardness. A consumer who prefers harder BGN may shorten the cooking time, which upon mastication may rupture some cells releasing starch granules that might increase the rate of starch digestion. On the contrary, a consumer who prefers a softer texture may extend the cooking time, resulting in starch encapsulation within cell walls, consequently limiting starch digestion. Hardness and interlinked pattern of microstructure determine starch digestion of BGN with distinct HTC levels. Chemical modification enhanced the stability of the BGN starches. Bambara groundnut starch showed a two-stage swelling pattern indicating a somewhat restricted swelling starch with good resistance to acid at the pH 4.6–7.0. Compared with BGN flour, the starch exhibited higher swelling power, breakdown and setback, but lower gelatinisation temperature, pasting temperature, water and oil absorption capacity.

References

Adebowale, K. O., Adeniyi Afolabi, T., & Lawal, O. S. (2002). Isolation, chemical modification and physicochemical characterisation of Bambarra groundnut (*Voandzeia subterranean*) starch and flour. *Food Chemistry, 78*, 305–311.

Afolabi, T. A. (2012). Synthesis and physicochemical properties of carboxymethylated Bambara groundnut (*Voandzeia subterranean*) starch. *International Journal of Food Science and Technology, 47*, 445–451.

Afolabi, T. A., Opara, A. O., Kareem, S. O., & Oladoyinbo, F. O. (2018). In vitro digestibility of hydrothermally modified Bambara groundnut (*Vigna subterranean* L.) starch and flour. *Food Science & Nutrition, 6*, 36–46.

Ashogbon, A. O. (2014). Physicochemical properties of Bambara groundnut starch and cassava starch blends. *African Journal of Food Science, 8*(6), 322–329.

Chigwedere, C. M., Olaoye, T. F., Kyomugasho, C., Kermani, Z. J., Pallares, A. P., Van Loey, A. M., Grauwet, T., & Hendrickx, M. E. (2018). Mechanistic insight into softening of Canadian wonder common beans (*Phaseolus vulgaris*) during cooking. *Food Research International, 106*, 522–531. ISSN 0963-9969. https://doi.org/10.1016/j.foodres.2018.01.016.

Falade, K. O., & Adebiyi, A. O. (2015). Effect of γ-irradiation on cooking, functional and pasting properties of Bambara groundnut (*Vigna subterranea* [L.] Verdc.) cultivars. *Journal of Food Process Engineering, 38*, 452–466.

Falade, K. O., & Nwajei, C. P. (2015). Physical, proximate, functional and pasting properties of four non- and γ-irradiated Bambara groundnut (*Vigna subterranean*) cultivars. *International Journal of Food Science and Technology, 50*, 640–651.

Gulu, N. B., Jideani, V. A., & Jacobs, A. (2019). Functional characteristics of Bambara groundnut starch-catechin complex formed using cyclodextrins as initiators. *Heliyon, 5*, e01562.

Gwala, S., Wainana, I., Pallares Pallares, A., Kyomugasho, C., Hendrickx, M., & Grauwet, T. (2019). Texture and interlinked post-process microstructures determine the in vitro starch digestibility of Bambara groundnuts with distinct hard-to-cook levels. *Food Research International, 120*, 1–11.

Gwala, S., Pallares Pallares, A., Pälchen, K., Hendrickx, M., & Grauwet, T. (2020). In vitro starch and protein digestion kinetics of cooked Bambara groundnuts depend on processing intensity and hardness sorting. *Food Research International, 137*, 109512.
Juliet, M., Vincenzo, F., Cathrine, C., & Anita, R. L. (2017). Hard-to-cook phenomenon in Bambara groundnut (*Vigna subterranea* (L.) Verdc.) processing: Options to improve its role in providing food security. *Food Reviews International, 33*, 167–194.
Ojimelukwe, P. C. (1999). Cooking characteristics of four cultivars of Bambara groundnuts seeds and starch isolate. *Journal of Food Biochemistry, 23*, 109–117.
Otegbayo, B., Oguniyan, D., & Akinwumi, O. (2013). Physicochemical and functional characterization of yam starch for potential industrial applications. *Starch-Starke, 65*, 1–16.
Oyeyinka, S. A., Singh, S., Adebola, P. O., Gerrano, A. S., & Amonsou, E. O. (2015). Physicochemical properties of starches with variable amylose contents extracted from Bambara groundnut genotypes. *Carbohydrate Polymers, 133*, 171–178.
Oyeyinka, S. A., Singh, S., Ma, Y., & Amonsou, E. O. (2016). Influence of high-pressure homogenization on the physicochemical properties of Bambara starch complexed with lysophosphatidylcholine. *LWT- Food Science and Technology, 74*, 120–127.
Oyeyinka, S. A., Singh, S., & Amonsou, E. O. (2017a). Physicochemical and mechanical properties of Bambara groundnut starch films modified with stearic acid. *Journal of Food Science, 82*(1), 118–123.
Oyeyinka, S. A., Singh, S., & Amonsou, E. O. (2017b). Physicochemical properties of starches extracted from Bambara groundnut landraces. *Starch/Staerke, 69*(3–4), 1600089.
Oyeyinka, S. A., Singh, S., Venter, S. L., & Amonsou, E. O. (2017). Effect of lipid types on complexation and some physicochemical properties of Bambara groundnut starch. *Starch/Staerke, 69*(3–4), 1600158.
Oyeyinka, S. A., Adegoke, R., Oyeyinka, A. T., Salami, K. O., Olagunju, O. F., Kolawole, F. L., Joseph, J. K., & Bolarinwa, I. F. (2018). Effect of annealing on the functionality of Bambara groundnut (*Vigna subterranea*) starch–palmitic acid complex. *International Journal of Food Science and Technology, 53*, 549–555.
Pallares Pallares, A. K., Loosveldt, B., Karimi, S. N., Hendrickx, M., & Grauwet, T. (2019). Effect of process-induced common bean hardness on structural properties of in vivo generated boluses and consequences for in vitro starch digestion kinetics. *British Journal of Nutrition, 122*, 388–399. ISSN: 0007-1145, 1475-2662. https://doi.org/10.1017/S0007114519001624.
Piyarat, S. (2008). Structure and functional properties of starch and flour from Bambarra groundnut. *Songklanakarin Journal of Science and Technology, 30*, 51–56.
Sirivongpaisal, P. (2008). Structure and functional properties of starch and flour from Bambara groundnut. *Songklanakarin Journal of Science and Technology, 30*, 51–56.
Yao, D. N., Kouassi, K. N., Erba, D., Scazzina, F., Pellegrini, N., & Casiraghi, M. C. (2015). Nutritive evaluation of the Bambara groundnut Ci12 landrace [*Vigna subterranea* (L.) Verdc. (Fabaceae)] produced in Côte d'Ivoire. *International Journal of Molecular Sciences, 16*, 21428–21441.
Zavareze, R. E., & Guerra Dias, A. R. (2011). Impact of heat-moisture treatment and annealing in starches: A review. *Carbohydrate Polymers, 83*, 317–328.

Chapter 9
Bambara Groundnut Proteins and Protein Isolates

9.1 Introduction

The knowledge of protein composition and structure is essential to explain legumes' functionality and application in food systems (Arise et al., 2016). Structure and composition may vary among legumes of the same genotypes, which will affect their functionality. Globulins (vicilin and legumin) and albumins are legume proteins that possess several functional and bioactive properties suitable as food ingredients (Kudre & Benjakul, 2014a). Legumes, including BGN have historically been used mainly as whole seeds. However, utilization of legumes in other forms, especially isolates or concentrates, has gained increasing attention because of their high protein content, beneficial effects on nutrition and health and potential for developing new foods (Kudre et al., 2013). Protein isolates from BGN (BGNPI) prepared by isoelectric precipitation exhibit excellent physicochemical properties and a well-balanced composition of essential amino acids in line with the FAO/WHO pattern. However, off-odour/flavour (e.g. beany, green, grassy) associated with oxidative degradation of residual lipids during either storage or processing is a major technical impediment to the application of BGN protein isolate (BGNPI) in foods (Kudre et al., 2013; Kudre & Benjakul, 2014a). A better understanding of BGNPI will lead to their practical uses. This chapter provides a preview of the knowledge about BGN protein isolate and concentrate.

9.2 Bambara Groundnut Flour and Protein

Bambara groundnut is an essential legume with a growing global appeal as a plant food (Duodu & Apea-Bah, 2017). BGN has high protein contents that could be utilized to formulate protein-rich diet or supplement diets low in protein content for the population dependent on plant-based protein (Adeleke et al., 2017). Delivering

V. A. Jideani, A. I. O. Jideani, *Bambara groundnut: Utilization and Future Prospects*, https://doi.org/10.1007/978-3-030-76077-9_9

the nutritional needs by food fortification of staples with underutilized plant species (Teye et al., 2020) meets food security in different parts of the world. The seed is relatively high in protein ranging from 15.88 to 24.91% (Adeleke et al., 2017; Afolabi et al., 2017; Atoyebi et al., 2017, 2018; Igbabul et al., 2013; Ikegwu et al., 2020; Kelechi-Ebisike et al., 2020; Ndidi et al., 2014; Temegne et al., 2018).

The properties of some protein fractions from BGN were evaluated by Poulter (1981). It is regarded as a good source of high-quality protein and micronutrients, like the common bean and soybean (Duodu & Apea-Bah, 2017) and characterised as gluten-free (Taylor & Awika, 2017). Atoyebi et al. (2017) reported that (i) the proteins of BGN are capable of combating the increased risk of venous thrombosis and the activation of protein c cofactor activity in the human plasma. Therefore, protein deficiency diseases such as marasmus, kwashiorkor and cachexia can be reduced, thus assisting in the overall prevention against protein-energy undernutrition; (ii) food supplementation with certain lines of BGN can be useful in the supply of essential amino acids (Table 9.1), thus making it to serve as a potential building block for the maintenance of body functions and metabolism and (iii) the essential amino acid contributes to cellular growth and protein synthesis. It has the

Table 9.1 Amino acid composition of Bambara groundnut, soybean and fava bean[a,c]

	Bambara groundnut			
Amino acid (g/100 g)	White	Brown	Soybean	Fava bean (*Vicia faba* L)
Threonine[b]	5.10	4.68	4.10	4.20
Lysine[b]	5.93	6.10	6.40	7.10
Methionine[b]	1.40	1.40	1.40	0.90
Valine[b]	5.10	4.98	5.00	5.10
Isoleucine[b]	3.98	4.10	4.00	4.50
Leucine[b]	6.70	6.48	7.80	8.40
Phenylalanine[b]	4.73	5.28	5.20	4.50
Arginine[b]	8.06	7.79	7.70	9.80
Histidine	2.42	2.62	2.80	2.80
Serine	5.56	5.71	5.60	6.10
Glycine	2.38	2.55	4.70	5.10
Tyrosine	7.12	7.42	3.80	3.50
Aspartic acid	12.34	11.90	11.60	12.00
Glutamic acid	15.34	15.08	19.10	14.90
Alanine	3.84	3.67	4.30	6.60
TAA	90.00	89.76	93.5	95.50
TEAA	41.00	40.81	41.60	44.50
TNEAA	49.00	48.95	51.9	51.00
%TEAA/TAA	45.56	45.47	44.49	46.60
%TNEAA/TAA	54.44	54.53	55.51	53.40

[a]Adapted from Adebowale et al. (2011a) and Sharan et al. (2021)
[b]Essential amino acid
[c]*TAA* Total amino acid, *TEAA* Total essential amino acid, *TNEAA* Total non-essential amino acid

potentials to serve as precursors for the body in producing the remaining non-essential amino acids.

Kissra is traditionally prepared by fermenting sorghum seed flour and then baking it into pancake-like leavened bread. Abdualrahman et al. (2019) studied incorporating BGN flour (BGNF) at 10, 20 and 30% into *kissra* bread. Protein, fat and energy of the control steadily increased with increasing BGNF. All minerals (except for Mg and P) and all essential amino acids (except for tryptophan and proline) increased significantly after fortification. Palmitic, stearic, oleic and linoleic acids were improved considerably while tannins, phytic acid and trypsin inhibitor activity were reduced significantly. The protein quality of *kissra* improved with BGNF fortification shown by improved nutritional index, essential amino acid index, biological value, predicted protein efficiency ratio and the in vitro protein digestibility.

Fortification increased the total phenolics and total flavonoids significantly and decreased the DPPH free radical scavenging activity. Overall, BGN flour could be used to upgrade the nutritional quality of cereal-based products.

9.2.1 Effect of Sprouting on Bambara Groundnut Proteins

Using SDS-PAGE Martınez et al. (2012) evaluated the proteins patterns of BGN seeds (Fig. 9.1) and observed (i) 12 bands in the original protein seeds and 1–9 days germinated samples proteins in the range of 28–49 and 49–80 kDa indicated an important reduction, but an increase in bands about 27 kDa, (ii) Phenolic compounds increased their concentration since the first day of germination until reaching 450% more than the original seed level, and (iii) the increase of phenolic compounds with antioxidant properties implies that germination results into legume foods for human consumption with better nutraceutical properties.

Okpuzor et al. (2010) identified the different types of proteins in the malted and dry seeds of BGN and their similarities to other *Vigna* species. A total of 10 different proteins were identified from the malted and 12 from the dry BGN seeds from 214 peptides isolated after searching 586 proteins of the genus *Vigna*.

Of the 22 different proteins, B storage protein and vicilin were dominant proteins common in malted and dry seeds and are similar to *Vigna luteola*. Others are beta and alpha isoforms of 8S globulin and 10 kDa protein precursors. The other proteins varied according to the condition of the seeds suggesting that each state of the seed is unique and develops the requirements that enable it to cope with biological demands. Some of the proteins have amino acid sequence homology to *Vigna radiata* and *Vigna unguiculata* species. Heat shock, BV1 and Bowman-Birk Inhibitor proteins are only in the malted state. The BV1 gene of the bipartite Begomovirus genome encodes a nuclear shuttle protein and is an avirulence determinant present in the common bean. The virus infection might have occurred during the malting process of *Vigna subterranea*.

Heat shock proteins had amino acid homology with that in *Phytophthora melonis*. Heat shock proteins are expressed in response to heat and other forms of stress.

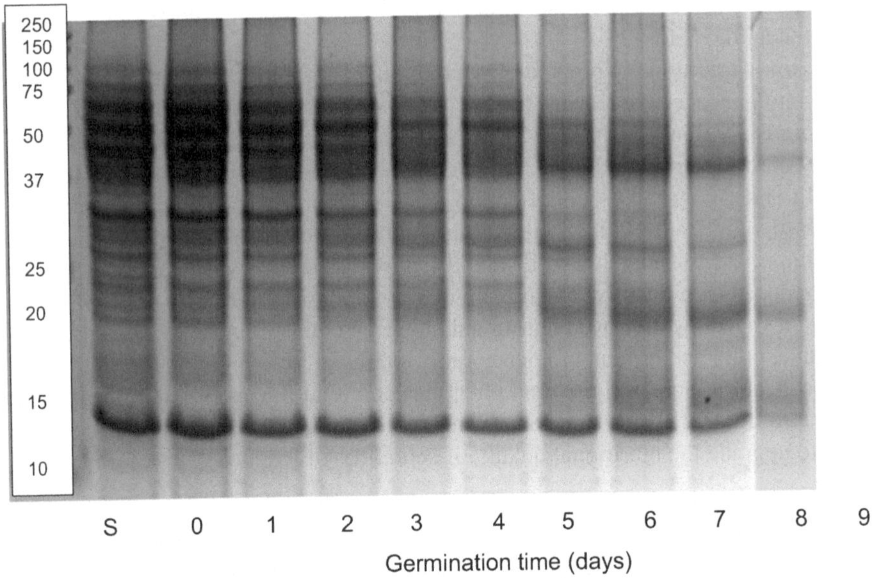

Fig. 9.1 Sodium dodecyl sulfate polyacrylamide gel electrophoresis (SDS-PAGE) of present proteins in the L. campestris germinated seed. S = standards (kDa)

They are responsible for protein folding, assembly, translocation and degradation in many normal cellular processes. They also stabilize proteins and membranes and under stress conditions, can assist in protein refolding. Okpuzor et al. (2010) postulated that during the malting process of BGN seeds, a form of stress in the form of excess or shortage of water and infection might have resulted in the appearance of heat shock proteins. This protein may be transient because it has a minimal percentage coverage and a low confidence level. Malted BGN seeds rather than increase or decrease seed nitrogen, produced other protein types such as Chain A, the crystal structure of adzuki bean 7S globulin-1 and Chain I, the crystal structure of the Bowman-Birk Inhibitor. Dehydrin present in the dry seeds may assist in withstanding stress due to lack of water. Seeds of higher plants accumulate large quantities of seed proteins including storage proteins and lectins. Seed storage proteins are deposited in protein bodies, specialised vacuoles known as protein storage vacuoles (PSVs) that serve as nitrogen, sulphur and carbon source for the developing embryo during seed germination. They are also significant sources of dietary protein for humans. Unique vesicles that are precursor-accumulating mediate the mass transport of storage proteins to PSVs and are known to contain a large amount of the precursors of major storage proteins of seeds. These types of precursor proteins are 8S globulin beta and alpha isoforms and 10 kDa protein precursors. BGN vicilin is a globulin storage protein found in plants. Vicilin (7S globulin subunit) protein of soybeans was reported to lower plasma lipids and upregulated liver (β)-VLDL receptors in rats fed with hypercholesterolemic diet. Seed proteins have been implicated in antihypertensive activities, regulation of food intake and body weight in humans and the improvement of plasma and liver lipid profile in rats fed with high cholesterol diet.

9.3 Effect of Cooking on Bambara Groundnut Protein Digestibility

The cooking of brown BGN did not significantly change the protein content but increased for brown BGN (Oyeyinka et al., 2019). The differences in the protein content of cooked grains depend on the kind of grain, the degree to which soluble solids are lost during soaking and cooking, and the incomplete elimination of certain amino acids and other nitrogenous compounds seeping during soaking. The amino acid content of BGN increased by cooking due to the degradation of some amino acids. Cooking resulted in increasing *in-vitro* protein digestibility of brown BGN than in the red BGN. The increase may be due to the decrease in the antinutrient content during thermal processing and also structural fragmentation of long polymer chains of native proteins. Heat treatment destroys heat-sensitive protease inhibitors and denatures proteins, consequently breaking the structure (especially globulins) making them less resilient to proteases.

There is an extensive variation in hardness in cooked BGN seeds at specific cooking times. Protein digestibility strongly depends on the hardness, making it a suitable food design variable to adjust starch and protein digestion of pulses. Low hardness seeds digest faster than high hardness seeds (Gwala et al., 2020).

9.4 Bambara Groundnut Protein Isolate and Vicilin Protein Fraction and Gelation

Arise et al. (2017b) reported similar protein content (90.7%) and protein yield (51.2%) across landraces for BGN protein isolate. The protein content is higher for pigeon pea (82.4%) and hemp (84.2%) seeds but similar to cowpea (89.5%). The yield is equivalent to 51% for cowpea protein isolate. The primary storage proteins in legume seeds known for their ability to form gels is the vicilin. Depending on their source, descriptive names include β-conglycinin from soybean (*Glycine max*) and phaseolin from kidney bean (*Phaseolus vulgaris*). Those from emerging crops are referred to as vicilins. BGN protein consists of broadband at 55 kDa, two medium bands at 62 kDa, 80 kDa and a high molecular weight (HMW) protein at 141 kDa, which is stabilized by disulphide bond. The protein bands at 55 and 62 kDa correspond to 7S vicilin subunits, the main polypeptides of BGN storage proteins (Arise et al., 2016; Diedericks et al., 2019).

Vicillin is the primary storage protein fraction in protein isolates. BGN vicilin contains 91.4% protein, fat (0.4%), moisture (0.2%), ash (2.2%) and carbohydrates (5.8%), being comparable to pea vicilin with a protein content of 90.0% and carbohydrate 6.8%. BGN vicilin presents two major bands 170 and 385 kDa in native-PAGE profiles. The broadband at 170 kDa corresponds to the trimeric structure of vicilin varying between 150 and 190 kDa, while that at 385 kDa corresponds to the hexamer legumin (Diedericks et al., 2019). The authors reported three major polypeptide bands at approximately 53, 65 and 118 kDa similar to the previous report for BGN protein isolates (BGNPIs) and vicilin-like protein (Arise et al., 2016; Busu & Amonsou, 2019; Diedericks et al., 2019). The band at 53 kDa is denoted as the β-subunit in legume seed vicilins dominant in BGN vicilin. The insolubility of BGN vicilin at low ionic strength further confirms the β-subunit composition. The band at 65 kDa common in other vicilin proteins and denoted at the α-subunit is also known as convicilin (Diedericks et al., 2019). The band at 118 kDa is similar to the bands identified for BGN protein isolate under non-reducing conditions and to the high molecular weight bands in pea vicilin and red kidney bean vicilin-rich protein isolate (Adebowale et al., 2011a; Diedericks et al., 2019). Mune Mune et al. (2018) reported three major bands of 62, 46 and 21 kDa under reducing conditions for BGNPI. Micellised and isoelectric BGN protein isolates had three major bands at 35, 43 and 112 kDa, but the isoelectric isolate bands were well defined compared to the micellised. All the BGNPIs were not dissociated by 1, 4-Dithiothreitol (DTT) suggesting that they do not contain subunits linked by a disulphide bond. Hence, 7S vicilin may be the primary storage protein in BGN isolates (Adebowale et al., 2011b). Alabi et al. (2020) also reported vicilin as the significant fraction of the BGN storage protein isolate with an intense band around 50 kDa being significantly different in composition and functionality than those of the whole storage protein isolate. Three major polypeptides (50, 70 and 80 kDa) in the vicilin fraction with a disulfide bond(s) in one strand. BGN concentrates from acid precipitation gave higher protein contents and yields 79, 52%, respectively, compared to salt solubilisation (Arise et al., 2015). Protein concentrates prepared through salt solubilisation exhibited better functional properties in water absorption capacity, oil absorption capacity, foaming capacity, foaming stability, and emulsion activities compared to concentrates obtained through acid precipitation. Hence, salt solubilisation might be the most appropriate method for the enhanced functionality and utilisation of BGNs protein concentrates.

Diedericks et al. (2020) investigated the structural changes in BGN protein due to soaking and roasting pre-treatments. Roasting BGN seeds changed the molecular composition of BGNPI while soaking resulted in minimal change comparing favourably to the control. Thermal aggregation of BGN proteins was evident in the roasted and soaked-roasted seeds supported by the formation of soluble and insoluble aggregates. Yet, BGNPIs are thermally stable since the same elution patterns for the control were present in the roasted seeds. Microstructurally, there were no marked differences between the red and black-eye varieties.

Kudre et al. (2013) reported a comparative study on chemical compositions and properties of protein isolates(PIs) from mung bean (MBPI), black bean (BBPI) and

BGN (BGNPI). All the protein isolates had a protein content in the range of 85.2–88.2%. The highest trypsin inhibitory activity was in BGPI. All PIs exhibited satisfactorily balanced amino acids in agreement with the FAO/WHO pattern. MBPI and BBPI had three predominant proteins with a molecular weight (MW) range of 42–54 kDa, whereas BGNPI had two dominant proteins with MW of 52 and 62 kDa. Based on differential scanning calorimetric analysis, MBPI and BGNPI had two endothermic peaks, whereas three peaks for BBPI. All PIs exhibited similar FTIR spectra, indicating similarity in functional group and structure. All protein isolates showed a minimum protein solubility at around pH 4–5 but 0.45 mol L^{-1} NaCl could increase their solubility. Comparatively, BGNPI exhibited superior solubility to BBPI and MBPI. All the PIs are good sources of lysine, had high thermal stability and could be used as ingredients in heat-treated food systems.

Glutamic and aspartic acids which may include glutamine and asparagine are the primary amino acids in BGNPI, similar to other legume seeds such as soybean, mung bean and black bean (Table 9.2). BGN protein is high in lysine (6.0 g/100 g protein) and identical among landraces. Hence, BGN is a food source of lysine and could be used in food formulation where it could complement cereal proteins (Arise

Table 9.2 Amino acid composition of Bambara groundnut protein isolates[a]

Amino acid (g/100 g)	Ruzengwe et al. (2020a)	Kudre et al. (2013)	Arise et al. (2017b) _Red	Arise et al. (2017b) _Maroon	Arise et al. (2017b) _Cream
Threonine[b]	5.00	2.76	5.11	5.13	4.61
Lysine[b]	6.10	6.30	5.95	6.03	5.81
Methionine[b]	1.20	1.27	1.26	1.26	1.1
Valine[b]	4.40	4.32	4.09	4.45	3.82
Isoleucine[b]	3.20	3.76	2.97	3.16	2.83
Leucine[b]	6.90	7.32	6.71	6.96	6.69
Phenylalanine[b]	6.10	5.33	5.35	5.97	4.83
Arginine[b]	7.20	5.90	6.61	7.31	5.9
Histidine	2.50	2.99	2.41	2.43	2.39
Serine	4.30	3.27	3.99	4.48	3.17
Glycine	2.80	3.09	2.63	2.81	1.87
Tyrosine	2.90	3.28	2.73	2.99	2.56
Aspartic acid	10.50	9.56	9.32	10.72	8.49
Glutamic acid	15.50	15.41	15.66	15.91	14.93
Alanine	3.80	3.51	3.57	3.92	3.23
Proline	3.30	2.68	3.08	3.41	2.42
TAA	85.70	80.75	81.44	86.94	74.65
TEAA	40.10	36.96	38.05	40.27	35.59
TNEAA	45.60	43.79	43.39	46.67	39.06
%TEAA/TAA	46.79	45.77	46.72	46.32	47.68
%TNEAA/TAA	53.21	54.23	53.28	53.68	52.32

[a]Essential amino acid

[b]*TAA* Total amino acid, *TEAA* Total essential amino acid, *TNEAA* Total non-essential amino acid

et al., 2017b). The total essential amino acid and its ratio with the total amino acid of BGN protein isolate is similar to that of soybean (Table 9.2). BGN protein isolate's thermal properties showed two endothermic peaks, the denaturation of the 7S protein at 68–71 °C and HMW protein at 78–90 °C. Higher denaturation temperatures (83–84 °C and 91–93 °C) are reported for cowpea proteins linked to a high proportion of β-sheet structure.

9.4.1 Effect of Extraction Methods on Bambara Groundnut Proteins

Arise et al. (2015) reported on the functional properties of protein concentrates prepared from three BGN landraces using acid precipitation and salt solubilisation methods. The protein content of BGN seeds (26–27%) was similar for the three landraces. The acid precipitation gave a much higher yield of protein concentrates (52%), which were also high in protein (79%) compared to the salt solubilisation method (yield: 25%, protein content: 57%). Functional properties of proteins were more influenced by the methods of extraction rather than the landraces. Protein concentrate prepared by salt solubilisation method showed higher emulsifying (63–66%), foaming (53–57%), water (1.4–2.0 mg/mL) and oil absorption properties (2.2–2.6 mg/mL) than the acid-precipitated concentrates (53–57%, 63–66%, 2.0–2.7 mg/mL, 1.4–1.7 mg/mL). The foaming capacity and stability of the protein concentrate decreased with increasing pH from 3 to 8. Salt solubilisation may be the most appropriate method for the enhanced functionality and utilisation of BGN protein concentrate.

Adebowale et al. (2011b) reported that the micellised BGNPIs isolates have a higher denaturation temperature (97.9–108.4 °C) than the corresponding isoelectric isolates (89.5–90.6 °C). BGN protein isolate has a strong secondary structure dominated by α-helix conformation at pH 3, decreasing slightly at higher pH. The foaming capacity of BGN proteins is pH-dependent with the maximum at pH 3, while the emulsion activity increases with pH. The minimum solubility of the isolates was at pH 5, with micellised isolates exhibiting higher solubility than the corresponding isoelectric isolates overall pH. Micellised isolate was superior in functional characteristics (foam capacity and stability, oil, water absorption, emulsifying, gelation) than the isoelectric isolate. However, the isoelectric isolate had a higher protein yield (56.3–58.2 g/100 g) than the micellised protein (14.2–15.6 g/100 g). The isolates contained higher levels of lysine, arginine and glutamic acid compared to soy protein. However, Ruzengwe et al. (2020a) reported that BGNPI is a good source of aspartic acid, lysine and histidine with amounts similar to those in soy protein. Although the glutamic acid content is high, it is lower than the soy protein amount (Ruzengwe et al., 2020a).

9.4.2 Protein Isolation Methods

The isolelectric precipitation involves dispersing defatted BGN flour in deionised water (1:10 w/v) and adjusting to pH 9.5 with 1 M NaOH to solubilise the proteins before isoelectric precipitation at pH 4 (adjusted with 1 M HCl). The precipitated protein is separated by centrifugation (4000 g, 30 min) to obtain the protein-rich residue. This residue is adjusted to pH 7 before being freeze-dried as the protein isolate (Diedericks et al., 2020). Modified isoelectric precipitation adjusts the pH to 9.0 with 1 M NaOH to facilitate protein solubilisation. The suspension was stirred for 1 h at 1000 rpm and centrifuged at 10000 × g for 15 min at 4 °C. The supernatant was collected, and the pH was adjusted to 4.6 with 1 M HCl to precipitate the protein, which was then recovered by centrifugation at 10000 × g for 15 min at 4 °C. The recovered protein isolate was washed with 25 ml of water, and pH adjusted to 7.0 using NaOH. The water was removed through centrifugation at 4000 × g for 5 min at 4 °C, and the protein isolate frozen at −80 °C, freeze-dried and kept at 4 °C until required (Ruzengwe et al., 2020a).

Vicilin extraction method by Osborne fractionation method: Defatted flour (1:10 1/v) was dispersed in 0.5 M NaCl (pH 6.1, unadjusted) stirred at 4 °C for 1 h and centrifuged at 4 °C (10,000 g, 20 min). This procedure was repeated twice (final extraction with deionised water) to obtain three supernatants containing the globulins, albumins, sugars and soluble non-starch polysaccharides. The pooled supernatants were adjusted to pH 4.6 with glacial acetic acid (precipitation of legumin) in an ice bath, stirred at 4 °C for 30 min and centrifuged. The supernatant (containing vicilin, albumin and soluble sugars/non-starch polysaccharides) was dialysed at 4 °C against several distilled water changes until the conductivity of the water remained constant. The retentate was centrifuged, the residue freeze-dried and designated as the vicilin fraction (Diedericks et al., 2019).

Defatted flour (1:10 w/v) was extracted with 0.2 M borate buffer (pH 9) containing 0.5 M NaCl and centrifuged (20,100 g, 30 min, 4 °C). The supernatant was retained, and the insoluble borate sediment was re-extracted as described above and centrifuged. The pooled supernatants were adjusted to pH 4.6 with glacial acetic acid in the cold, stirred for 3 min and centrifuged (20,100 g, 30 min, 4 °C). The supernatant was dialysed for 4 days against distilled water and centrifuged under the same condition. The sediment was freeze-dried as the 7S vicilin.

Protein concentrate: For acid precipitation, defatted flour was suspended in water at 1:10 (flour to water ratio), and pH was adjusted to 8.0 with 1 M NaOH, to facilitate protein solubilisation. The suspension was stirred for 4 h at 32 °C in a shaking water bath. The centrifuge was set at 4000 g for 30 min at 4 °C. The precipitate was resuspended in half the initial water volume, and extraction was carried out as described above. The supernatants were pooled and pH adjusted to 4.0 with 0.5 M HCl to precipitate the protein concentrates, which were recovered by centrifugation at 5000 g for 30 min at 4 °C. The protein concentrate was freeze-dried and kept at 4 °C until required (Arise et al., 2015).

Protein concentrates were prepared using the salt solubilisation method. Defatted flours were suspended in 0.5 M NaCl (1:10 w/v). The suspension was stirred in a shaking water bath for 4 h at 32 °C. The suspension was centrifuged at 12000 g for 20 min at 4 °C. The clear supernatant was dialysed (cut-off 10 kDa) against distilled water for 48 h. The dialysed extract was freeze-dried and stored at 4 °C until required (Arise et al., 2015).

The BGPI protein content and yield depend on the method of extraction, as illustrated in Table 9.3. It appeared the authors expressed the yield differently. For instance, Diedericks et al. (2020) expressed the yield as the dry weight of BGNPI per weight of defatted flour. However, it is not clear what the other authors based the yield on. It seems the yields on the protein content-based, hence the higher yield values. While the BGNPI content ranged from 89 to 91%, Diedericks et al. (2020) reported a lower range of 68.3–72%. Adebowale et al. (2011b) and Arise et al. (2017a, b) both used a ratio of 1:20 for flour:water for the extraction with isoelectric precipitation of the proteins at pH 5. However, Diedericks et al. (2020) precipitated the proteins at pH 4.

9.5 Peptides from Enzyme Hydrolysis of Bambara Groundnut Protein Isolates

Food protein hydrolysates and peptides possess multiple bioactive properties that could find application in functional foods. Bioactive properties of peptides produced from plant proteins include antioxidant activity, cholesterol-lowering ability, blood pressure-lowering (ACE inhibitory) effect and anti-hyperglycemic (DPPIV inhibitory) effect (Mune Mune et al., 2018). Limited enzymatic hydrolysis *in vitro* is widely used to release bioactive peptides from parent proteins, and some of these peptides have been recommended as a component of functional foods or new drugs for the prevention of cardiovascular diseases (CVD) and type-2 diabetes (T2D) (Mune Mune et al., 2018). Some important risk factors for developing CVD are raised blood pressure, vascular oxidative stress and associated endothelial

Table 9.3 Bambara groundnut protein isolate content and yield

Bambara groundnut	BGNPI protein content (%)	BGNPI yield (%)	References
Black-eye	68.3	11.5	Diedericks et al. (2020)
Red	72.0	14.7	Diedericks et al. (2020)
Red	91.7	49.60	Arise et al. (2017b)
Maroon	89.8	51.50	Arise et al. (2017b)
Cream	90.8	52.50	Arise et al. (2017b)
	90.7		Mune Mune et al. (2018)
White	90.0	56.30	Adebowale et al. (2011b)
Brown	91.0	58.b0	Adebowale et al. (2011b)

BGNPI Bambara groundnut protein isolate

dysfunction. Angiotensin-converting enzyme (ACE) is a crucial target for the treatment of hypertension. The inhibition of ACE causes a reduction of blood pressure by increasing the level of angiotensin II and decreasing the level of bradykinin, a vasodilatory peptide. Dipeptidyl peptidase IV (DPP-IV) is an essential enzyme for the prevention and treatment of T2D. Inhibition of DPP-IV, a serine protease, enhances insulin secretion by modulating the insulinotropic hormone's biological activity, glucagon-like peptide-I (GLP-1). Inhibitors of DPP-IV have been used as drugs in type-2 diabetes. Many pathological conditions such as diabetes, cardiovascular disease, cancer and other chronic and inflammatory diseases were closely related to oxidative damage. Natural antioxidants and more recently plant protein hydrolysates effectively prevent and treat oxidative stress-related conditions (Mune Mune et al., 2018).

Mune Mune et al. (2018) hydrolysed BGN protein isolate (BGNPI) with alcalase, thermolysin and trypsin. Bambara groundnut protein hydrolysate (BGNPH) produced by alcalase showed the highest angiotensin-converting enzyme (ACE) inhibitory properties (IC50: 52μg/mL). Hydrolysates produced by alcalase and thermolysin exhibited similar dipeptidyl peptidase-IV (DPP-IV) inhibitory activity (IC50: 1.73 mg/mL), while low inhibitory activity was observed for hydrolysate produced by trypsin. BGNPH also showed a protective effect against oxidative stress with significant 2,2-diphenyl-1-picrylhydrazyl radical scavenging and ferrous chelating activity. Bioactive peptides of BGNPH produced by thermolysin showed better resistance to simulated gastrointestinal digestion (SGID), while the DPP-IV and ACE inhibitory properties were reduced significantly. Molecular weight distribution showed a significant reduction in peptides of the molecular weight range 200–400 Da in BGNPH produced by alcalase, after SGID. LC-ESI-TOF-MS and in silico analysis revealed potential peptides with both ACE and DPP-IV inhibitory properties in BGNPH produced by thermolysin.

Arise et al. (2017a) studied the antihypertensive and antioxidant activities of BGN protein hydrolysates from different proteases (alcalase, trypsin and pepsin) and their peptide fractions (10, 5, 3 and 1 kDa). Alcalase hydrolysate contained the highest low molecular weight (LMW) peptides amount than pepsin and trypsin hydrolysates. LMW peptides fractions (<1 kDa) exhibited the most increased inhibitory activity against angiotensin-converting enzyme (ACE) for all the enzyme hydrolysates, with 93.4% by alcalase peptide fraction. This value is greater than that reported for alcalase kidney bean and rapeseed protein hydrolysates (Arise et al., 2017a). Alcalase hydrolysate showed the highest renin inhibition at 59% than other hydrolysates and their corresponding membrane fractions. The antioxidant power of BGN protein hydrolysates and peptide fractions was evaluated through the inhibition of linoleic acid peroxidation and ABTS scavenging activity. Among the hydrolysates, alcalase exhibited the highest inhibition of linoleic acid oxidation. Furthermore, all BGN hydrolysates scavenged $ABTS^+$ to a threefold greater extent than the isolate. Hydrolysates obtained from alcalase and trypsin contained higher hydrophobic amino acid (HAA) than pepsin hydrolysate. HAA has been reported to act as antioxidants by increasing the solubility of peptides in lipids, facilitating better interaction with free radicals. The amino acid profile of all the hydrolysates revealed high levels of glutamine, asparagine, arginine and leucine. Acidic amino acids such as glutamine and asparagine have potent antioxidant effects due to the presence of excess electrons donated during interaction with free radicals. The authors concluded that the BGN protein hydrolysates and LMW peptide fractions are potential useful ingredients in the formulation of functional foods and nutraceuticals against high blood pressure and oxidative stress. BGN is a promising source of protein for the production of bioactive peptides due to their low cost, low allergenicity and adequate nutritional and functional values.

9.6 Bambara Groundnut Protein Isolate for Food

BGN is consumed by a large part of Indonesia's population because of its palatability and demonstrated nutritional values. Much of Indonesia's use is limited to boiling and frying of the seeds (Astuti et al., 2016). Food allergies are adverse immune reactions by some individuals to food proteins (allergens), resulting in various symptoms, ranging from mild rashes to life-threatening systemic anaphylaxis. The four main types of symptoms are dermatological (hives, local swelling, dermatitis and eczema), gastrointestinal (nausea, vomiting, diarrhoea and abdominal pain), respiratory (runny nose, asthma and tightening of the throat) and systemic (anaphylactic shock, organ failure, cardiac arrhythmia and death) (Astuti et al., 2016). BGN protein molecular mass ranges from 19.2 to 66.1 kDa with a high-intensity protein band with a molecular mass of approximately 55.7 kDa (Astuti et al., 2016). The authors reported that BGN contains allergenic proteins for some people with cross-reactivity to peanut and soybean. Individuals allergic to peanut, tree nut, or both have IgE antibody to more than one nut hence sequence analysis, carbohydrate moieties, 3-dimensional structures, and epitope mapping of the BGN allergen homologs in peanut and soybean are necessary to determine the degree of sequence similarity with BGN allergens (de Leon et al., 2007; Traidl-Hoffmann et al., 2009). However, there is still a need for evidence of an associated allergic reaction in at least three patients established by double-blind placebo-controlled challenge with BGN or a convincing description of food-induced fatal anaphylaxis cases (Mills et al., 2003).

9.7 Bambara Groundnut Protein Interaction in the Food System

The interactions of food components are of a physicochemical, biochemical, or chemical nature. The physicochemical interactions are responsible for the formation of emulsions, foams, and also partially gels. Thus, they significantly impact the sensory quality of various food commodities and the shearing resistance and the flow of the material in the processing equipment. Chemical and biochemical reactions in fresh and processed foodstuffs may affect all quality attributes, including colour, appearance, flavour, aroma, texture, biological value, nutritional properties, safety, and food processing suitability. Thus they may improve the edible characteristics or in other cases delimit storage life. Chemical interactions are possible among constituent molecules of the food such as proteins, lipids, carbohydrates, minerals and vitamins. The macronutrients of food include water, proteins, lipids and carbohydrates. The micronutrients include minerals, vitamins, enzymes and nutraceuticals. Interactions can occur between the macronutrients resulting in protein-protein, protein-lipid or protein-carbohydrate, and between the macro- and micronutrients leading to protein, lipid-enzyme or carbohydrate-enzyme. Vitamins can play a role as cofactors for the enzyme interaction in some instances. Similarly, some enzymes require minerals to maintain their conformational state that leads to catalysis. Physical conditions of temperature, moisture, pH and time affect the intensity of these interactions.

9.7.1 Protein-Water Interaction

Functional properties of proteins are those physicochemical attributes that make them useful in foods that result from the interaction of proteins with water. These functional properties include solubility, water absorption water binding, viscosity, gelation, cohesion, adhesion, elasticity, emulsification and foaming all depend on the interaction of proteins with water (Table 9.4). Solubility is an essential attribute for the use of proteins in beverages. The mechanism involves hydrophilicity, and the presence of ions influences solubility. Both the type of ions and their concentrations affect solubility. This critical attribute of proteins also is a prerequisite for other functionalities such as viscosity, foaming, emulsification, and gelation. Since the mechanism for solubility involves hydrophilicity, the amino acid composition of a protein plays an integral role in determining proteins' interaction with water. Protein solubility is pH-dependent. For most proteins, the point of minimum solubility occurs near its isoelectric point. Above and below the isoelectric point, pH values alter the charge on the molecule and, therefore, its interaction potential with water.

Some proteins, such as whey proteins, retain their solubility at their isoelectric point (Hu et al., 2019). Increasing temperature leads to disorder in the structure of

Table 9.4 General classes of functional properties of proteins important in food applications

General property	Specific functional term
Organoleptic/ kinesthetic	Color, flavor, odor, texture, mouthfeel, smoothness, grittiness, turbidity, etc.
Hydration	Solubility, dispersibility, wettability, water absorption, swelling, thickening, gelling, water holding capacity, syneresis, viscosity, dough formation, etc
Surface	Emulsion, foaming, aeration, whipping, protein/lipid film formation, lipid binding, flavor binding, stabilization, etc.
Structural/textural/ rheological	Elasticity, grittiness, cohesion, chewiness, viscosity, adhesion, network cross-binding, aggregation, stickiness, gelation, dough formation, texturability, fiber formation, extrudability, elasticity, etc
Other	Compatibility with additives, enzymatic, inertness, modification properties

Adapted from Akharume et al. (2021)

the protein and also a loss in solubility. The desirable sensory properties of such foods as gravies, soups, and beverages are derived from their viscosity. Viscosity is highly dependent on solute type and concentration. Macromolecules increase viscosity at low concentrations depending on the size, shape, flexibility, and hydration of the protein in question. The viscosity of protein solutions increases exponentially with concentration. An increase in hydrodynamic radius caused by heating proteins also tends to increase viscosity. Environmental factors such as pH, ionic strength, and temperature also affect the viscosity of protein solutions. The viscosity of protein solutions decreases as the isoelectric point is approached.

Bambara groundnut vicilin displays substantially high solubility with better foaming and emulsifying activities than the storage protein due to its high proline content and less folded structure. In contrast to the storage protein, which had less than 40% protein solubility at pH 2–9, vicilin has a significantly higher solubility with a minimum of 50% at pH 5 and a maximum of 82% at pH 3. The emulsifying stability of the proteins significantly differed at pH 7. The vicilin fraction's foaming capacity was considerably higher than that of the storage protein at pH 3, 7, and 9. The high solubility of BGN vicilin in the acidic region establishes an opportunity for its application in acidified beverages (Alabi et al., 2020). The isoelectric point (pH 5.3), 86% solubility at NaCl concentration above 200 mM and thermal denaturation of 92 °C for BGN vicilin are in the range reported for other legume vicilins (Diedericks et al., 2019).

Gelation is a phenomenon in which protein solutions transform into a solid structure capable of holding large amounts of water. Proteins can form either coagula or transparent gels. Amino acid composition and the environmental conditions dictate the nature of the structure formed (Damodaran, 1997). Proteins with a high frequency of nonpolar amino acid residues tend to form coagula (Shimada & Matsushita, 1980). When the sum of Val, Pro, Leu, Ile, Phe, and Trp residues in the protein is above 31.5 mol%, coagulum-type gels are likely to form. Translucent gels hold more water than coagulum-type gels. Gelation of proteins can be induced by heating, altering the pH, adding salts to protein solutions, and enzymes' action.

The general principle is to modify the protein structure by adding one of the agents mentioned above. The interaction of the altered protein with water and with other protein molecules results in gelation. In all instances concentration of the reactive species is an important variable. Gelation is a process that involves not only protein-water interactions but also protein-protein interactions. Tofu, cheese, puddings, sausages, and cakes are also examples of foods in which gelation plays a key role. Gelation is vital in forming distinct texture characteristics of many foods such as cheeses, boiled eggs and jams. In particular, protein gelation provides a desirable texture, structure or stability to food products. It makes it possible to entrap other food ingredients such as oil droplets in its matrix, contributing to sensory properties. Understanding gelation processes provides an understanding of structure formation at the molecular scale, which determines functionality at the macroscopic scale enabling the use of individual components and mixtures in novel ways (Diedericks et al., 2019). The minimum gelling concentration for BGN protein isolate is 7.2–7.3%, whereas BGN vicilin is 5.5%. BGN vicilin can form a gel at a lower concentration compared to 12.5% for soybean β-conglycinin. BGN vicilin gels are characterised by weak-link fractal, which translates to rigid aggregate structures at 200 mM NaCl (Diedericks et al., 2019).

The interfacial properties of proteins are critical in foaming and emulsification. Foams are formed at the air-water interface, whereas emulsions form at oil-water interfaces. Whipped toppings, ice cream, cakes, and desserts are all examples of foods in which foams are involved. Foods in which emulsification plays an important role are salad dressings, mayonnaise, bologna, soups, gravies, and cakes. A prerequisite for foaming or emulsification is for the protein to be soluble in water. The soluble protein can then rearrange itself at the interface, thus lowering the surface and/or interfacial tension and facilitating the formation of foams or emulsions. As far as the protein-water interactions' molecular nature is concerned, these involve both polar and nonpolar amino acid residues on the protein molecules. The interaction itself may be mediated via dipole-dipole, charge-dipole, and dipole-induced dipole interactions. Charged amino acids can bind up to 6 mol of water per residue, whereas uncharged polar residues and nonpolar residues bind as much as 2 and 1 mol of water per residue, respectively. Therefore, the interaction of water with a protein is governed by the protein's amino acid composition. Typically, most nonpolar residues and a significant number of polar residues are buried in the protein's interior and thus are unavailable for interactions until the structure is perturbed. Such perturbations can occur by changing the environmental conditions (pH, ionic strength, and temperature) in food processing.

9.7.2 Protein-Protein Interaction

The molecular forces involved in protein-protein interactions are similar to those involved in the formation of the secondary and tertiary structure of the individual protein molecules. The quaternary structure of a protein is one example of protein-protein interaction. These forces are van der Waals interactions, hydrophobic interactions, electrostatic interactions, and hydrogen bonds. Kinetic or other constraints may prevent interactions from occurring. When solutions of globular proteins are heated, conformational changes occur. These changes can be followed by changes in viscosity, circular dichroism, surface hydrophobicity, and ultraviolet radiation absorption. Differential scanning calorimetry is a valuable tool in monitoring thermally induced changes in protein molecules. Conformational changes expose reactive groups, and covalent and noncovalent interactions within the same molecules and between different molecules can occur. If the protein concentration is appropriate, gelation may even occur.

Glutamine and asparagine are derivatives of glutamic acid and aspartic acid, respectively. The high content of glutamic acid (including glutamine) and lysine in the BGNPI suggests that its structure can be stabilised by transglutaminase (TGase) when used in hydrogel formation-mediated cross-linking. TGase catalyses the cross-linking reaction between the carboxamide group of glutamine (acyl donor) and the amino group of lysine and other primary amines (acyl acceptor) (Ruzengwe et al., 2020a). Thiol group-containing amino acids like cysteine are also present in BGNPI, and these are involved in disulfide bond formation during TGase cross-linking. Hence, the presence and abundance of these amino acids could favour TGase cross-linking of BGNPI, resulting in a structured network. TGase promoted the formation of an organised network with better hydrogel strength. TGase activity up to 15 U/g protein could be recommended as further increase had no significant effect on the hydrogel strength. The network structure is TGase activity-dependent up to 15 U/g protein with the treated hydrogel displaying numerous smaller-sized pores and a homogeneous network. Transglutaminase mediated the crosslinking of BGNPI hydrogel, as demonstrated by a reduction in amine and thiol groups. Transglutaminase crosslinking can, therefore, be used to improve the properties of BGNPI hydrogels. The approach used by Ruzengwe et al. (2020a) enabled gel formation at low temperatures; thus, the produced hydrogels is suitable for the encapsulation of heat-sensitive bioactive compounds.

Ruzengwe et al. (2020b) reported the rheological and microstructural properties of BGNPI gels, prepared under varying pH and salt conditions. BGNPI gels showed storage or elastic modulus (G′) > loss or viscous modulus(G″) over a frequency range of 0–100 rad s^{-1}. Although the gels displayed the characteristics of weak gels, slightly acidic conditions (pH 5.59–6) coupled with low NaCI concentration (0.5 mol/L) promoted the formation of more rigid gels. NaCl had a stabilising effect on the protein structure at slightly acidic conditions as demonstrated by the increase in denaturation temperature and enthalpy of denaturation with increasing NaCl concentration. Also, an increase in α-helix and β-turns contents and disappearance of random coils were observed when 0.5 mol/L NaCl was added at pH 6. Gels with the highest strength had the lowest water holding capacity and thiol content, suggesting more disulphide linkages during network formation. Microscopy images of gels prepared at the slightly acidic pH in the presence of NaCl showed porous homogeneous aggregates. The BGNPI gels have potential applications for the improvement of food texture.

Surimi is concentrated myofibrillar protein obtained from mechanically deboned fish flesh. Sardine (*Sardinella albella*) is a small coastal pelagic fish species, which can be used for surimi production in Thailand. Apart from the dark colour, surimi from sardine has poorer gel properties due to high proteolytic degradation of myofibrillar proteins by heat-stable endogenous proteases leading to less demand for sardine surimi. Plant-based protein additives containing protease inhibitors such as protein isolates from mung bean, black bean, and BGN are used to alleviate the soft texture problem caused by endogenous proteases. Due to relatively low cost and high nutrition value, protein isolate from legume seeds have gained much attention for their use in surimi-based food products. Kudre and Benjakul (2014b) and Kudre and Benjakul (2013) investigated the different BGNPIs at a level of 6% (w/w) in combination with microbial transglutaminase (MTGase) at a concentration of 0.6 U g^{-1} surimi on gel properties of sardine (*Sardinella albella*) surimi. In the absence of MTGase, all BGNPIs showed an adverse effect on surimi's gel-forming properties, as evidenced by the decrease in breaking force and deformation. When MTGase was incorporated, an increase in breaking force and deformation were observed. The water holding capacity of all gels was improved when BGNPIs were added in combination with MTGase. The whiteness of gels slightly decreased with BGNPIs; however, MTGase had no impact on whiteness. Surimi gel added with BGPI prepared from defatted flour with heat treatment in the presence of ethylenediaminetetraacetic acid (EDTA) and MTGase showed well-ordered network and exhibited the lowest peroxide value and thiobarbituric acid-reactive substances, negligible beany flavour, lower amount of volatile compounds after storage of 30 days at room temperature. Thus, the addition of 6% in conjunction with MTGase (0.6 U g − 1 surimi) in sardine surimi could improve gel property without beany flavour.

The impact of BGNPI on autolysis and gel properties of surimi from threadfin bream (*Nemipterus bleekeri*) was investigated by Oujifard et al. (2012). BGNPI with trypsin inhibitory activity of 6356.3 unit/g markedly increased the breaking force and deformation of modori gel as the levels (0–3 g/100 g) increased. Nevertheless, only 0.25 g/100 g BGNPI increased breaking force and deformation of kamaboko gel beyond which detrimental effect on gelation was observed. Myosin heavy chain (MHC) was more retained when BGNPI concentration increased, especially in modori gel. However, whiteness slightly decreased with increasing BGPI levels. The microstructure of kamaboko gel with 0.25 g/100 g BGNPI had finer and more ordered fibrillar structure than that without BGPI. Therefore, BGNPI at an appropriate level could be an alternative food-grade inhibitor to improve the gel properties of surimi.

9.7.3 Protein-Lipid Interaction

One aspect of such interactions mentioned briefly is interfacial behavior that leads to the formation of emulsions. In emulsions, proteins adsorb at the oil-water interface. They then undergo unfolding and rearrangement, allowing the hydrophobic parts of the protein to interact with the oil phase and the hydrophilic portion of the same molecule to anchor in the aqueous phase. Typically food emulsions are produced using combinations of small molecule surfactants and macromolecular surfactants. Although competitive adsorption may occur, there also can be interactions between proteins and the low-molecular-weight surfactants. Cooperative interactions lead to emulsion stability, and competitive or negative interactions result in macromolecular desorption. One common method of improving the stability of dairy emulsions is to add mono- and diglycerides to the system. These compounds adsorb onto the fat droplet surface and impart stability toward coalescence.

9.7.4 Protein-Carbohydrate Interaction

The most studied and common family of reactions involve those between reducing sugars and the amino groups of proteins known as the Maillard reaction. More recently, the reactivity between the amino group of asparagine and the reducing group of sugars has been implicated in acrylamide formation in starch-based foods subjected to high temperatures (Mottram et al., 2002). Although the presence of acrylamide in foods has been irrevocably established, its public health significance remains unclear. In most proteins, the amino groups are involved in the peptide linkage. Therefore, the only free amino groups available for reactions are glutamine, asparagine, lysine, and the N-terminal amino group. There are, however, free amino acids and peptides also present in foods. The peptide and free amino acid content of foods increase in the event of proteolysis. Proteolysis can be caused by endogenous proteases in fresh foods, exogenous proteases added to achieve a desirable property or subjection of the food substrate to fermentation. A Maillard reaction occurs during the processing and storage of foods. Its effects can be desirable in producing flavors and imparting color. It may be undesirable when nutritional properties are adversely affected, or the color and flavor may be undesirable. A Maillard reaction also leads to flavor generation, cross-linking of protein chains (in advanced stages), loss of nutritional value due to loss of lysine, the formation of various colors, and antioxidative and even mutagenic and carcinogenic effects. Protein and starch constitute the major components in many food systems. Cereals, legumes, and tubers, generally consumed after cooking, are examples of foods composed mainly of starch and protein. In products such as bakery foods, pasta, snack foods, breakfast cereals, and the like, starch and protein are shaped and flavored in various ways. The characteristic flavor and texture of these foods are due in part to the protein-starch interactions. The breakdown of starch by enzymes to yield high fructose corn syrups, maltodextrins, corn syrup solids, etc., are also specific instances of protein-starch interactions because enzymes are catalytic proteins. In the absence of heat, the mechanism of the protein-starch interaction is a charge-charge interaction. Hence, it is highly pH-dependent and on the proteins isoelectric point (Takeuchi, 1969; Dahle, 1971).

Heating increases the complexity of the reactions between starch and protein. Thermal changes in protein lead to denaturation (Kilara & Sharkasi, 1986). This denaturation is accelerated in the presence of moisture. Denaturation of cereal proteins leads to disulfide-sulfhydryl inter-change reactions resulting in the extensive cross-linking of proteins, i.e., protein-protein interactions. Starch undergoes a loss of crystallinity, swelling of granules, and leaching of amylose out of the granules leaving mostly amylopectin behind. The granules collapse and the matrix of amylose forms a part of the gel network (Remsen & Clark, 1978). When protein and starch are in contact, a stable protein-starch matrix involving hydrogen bonding, covalent bonding, and ionic linkages may form.

Hydrocolloids or large-molecular-weight polysaccharides are used as stabilizers in many processed foods. Proteins and hydrocolloids can also interact to form desirable structures in foods, a topic of considerable interest to the food industry. When protein and polysaccharide solutions are mixed, four different results can occur. In two cases, stable solution mixtures are formed. In the other two cases, phase-separated mixtures are formed. The protein-polysaccharide mixtures can be soluble or insoluble. Interpolymer complex formation occurs at pH values below the isoelectric point of the protein and low ionic strengths [<0.3] (Tolstoguzov, 1997). Proteins are positively charged below their isoelectric point and negatively charged above their isoelectric point. Carboxyl-containing polysaccharides at acidic to neutral pH range carry a net positive charge. Therefore, electrostatic attraction between a protein and a polysaccharide is possible, depending on the protein's pH and the isoelectric point. Such a reaction is reversible by changing either the ionic strength or pH values of the solutions. Sulfated polysaccharides form soluble complexes with proteins at pH values above the isoelectric point of the protein. Irreversible complexation between polysaccharides and proteins is nonelectrostatic, and the functional properties of the complex are different from those of either reactant. Limited compatibility of biopolymers leads to single-phase and biphasic mixtures and usually takes place at high ionic strength or pH values above the isoelectric point of the protein. Such conditions are commonly encountered in food systems. Understanding the phase behavior can lead to new food processes and new food products.

Kaewiad et al. (2017) reported the development of BGNPI-alginate encapsulated capsules containing *Lactobacillus rhamnosus* GG using 8.66% BGNPI and 1.85% w/v alginate. The ratio resulted in greater cell-encapsulation efficiency, significantly higher survival rates in both acid conditions and after the freeze-drying process. The entrapped cells were protected when exposed to simulated gastric fluid and efficiently released in simulated intestinal fluid. The BGNPI-alginate wall material provided better protection, significantly extending *L. rhamnosus* GG's survival relative to free cells over 6 months.

The BGN protein isolate extracted at pH 2, 7 and 9 following the salt solubilisation method were complexed with gum Arabic (Busu & Amonsou, 2019). The complexes were aggregated spherical particles ranging from 100 to 200 nm. The optimal pH for the interaction decreased from 4.8 to 2.9 at more acidic extraction pH. The lowest coacervate yield (42%) was observed for pH 2 fraction with a high molecular weight protein of 100 kDa that was absent in the other fractions. It also contained the highest content of basic amino acids. The low coacervate is attributed to protonation of carboxyl groups on gum Arabic at the optimal coacervation (pH = 2.91), consequently hindering protein-polysaccharide interactions. The high coacervate yield at pH 9 is due to the charge configuration of the biopolymers at the pH of optimum interaction (4.05). The authors concluded that extraction pH could be manipulated to produce better acid-stable structures when producing complexes from protein-polysaccharide interaction. Protein-polysaccharide interactions are important in structure formation and controlling the physical and functional properties of foods such as stabilizing emulsions, fat mimics and meat replacers.

9.8 Protein Modification

The nature of plant protein and the associated physical, chemical, and biological processing techniques (traditional and emerging technologies) that grains undergo has a way of affecting the functionality of some plant protein ingredients (Fig. 9.2). Some of the common functional properties of proteins include solubility, gelation,

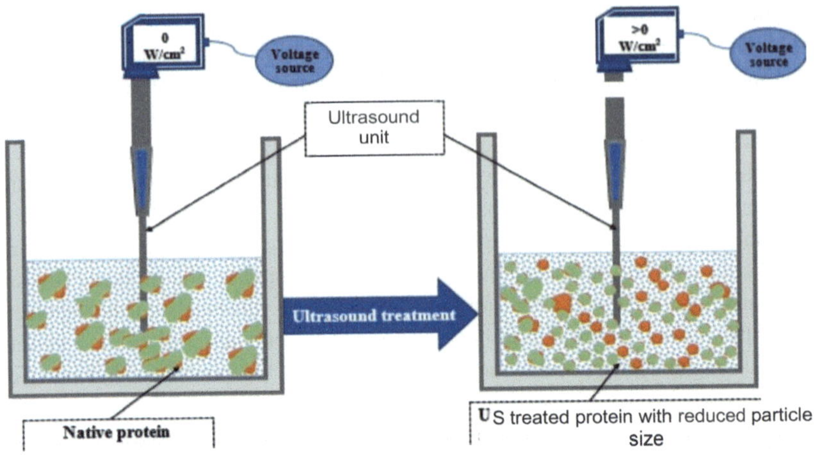

Fig. 9.2 Schematic of a typical ultrasound system showing protein in solution before (left) and after (right) ultrasound treatment. Ultrasound treatment reduces particle size through hydrodynamic shearing and exposes hydrophobic residue through localized heating (Akharume et al., 2021)

emulsification, foam formation, water, and fat binding, viscosity, film formation, and others. The general classes of functional properties of proteins important in food applications and the chemistry behind processing techniques and how they modify the protein structure for improved functionality have been reviewed by Akharume et al. (2021).

References

Abdualrahman, M. A. Y., Ma, H., Yagoub, A. E. G. A., Zhou, C., Ali, A. O., & Yang, W. (2019). Nutritional value, protein quality and antioxidant activity of Sudanese sorghum-based kissra bread fortified with Bambara groundnut (*Voandzeia subterranea*) seed flour. *Journal of the Saudi Society of Agricultural Sciences, 18*, 32–40.

Adebowale, Y. A., Schwarzenbolz, U., & Henle, T. (2011a). Protein isolates from Bambara groundnut (*Voandzeia subterranean* L.): Chemical characterization and functional properties. *International Journal of Food Properties, 14*, 758–775.

Adebowale, Y. A., Schwarzenbolz, U., & Henle, T. (2011b). Protein isolates from Bambara groundnut (*Voandzeia subterranea* L.): Chemical characterization and functional properties. *International Journal of Food Properties, 14*(4), 758–775.

Adeleke, O. R., Adiamo, O. Q., & Fawale, O. S. (2017). Nutritional, physicochemical, and functional properties of protein concentrate and isolate of newly-developed Bambara groundnut (*Vigna subterrenea* L.) cultivars. *Food Science & Nutrition, 6*(1), 229–242. https://doi.org/10.1002/fsn3.552.

Afolabi, A. T., Opara, A. O., Kareem, S. O., & Oladoyinbo, F. O. (2017). In vitro digestibility of hydrothermally modified Bambara groundnut (*Vigna subterranean* L.) starch and flour. *Food Science & Nutrition, 6*, 36–46. https://doi.org/10.1002/fsn3.510.

Akharume, F. U., Aluko, R. E., & Adedeji, A. A. (2021). Modification of plant proteins for improved functionality: A review. *Comprehensive Reviews in Food Science and Food Safety, 20*, 198–224. https://doi.org/10.1111/1541-4337.12688.

Alabi, O. O., Ali, N., Nwachukwu, I. D., Aluko, R. E., & Amonsou, E. O. (2020). Composition and some functional properties of Bambara groundnuts vicilin fraction. *LWT - Food Science and Technology, 125*, 109256.

Arise, A. K., Alashi, A. M., Nwachukwu, I. D., Ijabadeniyi, O. A., Aluko, R. E. & Amonsou, E. O. (2016). Antioxidant activities of Bambara groundnut (Vigna subterranea) protein hydrolysates and their membrane ultrafiltration fractions. *Food and Function, 7*, 2431–2437

Arise, A. K., Amonsou, E. O., & Ijabadeniyi, O. A. (2015). Influence of extraction methods on functional properties of protein concentrates prepared from South African Bambara groundnut landraces. *International Journal of Food Science and Technology, 50*, 1095–1101.

Arise, A. K., Alashi, A. M., Nwachukwu, I. D., Malomo, S. A., Aluko, R. E., & Amonsou, E. O. (2017a). Inhibitory properties of Bambara groundnut protein hydrolysate and peptide fractions against angiotensin-converting enzymes, renin and free radicals. *Journal of the Science of Food and Agriculture, 97*, 2834–2841.

Arise, A. K., Nwachukwu, I. D., Aluko, R. E., & Amonsou, E. O. (2017b). Structure, composition and functional properties of storage proteins extracted from Bambara groundnut (*Vigna subterranea*) landraces. *International Journal of Food Science and Technology, 52*, 1211–1220.

Astuti, R. M., Palupi, N. S., & Zakaria, F. R. (2016). Allergic reactivity of Bambara groundnut (*Vigna subterranea*) proteins. *Food and Agricultural Immunology, 27*(4), 535–546.

Atoyebi, O. J., Osilesi, O., Adebawo, O., & Abberton, M. (2017). Evaluation of nutrient parameters of selected African accessions of Bambara groundnut (*Vigna subterranea* (L.) Verdc.). *American Journal of Food and Nutrition (SCIEP Series), 5*(3), 83–89. https://doi.org/10.12691/ajfn-5-3-1.

Atoyebi, J. O., Osilesi, O., Abberton, M., Adebawo, O., & Oyatomi, O. (2018). Quantification of selected anti-nutrients and bioactive compounds in African Bambara groundnut (*Vigna subterranea* (L.) Verdc.). *American Journal of Food and Nutrition, 6*(3), 88–95. https://doi.org/10.12691/ajfn-6-3-5.

Busu, N. M., & Amonsou, E. O. (2019). Fractionation pH of Bambara groundnut (*Vigna subterranea*) protein impacts the degree of complexation with gum Arabic. *Food Hydrocolloids, 87*, 653–660.

Dahle, L. K. (1971). Wheat protein-starch interaction. I. Some starch-binding effects of wheat flour proteins. *Cereal Chemistry, 48*, 706–713.

Damodaran, S. (1997). *Food proteins and their applications* (1st ed.). CRC Press. 694 Pages. ISBN 9780367401047.

de Leon, M. P., Drew, A. C., Glaspole, I. N., Suphioglu, C., O'Hehir, R. E., & Rolland, J. M. (2007). IgE cross-reactivity between the major peanut allergen Ara h 2 and tree nut allergens. *Molecular Immunology, 44*(4), 463–471. ISSN 0161-5890. https://doi.org/10.1016/j.molimm.2006.02.016.

Diedericks, C. F., de Koning, L., Jideani, V. A., Venema, P., & van der Linden, E. (2019). Extraction, gelation and microstructure of Bambara groundnut vicilins. *Food Hydrocolloids, 97*, 105226.

Diedericks, C. F., Venema, P., Mubaiwa, J., Jideani, V. A., & van der Linden, E. (2020). Effect of processing on the microstructure and composition of Bambara groundnut (*Vigna subterranea* (L.) Verdc.) seeds, flour and protein isolates. *Food Hydrocolloids, 108*, 1–11.

Duodu, K. G., & Apea-Bah, F. B. (2017). African legumes: Nutritional and health-promoting attributes (chapter 9). In J. Taylor & J. Awika (Eds.), *Gluten-free ancient grains: Cereals, pseudocereals, and legumes: Sustainable, nutritious, and health-promoting foods for the 21st century, 1st Edition, pp. 356* (pp. 223–269). Woodhead Publishing. eBook ISBN: 9780081008911. https://www.elsevier.com/search-results?query=Bambara%20groundnut&labels=books&page=1

Gwala, S., Pallares Pallares, A., Pälchen, K., Hendrickx, M., & Grauwet, T. (2020). In vitro starch and protein digestion kinetics of cooked Bambara groundnuts depend on processing intensity and hardness sorting. *Food Research International, 137*, 109512.

Hu, J., Zhao, T., Li, S., Wang, Z., Wen, C., Wang, H., Yu, C., & Ji, C. (2019). Stability, microstructure, and digestibility of whey protein isolate—*Tremella fuciformis* polysaccharide complexes. *Food Hydrocolloids, 89*, 379–385.

Igbabul, B., Adole, D., & Sule, S. (2013). Proximate composition, functional and sensory properties of Bambara nut (*Voandzeia subterranean*), cassava (*Manihot esculentus*) and soybean (*Glycine max*) flour blends for "Akpekpa" production. *Current Research in Nutrition and Food Science Journal, 1*(2), 147–155. https://doi.org/10.12944/CRNFSJ.1.2.06.

Ikegwu, T. M., Nkama, I., & Okafor, I. G. (2020). Comparative studies of the proximate, microscopic and thermal properties of processed maize, wheat, millet, cassava and Bambara nut flours. *Acta Scientific Nutritional Health, 4*(8). ISSN:2582-1423.

Kaewiad, K., Kaewnopparat, N., Faroongsarng, D., Wungsintaweekul, J., & Kaewnopparat, S. (2017). Statistical_optimization_of_zi.PDF. *AIMS Microbiology, 3*, 713–732.

Kelechi-Ebisike, V. O., Solomon, M. D., Jaryum, K. H., Ebisike, K. J., Mshelia, P., & Danjin, M. (2020). Energy/macronutrient density and sensory evaluation of formulated complementary foods for use in low-income populations in northern Nigeria. *Acta Scientific Nutritional Health, 4*(3), 01–06.

Kilara, A., & Sharkasi, T. Y. (1986). Effects of temperature on food proteins and its implications on functional properties. *C R C Critical Reviews in Food Science and Nutrition, 23*(4), 323–395. https://doi.org/10.1080/10408398609527429.

Kudre, T. G., & Benjakul, S. (2013). Combining effect of microbial transglutaminase and Bambara groundnut protein isolate on gel properties of surimi from sardine (*Sardinella albella*). *Food Biophysics, 8*, 240–249.

Kudre, T. G., & Benjakul, S. (2014a). Physicochemical and functional properties of beany flavour-free Bambara groundnut protein isolate. *Journal of the Science of Food and Agriculture, 94*, 1238–1247.

Kudre, T. G., & Benjakul, S. (2014b). Effects of Bambara groundnut protein isolates and microbial transglutaminase on textural and sensorial properties of surmi gel from sardine (*Sardinella albella*). *Food and Bioprocess Technology, 7*, 1570–1580.

Kudre, T. G., Benjakul, S., & Kishimura, H. (2013). Comparative study on chemical compositions and properties of protein isolates from mung bean, black bean and Bambara groundnut. *Journal of the Science of Food and Agriculture, 93*(10), 2429–2436. https://doi.org/10.1002/jsfa.6052.

Martınez, C. J., Martınez, A. C., Ayala, A. L. M., Muzquiz, M., Pedrosa, M. M., & Davila-Ortiz, G. (2012). Changes in protein, nonnutritional factors, and antioxidant capacity during germination of *L. campestris* seeds. *International Journal of Agronomy, 2012*. Article ID 387407. https://doi.org/10.1155/2012/387407.

Mills, E. N. C., Madsen, C., Shewry, P. R., & Wichers, H. J. (2003). Food allergens of plant origin-their molecular and evolutionary relationships. *Trends in Food Science & Technology, 14*, 145–156.

Mottram, D., Wedzicha, B., & Dodson, A. (2002). Acrylamide is formed in the Maillard reaction. *Nature, 419*, 448–449. https://doi.org/10.1038/419448a.

Mune Mune, M. A., Minka, S. R., & Henle, T. (2018). Investigation on antioxidant, angiotensin converting enzyme and dipeptidyl peptidase IV inhibitory activity of Bambara bean protein hydrolysates. *Food Chemistry, 250*, 162–169.

Ndidi, U. S., Ndidi, C. U., Aimola, I. A., Bassa, O. Y., Mankilik, M., & Adamu, Z. (2014). Effects of processing (boiling and roasting) on the nutritional and antinutritional properties of Bambara groundnuts (*Vigna subterranea* [L.] Verdc.) from southern Kaduna, Nigeria. *Journal of Food Processing*. https://doi.org/10.1155/2014/472129.

Okpuzor, J., Ogbunugafor, H. A., Okafor, U., & Sofidiya, M. O. (2010). Identification of protein types in Bambara nut seeds: Perspectives for dietary protein supply in developing countries. *Experimental and Clinical Sciences Journal, 9*, 17–28.

Oujifard, A., Benjakul, S., Ahmad, M., & Seyfabadi, J. (2012). Effect of Bambara groundnut protein isolate on autolysis and gel properties of surimi from threadfin bream (*Nemipterus bleekeri*). *LWT – Food Science and Technology, 47*, 261–266.

Oyeyinka, A. T., Pillay, K., & Siwela, M. (2019). In vitro digestibility, amino acid profile and antioxidant activity of cooked Bambara groundnut grain. *Food Bioscience, 31*, 100428.

Poulter, N. H. (1981). Properties of some protein fractions from Bambara groundnut. *Journal of Science and Food Agriculture, 32*, 44–50.

Remsen, C. H., & Clark, J. P. (1978). A viscosity model for a cooking dough. *Journal of Food Process Engineering, 2*, 39–64.

Ruzengwe, F. M., Amonsou, E. O., & Kudanga, T. (2020a). Transglutaminase-mediated crosslinking of Bambara groundnut protein hydrogels: Implications on rheological, textural and microstructural properties. *Food Research International, 137*, 109734. https://doi.org/10.1016/j.foodres.2020.109734. PMID: 33233301.

Ruzengwe, F. M., Amonsou, E. O., & Kudanga, T. (2020b). Rheological and microstructural properties of Bambara groundnut protein gels. *LWT - Food Science and Technology, 123*, 109070.

Sharan, S., Zanghelini, G., Zotzel, J., Bonerz, D., Aschoff, J., Saint-Eve, A., & Maillard, M. (2021). Fava bean (*Vicia faba* L.) for food applications: From seed to ingredient processing and its effect on functional properties, antinutritional factors, flavor, and color. *Comprehensive Reviews in Food Science and Food Safety, 20*, 401–428. https://doi.org/10.1111/1541-4337.12687.

Shimada, K., & Matsushita, S. (1980). Relationship between thermocoagulation of proteins and amino acid compositions. *Journal of Agricultural and Food Chemistry, 28*(2), 413–417. https://doi.org/10.1021/jf60228a019.

Takeuchi, K. (1969). A note on the test for the location parameter of an exponential distribution. *Annals of Mathematical Statistics, 40*(5), 1838–1839.

Taylor, J., & Awika, J. (2017). *Gluten-free ancient grains: Cereals, pseudocereals, and legumes: Sustainable, nutritious, and health-promoting foods for the 21st century* (1st ed., p. 356). Woodhead Publishing. eBook ISBN: 9780081008911.

Temegne, N. C., Gouertoumbo, W. F., Wakam, G. A., Nkou, F. T. D., Youmbi, E., & Ntsomboh-Ntsefonng, G. (2018). Origin and ecology of Bambara groundnut (*Vigna subterranea* (L.) Verdc): A review. *Journal of Ecology & Natural Resources, 2*(4), 000140.

Teye, E., Deha, C. I., Dadzie, R., & MacArthur, R. L. (2020). Delivering the nutritional needs by food to food fortification of staples using underutilized plant species in Africa. *International Journal of Food Science*. Article ID 8826693, 8 pages. https://doi.org/10.1155/2020/8826693.

Tolstoguzov, V. (1997). Thermodynamic aspects of dough formation and functionality. *Food Hydrocolloids, 11*, 181–193.

Traidl-Hoffmann, C., Jakob, T., & Behrendt, H. (2009). Determinants of allergenicity. *Journal of Allergy and Clinical Immunology, 123*(3), 558–566. ISSN 0091-6749. https://doi.org/10.1016/j.jaci.2008.12.003.

Chapter 10
Phytonutrients and Antioxidant Activity of Bambara Groundnut

10.1 Introduction

Bioactive compounds in plants are secondary metabolites resulting from the biosynthesis of primary plant metabolites (protein, carbohydrate, fat). They are essential phytochemicals that form the basis of modern pharmacological effects and medical treatment because it has natural beneficial compounds. These plant compounds, having antioxidant properties are composed of a wide variety of different substances like ascorbic acid and tocopherols, polyphenolic compounds, or terpenoids (Sharan et al., 2021; Amandeep et al., 2010, Rabaneda et al., 2003, Gouveia & Castilho, 2011, Lin et al., 2008, Jideani et al., 2021).

Significant chemical substances of interest in plants have been the polyphenols and tannins with other groups of naturally occurring phytochemicals such as alkaloids and saponins. These plant compounds have complex activities and are investigated for chemical, toxicological and pharmacological features. The alkaloids are a diverse group of low molecular weight nitrogen-containing compounds derived mostly from amino acids. Alkaloids have a wide range of biological activities of alkaloids including emetic, anti-cholinergic, antitumor, diuretic, sympathomimetic, antiviral, antihypertensive, hypnoanalgesic, antidepressant, miorelaxant, antitussigen, antimicrobial and anti-inflammatory (Aberoumand, 2012; Mashau et al., 2020). This chapter details the phytonutrients and antioxidant activity of Bambara groundnut.

V. A. Jideani, A. I. O. Jideani, *Bambara groundnut: Utilization and Future Prospects*, https://doi.org/10.1007/978-3-030-76077-9_10

10.2 Phytochemical Profile of Bambara Groundnut Seeds

Tsamo et al. (2018) described the polar metabolome of the seed from 21 BGN landraces sourced from South Africa, Mozambique and Swaziland. Metabolites belonging to varied compound classes were detected and identified. The total phenolic, flavonoid and anthocyanin contents varied from 0.75 to 17.71 mg GAE.g – 1, 0.01 to 2.51 mg QUE.g – 1 and 0.03 to 1.31 mg CYE.g^{-1}, respectively. Caffeic and catechin conjugates are the most decisive marker compounds discriminating the landraces. The authors highlighted the potential of BGN groundnut as an economical source of natural antioxidants for human consumption and food industries. Therefore, it highlights the industrial use of BGN flours in developing functional food and feed products. Significant differences exist in phenolic content among the 21 landraces. The cream-colored landrace BG12 contained the highest total phenolic content (TPC) of 17.71 mg g^{-1}, followed by the black-colored landrace BG20 (11.27 mg g^{-1}). The mottled cream BG18 (0.75 mg g^{-1}) has the lowest content. TPC average was 5.98 mg g^{-1} and decreased based on average values of seed color in all the landraces in the order black > red > brown >cream > white. The seed of grain legumes' total phenolic content varies depending on the darkness of the seed coat, increasing with the darkness of the seed coat color and decreases with its lightness (Diedericks & Jideani, 2015; Maphosa & Jideani, 2016; Harris et al., 2018). The total flavonoid content (TFC) varied significantly among samples ranging from 0.01 to 2.51 mg g^{-1}. Landrace BG20 had the highest TFC (2.51 mg g^{-1}) followed by BG12 (1.75 mg g^{-1}) while the white-colored landrace BG13 (0.01 mg g^{-1}) had the lowest TFC followed by the red BG8 (0.02 mg g^{-1}). There was significant variation in total anthocyanin content (TAC) among the samples. The TAC ranged from 0.03 to 1.31 mg g^{-1} with an average of 0.22 mg g^{-1} (Tsamo et al., 2018) The highest anthocyanin content was in the black seed coated landraces BG16 (1.31 mg g^{-1}), BG6 (0.80 mg g^{-1}), BG19 (0.56 mg g^{-1}) and BG20 (0.25 mg g^{-1}), while the other landraces showed traces to low anthocyanin content (Harris et al., 2018; Tsamo et al., 2018). The high TPC, TFC and TAC of BGN seeds are comparable to their levels in other grain legume crops such as black beans, common beans, adzuki beans cowpea and peanuts (Agostini-Costa et al., 2015; Akond et al., 2011; Chon, 2013). The presence of catechin, quercetin, kaempferol, apigenin, phenolic acid, saponins, sphingolipids and fatty acids in BGN seeds elevates the crop for human consumption, medicinal and pharmaceutical purposes. Caffeic acid and catechin conjugates contributed most to the discrimination of all landraces (Tsamo et al., 2018).

The beneficial health-promoting properties present in Bambara groundnut are attributed to their phytochemical composition, especially polyphenolics which act as free radical scavengers or primary antioxidants (Tsamo et al., 2018). Some reported BGN phytochemicals, including anthocyanin and flavonoids, impart many beneficial health benefits and reduce the risk of many chronic illnesses (Pale et al., 1997). Phenolic phytochemicals, as health-promoting constituents, have antioxidant properties in human nutrition leading to possible protection against most non-communicable diseases (cardiovascular and cancer) involving reactive oxygen species in the category superoxide anion, peroxy and hydroxyl radicals (Duthie & Crozier, 2000). Some phytonutrients of BGN are detailed in Fig. 10.1.

Generally, a diversity of phenolic compounds (phenolic acids and flavonoids - condensed tannins, flavanols, flavonones, flavonols, flavan-3-ol and flavonoids) found in BGN are mainly hydroxybenzoic and hydroxycinnamic acids, flavonols, and monomeric flavan-3-ols. Also found are protocatechuic acid, p-coumaric aldaric acid isomers, procyanidin (dimer and trimers), myricetin glucuronide, quercetin 3-O-glucoside, naringenin, naringenin derivative and kaempferol 3-O-glucoside, gallic acid, procyanidin dimers, quercetin, kaempferol derivative and kaempferol glucoside, catechin, kaempferol and ferulic acid (Kumar & Pruthi, 2014; Randjelović et al., 2015). These phenolic compounds are of great relevance because most have anticancer potential (Bernatoniene & Kopustinskiene, 2018; Schwarz et al., 2018; Zhang et al., 2015; Kakkar & Bais, 2014; Huang et al., 2015; Ahamad et al., 2014; Blade et al., 2016).

Fig. 10.1 Phytonutrients in Bambara groundnut seeds

10.3 Phenolic-Rich Bambara Groundnut Extracts and Peptide Hydrolysates Inhibition of Amylase, Glucosidase and Angiostensin-Converting Enzymes

Bambara protein isolate (BPI) was digested with three proteases (alcalase, trypsin and pepsin), to produce BGN protein hydrolysates (BPHs) (Arise et al., 2016). These hydrolysates were passed through ultrafiltration membranes to obtain peptide fractions of different sizes (<1, 1–3, 3–5 and 5–10 kDa). The hydrolysates and their peptide fractions were investigated for antioxidant activities. The membrane fractions showed that peptides with sizes <3 kDa had significantly reduced surface hydrophobicity when compared with peptides >3 kDa. This is in agreement with the result obtained for the ferric reducing power, metal chelating and hydroxyl radical scavenging activities where higher molecular weight peptides exhibited better activity than low molecular weight peptide fractions. However, the low molecular weight peptides were more effective diphenyl-1-picrylhydrazyl (DPPH) radical scavengers but not superoxide radicals than the bigger peptides. Compared with glutathione (GSH), BPHs and their membrane fractions had better reducing power and ability to chelate metal ions except for the pepsin hydrolysate and its membrane fractions did not show any metal chelating activity. However, the 5–10 kDa pepsin hydrolysate peptide fractions had greater (88%) hydroxyl scavenging activity than GSH, alcalase and trypsin hydrolysates (82%). These findings show the potential use of BPHs and their peptide fraction as antioxidants in reducing food spoilage or management of oxidative stress-related metabolic disorders.

Ademiluyi and Oboh (2012a) investigated the modulatory effect of fermented legumes condiment diet on oxidative stress in streptozotocin (STZ) induced diabetic rats. Adult male Wistar rats were randomly divided into six groups, with six animals in each group. Diabetes was induced by intraperitoneal injection of STZ (35 mg/kg b.w.). The diabetic rats were fed with fermented BGN, locust bean and soybean diets for 14 days. The plasma was obtained after 14-day treatment and analyzed for hepatic damage marker enzymes (AST, ALT and ALP) and in vivo antioxidant indices. The predominant phenolic compound in fermented BGN includes carvacrol, p-coumaric acid, vnillic acid, p-hydroxybenzoic acid, caffeic acid, genistein, apigenin, shagaol, glycitein, kaempferol, luteolin, capsaicin, isorhamnetin, myricetin and romarinio acid. The fermented legume condiments demonstrated intense free radical scavenging activities by scavenging moderately stable ABTS radicals *in vitro*. Free radicals may play an essential role in the causation and complications of diabetes mellitus. Alterations in the endogenous free radical scavenging defense mechanisms could lead to ineffective scavenging of reactive oxygen species resulting in oxidative damage and tissue injury. Hence, a steady supply of dietary antioxidants to augment or boost the endogenous antioxidant defense mechanisms could be one practical approach through which free radical-mediated oxidative stress in diabetes mellitus may be restrained. Phenolic compounds due to their structure are known to be involved in the healing process of free radical-mediated diseases, including diabetes. Thus the observed attenuation of oxidative stress in the diabetic rats treated with fermented legume condiment diets could be due to their phenolic constituents. Consumption of phenolic-rich fermented tropical legume condiments could be a practical dietary approach through which oxidative stress arising from diabetes or its complications can be managed. The diabetic untreated rats had significantly elevated levels of AST, ALT, ALP and malondialdehyde with reduced activities of glutathione-S-transferase, catalase, and plasma reduced glutathione, vitamin C and total protein content. Feeding the diabetic rats with fermented legume condiments diets for 14 days significantly reversed to normal the above parameters, suggesting their modulation of oxidative stress, which may be due to their high phenolic content and antioxidant capacity. The attenuation of oxidative stress and protection of hepatic tissue damage by the legume condiment diets in STZ induced diabetic rats compared favourably with that of metformin, a well-known oral hypoglycemic drug.

Interaction of free and bound phenolic extracts from selected tropical underutilized legumes (popular folklore for the management of diabetes and hypertension) with crucial enzymes linked to type 2 diabetes (α-glucosidase and α-amylase) and hypertension (angiotensin I converting enzyme, ACE) was assessed by (Ademiluyi & Oboh, 2012b). The free phenolic extracts of the soybean were obtained by extracting powdered legume seeds with 80% acetone. In contrast, the residue was subjected to alkaline and acid hydrolysis before extraction with ethyl acetate to obtain the bound extracts. After that, the enzyme (α-amylase, α-glucosidase, and ACE) inhibitory studies and antioxidant capacity of the extracts were investigated. The free and bound phenolic extracts inhibited α-amylase, α-glucosidase, and ACE activities in a dose-dependent manner. However, the free phenolic extracts of all the legumes were stronger inhibitors of α-glucosidase than their corresponding bound phenolic extracts. The free phenolic extracts (except melon) exhibited more potent ACE inhibition than their corresponding bound phenolic extracts. All the phenolic extracts significantly inhibited $Fe2^{+}$-induced oxidative stress in the pancreas (in vitro) and exhibited intense antioxidant activities. The authors concluded that the inhibition of the α-amylase, α-glucosidase, ACE, and oxidative stress by the phenolic extracts could be a part of the mechanism through which these underutilized legumes manage/prevent type-2 diabetes and hypertension. Of all the legumes BGN had the best enzyme inhibitory potentials coupled with potent antioxidant properties ranked highest.

10.4 Bambara Groundnut Seeds Antioxidant and Antimicrobial Activity

Bambara groundnut (BGN), originally from Africa, is widely distributed in Asian countries, especially in southern Thailand, and is used for food and functional foods. Chutoprapat et al. (2020) reported the phytochemicals and the potential efficacy of BGN in protecting against free radicals, metal ions, lipid peroxidation, enzyme tyrosinase, as well as anti-melanogenesis and collagen biosynthesis simulation. The powdered hulls (H) and seeds (S) of BGN were collected from Trang province, Thailand and extracted by Soxhlet (S) [95% v/v ethanol at the ratio of 1:10 at 60 °C for 24 h] and maceration (M) using ethanol [soaked in 95% ethanol (v/v) (1:10) at 37 °C for 48 h] in a shaker and powders boiled at 90 °C with distilled water (1:10) for 2 h (B). The extracts were filtered through filter paper no.1, evaporated, the dried extracts stored in an amber bottle at 4 °C. The seed extract prepared by Soxhlet with ethanol (BGN-SS) gave the highest TPC and TFC. The hull extract prepared by maceration with ethanol (BGN-HM) showed the highest antioxidant activities (SC_{50} of 0.87 ± 0.02 mg/mL, MC_{50} of 1.83 ± 0.09 mg/mL and LC_{50} of 0.70 ± 0.06 mg/mL), tyrosinase inhibition activity (IC_{50} of 0.45 ± 0.23 mg/mL), and anti-melanogenesis activities (72.9 ± 0.08%), whereas the seed extract (BG-SB) exhibited the highest stimulation of collagen biosynthesis (18.04 ± 0.03%). All BGN extracts at 0.1 mg/mL showed no cytotoxicity on human dermal fibroblasts. The biological activities of BGN extracts might be from their phytochemicals, especially phenolic and flavonoid contents. The authors concluded that the BGN-HB and BGN-HM extracts might be promising novel active sources for anti-ageing and whitening cosmeceuticals. Although the extracts were cosmeceutical, they are also applicable in functional foods. The effectiveness of the BGN hulls attributed to their phytochemical, phenolic and flavonoid contents as reported by Chutoprapat et al. (2020) is in agreement with that of Harris et al. (2018). In 70% methanolic extract flavonoids and tannins were highly concentrated in the red and brown BGN hulls. Among the flavonoid compounds, rutin was highest in the brown hull (24.46 mg/g) and myricetin (1.80 mg/g). The tannin compounds chlorogenic acid concentration was highest (0.115 mg/g) and ellagic acid in the red hull (0.105 mg/g). The hulls of the BGN are the optimum source of flavonoids and tannins. Darker coloured BGN hulls such as red and brown have a higher concentration of flavonoids than the black-eye and brown-eye seeds.

The increased rate of high blood pressure has led to critical human hypertensive conditions in most nations. Arise et al. (2017) reported on the antihypertensive and antioxidant activities of BGN protein hydrolysates (BPHs) and their peptide fractions (molecular weight: 10, 5, 3 and 1 kDa) obtained from three different proteases (alcalase, trypsin and pepsin). Alcalase hydrolysate contained the highest amount of low molecular weight (LMW) peptides compared to pepsin and trypsin hydrolysates. LMW peptides fractions (<1 kDa) exhibited the highest inhibitory activity against angiotensin-converting enzyme (ACE) for all the enzyme hydrolysates. For renin inhibition, alcalase hydrolysate showed the highest inhibition at 59% than other hydrolysates and their corresponding membrane fractions. The antioxidant power of BGN protein hydrolysates and peptide fractions was evaluated through the inhibition of linoleic acid peroxidation and ABTS scavenging activity. Among the hydrolysates, alcalase exhibited the highest inhibition of linoleic acid oxidation. Furthermore, all BPHs scavenged $ABTS^+$ to a threefold greater extent compared to the isolate. BPH and LMW peptide fractions could potentially serve as useful ingredients in the formulation of functional foods and nutraceuticals against high blood pressure and oxidative stress.

Chinnapun (2018) investigated the effects of processing by dry heating, boiling and steaming on the antioxidant activity and DNA protection against oxidative damage of BGN seeds. Comparing raw and processed BGN seeds, dry heating caused a significant reduction of ferric-reducing antioxidant power (FRAP), metal chelating activity, DPPH and $ABTS^+$ radical scavenging activity. The boiling process did not cause a considerable difference in FRAP and metal chelating activity, only smaller losses in DPPH and $ABTS^+$ radical scavenging activity than the dry heating and steaming processes. The steaming process caused a significant reduction of FRAP, DPPH and $ABTS^+$ radical scavenging activity. For DNA protection against oxidative damage, boiled and steamed BGN seeds were more effective with a lower minimum concentration of 50μg/mL than raw and dry heated seeds. Consequently, the boiling process caused smaller losses in antioxidant activity than dry heating and steaming. Therefore, boiling is recommended as a processing method for BGN seeds to preserve antioxidant- components and activity.

The antioxidant mode of action includes (i) quenching of singlet oxygen, (ii) hydrogen transfer and (iii) electron transfer among others (Bernhoft, 2010). Atoyebi et al. (2018) quantified certain bioactive compounds in BGN. They observed that selected African accession had varying but high ascorbic acid (29.90 mg/100 g) and oxalic acid (0.0049 g/g).

Clinically, the extracts of BGN have been reported to exhibit antimicrobial activities against bacteria such as *Klebsiella pneumoniae* subsp. *pneumoniae* ATCC 700603, *Pseudomonas aeruginosa* ATCC 27853, *Staphylococcus aureus* subsp. *aureus* ATCC 33591, *Escherichia coli*, *Bacillus cereus*, yeast (*Candida albicans*) and mold (*Aspergillus niger*) (Udeh et al., 2020).

Bioactive compounds and antioxidant activities have been reported in BGN (Bamishaiye et al., 2011; Nyau et al., 2015; Atoyebi et al., 2018; Mashau et al., 2020). Table 10.1 shows the DDPH free radical scavenging activities of BGN (Nyau et al., 2015). Table 10.2 shows polyphenolic compounds and antioxidant activity of maize-Bambara groundnut (BGN) flour blends in tortillas production (Mashau et al., 2020). These reports suggest that it could be a useful source of antioxidants compared to commonly consumed legumes such as lentils, common beans and chickpeas. Nyau et al. (2015) investigated the nutraceutical potential of BGNs (red and brown) from Zambia based on the antioxidant properties, using 1,1-diphenyl-2-picrylhydrazyl (DPPH) and ferric reducing antioxidant power (FRAP) *in vitro*

Table 10.1 DDPH Free radical scavenging activities of Bambara groundnut (BGN): Pseudo-first order rate constant of antiradical (Y-H) in BGNs extracts and the amount of DPPH scavenged and EC50 values for DPPH free radical scavenging by BGN extracts after 30 min of incubation

Market class	Amount DPPH quenched [%] after 80 min incubation		DPPH radical scavenging EC50 (μg dried extract/ml)	
	Aqueous extract	70% Methanol extract	70% Methanol extract	Aqueous extract
Brown	57.1 ± 3.1	53.7 ± 2.6	477.5 ± 3.5	347.0 ± 4.2
Red	38.9 ± 1.8	43.3 ± 4.6	525.5 ± 7.8	495.5 ± 12.0

Source: Nyau et al. (2015)

Table 10.2 Thermal properties, polyphenolic compounds and antioxidant activity of control and maize-Bambara groundnut (BGN) flour blends in tortillas production

Thermal properties				
Flours	Onset temperature To (°C)	Peak temperature Tp (°C)	End temperature Tc (°C)	Enthalpy of gelatinisation ΔH (J/G)
Maize flour (control)	57.50 ± 0.58[a]	74.94 ± 0.61[a]	91.58 ± 1.49[b]	5.57 ± 1.39[a]
BGN:maize (5:95)	61.70 ± 0.72[b]	75.86 ± 0.87[a]	89.93 ± 1.86[b]	6.31 ± 0.83[b]
BGN:maize (10:90)	66.10 ± 0.70[c]	76.74 ± 0.71[ab]	87.70 ± 0.81[c]	8.59 ± 1.99[c]
BGN:maize (15:85)	67.38 ± 0.44[d]	77.95 ± 0.32[b]	84.34 ± 0.45[a]	11.31 ± 0.29[d]
BGN:maize (20:80)	71.95 ± 0.62[e]	78.45 ± 0.70[b]	81.72 ± 0.69[a]	14.81 ± 0.49[e]
Polyphenolic compounds and antioxidant activity				
Flour	Total phenolic content TPC (mg GAE/g)	Total flavonoids content TFC (mg CE/g)	FRAP (mg GAE/g)	DPPH (%)
Control	43.82 ± 2.00[a]	10.83 ± 1.10[a]	0.071 ± 0.01[a]	43.09 ± 1.92[a]
BGN1	55.05 ± 3.01[b]	13.21 ± 1.23[b]	0.112 ± 0.01[b]	55.87 ± 3.07[b]
BGN2	60.21 ± 3.52[d]	17.34 ± 1.50[c]	0.124 ± 0.01[b]	65.46 ± 3.48[d]
BGN3	66.04 ± 3.88[e]	20.44 ± 1.68[d]	0.141 ± 0.03[c]	75.19 ± 5.60[e]
BGN4	72.05 ± 4.22[f]	23.58 ± 1.98[e]	0.156 ± 0.04[d]	80.54 ± 6.17[f]
Tortillas				
Control	46.61 ± 2.10[b]	10.03 ± 1.15[a]	0.106 ± 0.01[b]	60.70 ± 3.50[c]
BGN1	60.11 ± 3.49[d]	10.87 ± 1.30[a]	0.118 ± 0.02[b]	77.87 ± 4.96[b]
BGN2	62.11 ± 3.92[d]	13.31 ± 1.63[b]	0.126 ± 0.02[b]	82.38 ± 5.93[f]
BGN3	72.86 ± 4.20[f]	16.43 ± 1.70[c]	0.131 ± 0.02[c]	87.21 ± 6.32[g]
BGN4	79.31 ± 5.60[g]	19.87 ± 1.99[d]	0.183 ± 0.01[d]	90.99 ± 6.66[h]

BGN1, BGN2, BGN3 and BGN4 refer to the composite flours and tortillas prepared from BGN and maize flour (BGN: maize) in the percentages of 5:95, 10:90, 15:85 and 20:80 and maize flour as control. Values are mean ± standard deviation (n = 3). Values followed by the same superscript(s) within the same column are not significantly different at $p < 0.05$. Source: Mashau et al. (2020)

assays and found that (i) BGNs possess antioxidant activities with the brown BGNs having the highest DPPH free radical scavenging activity (347 ± 4.2μg dried extract/ml) compared to 495 ± 12μg dried extract/ml for the red BGNs, (ii) FRAP derived total antioxidant power was higher in the brown (6.00 ± 0.21 mmole $Fe2+$ / 100 g DW) compared to (5.00 ± 0.13 mmole $Fe2+$ / 100 g DW) in the red.

10.5 Sprouting and Polyphenol Content/Phenolics Compounds in Sprouted Bambara Groundnut

Consumption of sprouted legumes is a popular global trend. Plant-derived phenolic compounds are important as antioxidants in our diet. Brajdes and Vizireanu (2012) stated that sprouted buckwheat is an important vegetable source of antioxidants. Nyau et al. (2017a, b) investigated the effect of sprouting on the phenolic phytochemicals of the red BGN using HPLC-PDA-ESI-MS and Folin Ciocalteu assay and observed (i) noticeable changes in the phenolic profiles due to sprouting, (ii) total polyphenol content increased by 1.3-fold after sprouting in red BGN (135.9 ± 2.0 to 174.2 ± 4.8 mg GAE/100 g DW), (iii) the HPLC-PDA-ESI-MS profiling of the methanolic extracts of the sprouts revealed the following eleven new emerging compounds in red BGN: caffeic acid hexoside, resveratrol glucoside, a caffeic acid derivative, naringenin and kaempferol glucoside, (iv) sprouting enhances the polyphenolic profiles of BGN.

Nyau et al. (2017a, b) attributed the total polyphenol concentration changes during sprouting to the enzymatic activities in the seed during the germination process. Polyphenols in the cell are bound to the cellular components (carbohydrates, pectin, lignin and proteins). During the enzymatic degradation of the cellular components, bound phenolics may be released and made available for quantification.

Brajdes and Vizireanu (2012) and Martınez et al. (2012) explained the increase of phenolic compounds to an increase in the number of free forms occurring due to hydrolytic enzyme activity, that breakdown the cell wall during germination. Conjugated phenolic acids are released from the breakdown of cell walls, to protect the inner parts of the caryopsis which is still needed to support the developing germ (Engert et al., 2011). Condensed tannins or proanthocyanidins that exist as oligomer and polymers may disintegrate into simple phenolics due to the activities of the hydrolytic enzymes.

References

Aberoumand, A. (2012). Screening of phytochemical compounds and toxic proteinaceous protease inhibitor in some lesser-known food based plants and their effects and potential applications in food. *International Journal of Food Science and Nutrition Engineering, 2*(3), 16–20. https://doi.org/10.5923/j.food.20120203.01. ISSN: 2166-5168 e-ISSN: 2166-5192.

Ademiluyi, A. O., & Oboh, G. (2012a). Attenuation of oxidative stress and hepatic damage by some fermented tropical legume condiment diets in streptozotocin-induced diabetes in rats. *Asian Pacific Journal of Tropical Medicine, 5*, 692–697.

Ademiluyi, A. O., & Oboh, G. (2012b). Phenolic-rich extracts from selected tropical underutilized legumes inhibit α-amylase, α-glucosidase, and angiotensin I converting enzyme in vitro. *Journal of Basic and Clinical Physiology and Pharmacology, 23*, 17–25.

Agostini-Costa, T. d.-S., Teodoro, A. F. P., Alves, R. d. B. d.-N., Braga, L. R., Ribeiro, I. F., Silva, J. P., Quintana, L. G., & Burle, M. L. (2015). Total phenolics, flavonoids, tannins and antioxidant activity of lima beans conserved in a Brazilian Genebank. *Ciência Rural, 45*(2), 335–341. Epub October 14, 2014. https://doi.org/10.1590/0103-8478cr20140030.

Ahamad, M. S., Siddiqui, S., Jafri, A., Ahmad, S., Afzal, M., & Arshad, M. (2014). Induction of apoptosis and antiproliferative activity of naringenin in human epidermoid carcinoma cell through ROS generation and cell cycle arrest. *PLoS One, 9*(10), e110003.

Akond, A. S. M. G. M., Khandaker, L., Berthold, J., Gates, L., Peters, K., Delong, H., & Hossain, K. (2011). Anthocyanin, total polyphenols and antioxidant activity of common bean. *American Journal of Food Technology, 6*, 385–394. https://doi.org/10.3923/ajft.2011.385.394.

Amandeep, K., Sandhu, & Liwei, G. (2010). Antioxidant capacity, phenolic content, and profiling of phenolic compounds in the seeds, skin, and pulp of *Vitis rotundifolia* (Muscadine grapes) as determined by HPLC-DAD-ESI-MSn. *Journal of Agricultural and Food Chemistry, 58*, 4681–4692. https://doi.org/10.1021/jf904211q.

Arise, A. K., Alashi, A. M., Nwachukwu, I. D., Ijabadeniyi, O. A., Aluko, R. E., & Amonsou, E. O. (2016). Antioxidant activities of Bambara groundnut (*Vigna subterranea*) protein hydrolysates and their membrane ultrafiltration fractions. *Food & Function, 7*, 2431–2437.

Arise, A. K., Alashi, A. M., Nwachukwu, I. D., Malomo, S. A., Aluko, R. E., & Amonsou, E. O. (2017). Inhibitory properties of Bambara groundnut protein hydrolysate and peptide fractions against angiotensin-converting enzymes, renin and free radicals. *Journal of the Science of Food and Agriculture, 97*, 2834–2841.

Atoyebi, J. O., Osilesi, O., Abberton, M., Adebawo, O., & Oyatomi, O. (2018). Quantification of selected anti-nutrients and bioactive compounds in African Bambara groundnut (*Vigna subterranea* (L.) Verdc.). *American Journal of Food and Nutrition, 6*(3), 88–95. https://doi.org/10.12691/ajfn-6-3-5.

Bamishaiye, O. M., Adegbola, J. A., & Bamishaiye, E. I. (2011). Bambara groundnut: An underutilized nut in Africa. *Advances in Agricultural Biotechnology, 1*, 60–72.

Bernatoniene, J., & Kopustinskiene, D. M. (2018). The role of catechins in cellular responses to oxidative stress. *Molecules, 23*(4), 965.

Bernhoft, A. (2010). A brief review on bioactive compounds in plants. In A. Bernhoft (Ed.), *Bioactive compounds in plants – Benefits and risks for man and animals. Proceedings from a Symposium held at the Norwegian Academy of Science and Letters, Oslo, Norway, 13–14 November 2008. pp. 255* (pp. 11–17). Novus forlag. ISBN 978-82-7099-583-7. Printed in Norway 2010 by AIT Otta AS. http://www.dnva.no/geomed

Blade, C., Aragones, G., Arola-Arnal, A., Muguerza, B., Bravo, F. I., Salvado, M. J., & Suárez, M. (2016). Proanthocyanidins in health and disease. *BioFactors, 42*(1), 5–12.

Brajdes, C., & Vizireanu, C. (2012). Sprouted buckwheat an important vegetable source of antioxidants. *The Annals of the University Dunarea de Jos of Galati Fascicle VI – Food Technology, 36*(1), 53–60.

Chinnapun, D. (2018). Antioxidant activity and DNA protection against oxidative damage of Bambara groundnut seeds (*Vigna subterranea* (L.) Verdc .) as affected by processing methods. *International Journal of Food Properties, 21*(1), 1661–1669.

Chon, S-U (2013) Total polyphenols and bioactivity of seeds and sprouts in several legumes. *Current Pharmaceutical Design 19*(34):6112-24. doi: 10.2174/1381612811319340005

Chutoprapat, R., Malilas, W., Rakkaew, R., Udompong, S., & Boonpisuttinant, K. (2020). Collagen biosynthesis stimulation and anti-melanogenesis of Bambara groundnut (*Vigna subterranea*) extracts. *Pharmaceutical Biology, 58*, 1023–1031.

Diedericks, C. F., & Jideani, V. A. (2015). Physicochemical and functional properties of insoluble dietary fiber isolated from Bambara groundnut (*Vigna subterranea* [L.] Verdc.). *Journal of Food Science, 80*, C1933–C1944.

Duthie, G., & Crozier, A. (2000). Plant-derived phenolic antioxidants. *Current Opinion in Clinical Nutrition and Metabolic Care, 3*(6), 447–451. https://doi.org/10.1097/00075197-200011000-00006.

Engert, N., John, A., Henning, W., & Honermeier, B. (2011). Effect of sprouting on the concentration of phenolic acids and antioxidative capacity in wheat cultivars (*Triticum aestivum* ssp. aestivum L.) in dependency of nitrogen fertilization. *Journal of Applied Botany and Food Quality, 84*(1), 111–118.

Gouveia, S., & Castilho, P. C. (2011). Characterisation of phenolic acid derivatives and flavonoids from different morphological parts of *Helichrysum obconicum*by a RP-HPLC-DAD-ESI-MSnmethod. *Food Chemistry, 129*, 333–344. https://doi.org/10.1016/j.foodchem.2011.04.078.

Harris, T., Jideani, V., & Roes-Hill, M. L. (2018). Flavonoids and tannin composition of Bambara groundnut (*Vigna subterranea*) of Mpumalanga, South Africa. *Heliyon, 4*, e00833.

Huang, H., Chen, A. Y., Ye, X., Li, B., Rojanasakul, Y., Rankin, G. O., & Chen, Y. C. (2015). Myricetin inhibits proliferation of cisplatin-resistant cancer cells through a p53-dependent apoptotic pathway. *International Journal of Oncology, 47*(4), 1494–1502. https://doi.org/10.3892/ijo.2015.3133.

Jideani, A. I. O., Silungwe, H., Takalani, T., Omolola, A. O., Udeh, H. O., & Anyasi, T. A. (2021). Antioxidant-rich natural fruit and vegetable products and human health. *International Journal of Food Properties, 24*(1), 41–67. https://doi.org/10.1080/10942912.2020.1866597.

Kakkar, S., & Bais, S. (2014). *A review on protocatechuic acid and its pharmacological potential*. ISRN Pharmacology, 952943. https://doi.org/10.1155/2014/952943.

Kumar, N., & Pruthi, V. (2014). Potential applications of ferulic acid from natural sources. *Biotechnology Reports, 4*, 86–93.

Lin, L. Z., Harnly, J. M., Pastor-Corrales, M. S., & Luthria, D. L. (2008). The polyphenolic profiles of common beans (*Phaseolus vulgaris* L.). *Food Chemistry, 107*, 399–410. https://doi.org/10.1016/j.foodchem.2007.08.038.

Maphosa, Y., & Jideani, V. A. (2016). Physicochemical characteristics of Bambara groundnut dietary fibers extracted using wet milling. *South African Journal of Science, 112*, 72–79.

Martınez, C. J., Martınez, A. C., Ayala, A. L. M., Muzquiz, M., Pedrosa, M. M., & Davila-Ortiz, G. (2012). Changes in protein, non-nutritional factors, and antioxidant capacity during germination of *L. campestris* seeds. *International Journal of Agronomy*. https://doi.org/10.1155/2012/387407.

Mashau, M., Mabodze, T., Tshiakhatho, O., Silungwe, H., & Ramashia, S. (2020). Evaluation of the content of polyphenols, antioxidant activity and physicochemical properties of tortillas added with Bambara groundnut flour. *Molecules, 25*, 3035.

Nyau, V., Prakash, S., Rodrigues, J., & Farrant, J. (2015). Antioxidant activities of Bambara groundnuts as assessed by FRAP and DPPH assays. *American Journal of Food and Nutrition, 3*(1), 7–11. https://doi.org/10.12691/ajfn-3-1-2.

Nyau, V., Prakash, S., Rodrigues, J., & Farrant, J. (2017a). Profiling of phenolic compounds in sprouted common beans and Bambara groundnuts. *Journal of Food Research, 6*, 74.

Nyau, V., Prakash, S., Rodrigues, J., & Farrant, J. (2017b). Domestic cooking effects of Bambara groundnuts and common beans in the antioxidant properties and polyphenol profiles. *Journal of Food Research, 6*(2), 24–37.

Pale, E., Kouda-Bonafos, M., Nacro, M., Vanhaelen, M., Vanhaelen-Fastré, R., & Ottinger, R. (1997). 7-O-methylapigeninidin, an anthocyanidin from *Sorghum caudatum*. *Phytochemistry, 45*(5), 1091–1092., ISSN 0031-9422. https://doi.org/10.1016/S0031-9422(97)00099-X.

Rabaneda, F. S., Jáuregui, O., Lamuela-Ravento, R. M., Bastida, J., Viladomat, F., & Codina, C. (2003). Identification of phenolic compounds in artichoke waste by high performance liquid chromatography-tandem mass spectrometry. *Journal of Chromatography, A, 1008*, 57–72. https://doi.org/10.1016/S0021-9673(03)00964-6.

Randjelović, P., Veljković, S., Stojiljković, N., Sokolović, D., Ilić, I., Laketic, D., Randjelović, D., & Randjelović, N. (2015). The beneficial biological properties of salicylic acid. *Acta facultatis medicae Naissensis, 32*(4), 259–265.

Schwarz, N. A., Blahnik, Z. J., Prahadeeswaran, S., McKinley-Barnard, S. K., Holden, S. L., & Waldhelm, A. (2018). (−)-Epicatechin supplementation inhibits aerobic adaptations to cycling exercise in humans. *Frontiers in Nutrition, 5*, 132.

Sharan, S., Zanghelini, G., Zotzel, J., Bonerz, D., Aschoff, J., Saint-Eve, A., & Maillard, M. (2021). Fava bean (*Vicia faba* L.) for food applications: From seed to ingredient processing and its effect on functional properties, antinutritional factors, flavor, and color. *Comprehensive Reviews in Food Science and Food Safety, 20*, 401–428. https://doi.org/10.1111/1541-4337.12687.

Tsamo, A. T., Ndibewu, P. P., & Dakora, F. D. (2018). Phytochemical profile of seeds from 21 Bambara groundnut landraces via UPLC-qTOF-MS. *Food Research International, 112*, 160–168. ISSN 0963-9969. https://doi.org/10.1016/j.foodres.2018.06.028.

Udeh, E. L., Nyila, M. A., & Kanu, S. A. (2020). Nutraceutical and antimicrobial potentials of Bambara groundnut (*Vigna subterranean*): A review. *Heliyon, 6*(10), e05205. https://doi.org/10.1016/j.heliyon.2020.e05205.

Zhang, X. A., Zhang, S., Yin, Q., & Zhang, J. (2015). Quercetin induces human colon cancer cells apoptosis by inhibiting the nuclear factor-kappa B pathway. *Pharmacognosy Magazine, 11*(42), 404–409. https://doi.org/10.4103/0973-1296.153096.

Chapter 11
Global Research to Expand Utilization of Bambara Groundnut for Food and Nutrition Security

11.1 Introduction

The knowledge on the value-addition of grain crops is fundamental to food security and part of promoting production strategies. Value addition results in various products from grain crops and this may increase utilization and profitability (Gregory et al., 2019; Hoh, 2020; Holt et al., 1992; Yang et al., 2020). The production of grain crops is vital to the food security and nutrition of many countries, and hence the knowledge on utilization strategically contributes to promoting production. Among the global grain crops like maize, rice, soybeans, wheat, sorghum, groundnuts, pearl millets, and cowpea, cereal grains make up the greater percentage of plant species that nourish humanity (Teixeira et al., 2016). Hence legumes are less known, especially Bambara groundnut (BGN). Nevertheless, food legumes are vital nutritional components for dietary protein, vitamin B, and diversified recipes in traditional food dishes, contributing to wellness. Legumes increasingly appear in centre-plate entrees, soups, salads and side dishes. New product applications for legume ingredients include nutritional products, sweet baked goods, bread, nutrition bars, extruded snacks, crackers, snack chips, battered and fried foods, beverages and healthy fitness products. The properties of BGN have received adequate attention such that its use in the formulation of composite flour, breakfast food, production of various bakery and confection products, noodle (Ho, 2020; Awang, 2020; Hampson et al., 2000; Kiin-Kabari & Banigo, 2015) production and ingredients contributes to the attainment of food security. Value addition can stimulate the production of BGN and profitability among farmers (Coltrain et al., 2000; Linnemann, 1988; Ndidi et al., 2014; Ja'afar-Furo et al., 2011; Chandiposha et al., 2013). Increasing value and consumer appeal of BGN would lead to various products based on the physico-functionality of the legume (Linnemann & Azam-Ali, 1993; Oguntunde, 1985; Tan et al., 2020).

The three processes in value addition, depending on the chemical constituent of a crop, are primary, secondary and high-end processing (Misheck et al., 2013). The

V. A. Jideani, A. I. O. Jideani, *Bambara groundnut: Utilization and Future Prospects*, https://doi.org/10.1007/978-3-030-76077-9_11

first, primary processing, involving cleaning, grading, drying, de-stoning, shelling, roasting, blanching, grinding, and packaging. Cleaning removes all extraneous material like weed seeds and soil thereby improving the appearance and this may attract a higher price. The second, secondary processing, encompasses basic processing like size reduction operation, packaging and branding. Basic processing is typically performed after cleaning and grading, involving grinding grain into meal or other products. Packaging and branding give the product uniqueness, which is vital in marketing. High-end processing consists of advanced processing technology in the commercial industry giving rise to secondary and derived products (John & Babynath, 2011) like canning and freezing processes to preserve (Hillocks et al., 2012). BGN is gradually entering this complex processing level like the canning in Ghana and Nigeria (Hillocks et al., 2012) and extraction of oil or fiber. Innovative technologies like decortication, fermentation and extrusion have promising benefits for BGN new product formulation and fortification. The meal from oil extraction has been reported to contain antinutritional substances like trypsin inhibitors, saponins, phytoestrogens, agglutinins, goitrogens, lectins, urease and other substances (Chandiposha et al., 2013). However, acid and heat treatment reduces the level of these substances.

Fig. 11.1 Some processed food products made from Bambara groundnut. (Muhammad et al., 2020)

Value addition (Fig. 11.1) of BGN is essential to increase its production and utilization (Muhammad et al., 2020). Jideani (2016) reported that value-added products from BGN seeds and its flour would provide functional foods with therapeutic benefits. This section explores scientific research towards utilising BGN seeds and flour in value-added products and sometimes compared with other legumes and crops like maize, soybeans, wheat, sorghum, groundnuts, pearl millets, and cowpea.

11.2 Bambara Groundnut Flour for Protein Enrichment in Foods

A considerable number of value-added products have been developed from BGN (Barends, 2019, Ijarotimi, 2008). Ijarotimi (2008; Awobusuyi & Siwela, 2019; Bultosa et al., 2020; Mubaiwa et al., 2018a) produced weaning food using a mixture of cooking banana fruit and BGN flours in 60:40 ratio. Nutrend (a commercial formula) and ogi (corn gruel, a traditional weaning food) were used as control. The red blood cell (RBC), white blood cell (WBC), pack cell volume (PCV) and erythrocyte sedimentation rate (ESR) were higher in the experimental food samples than the commercial food. The growth rate of animals fed with experimental food sample was lower than those fed with the Nutrend but higher than those fed with ogi. The author concluded that the nutritional quality of cooked banana and fermented BGN mix of 60:40 ratio was better than ogi; and comparable to the Nutrend. The amino acid composition increased as the BGN ratio increased. The mixture provided 42.4% of the recommended dietary allowance (RDA) for essential amino acids. The biological value (BV) was high (90.5%) and Nutrend (93.8%). Furthermore, the net protein utilization (NPU), total digestibility (TD), protein efficiency ratio (PER), feed efficiency ratio (FER) and nitrogen retention (NR) were like those of Nutrend. Hence, the 60:40 mixture of cooking banana fruit and BGN flours contained the essential amino acids needed to support infant growth and development (Ijarotimi & Olopade, 2009).

Anyika (2003) studied the effect of dehulled and soaked BGN combined with sorghum or crayfish on iron and phosphorus balances of adult Wistar albino male rats with casein as a reference. Rats fed the casein diet significantly decreased in food intake, maintenance body weight, phosphorus intake and phosphorus balance than those of the test groups. Dehulled brown BGN was superior to the white variety in terms of iron. Soaking affected iron utilization in the brown variety. Dehulling produced significant increases in phosphorus balance in animals fed the white variety of BGN than the brown variety. However, soaking had a positive phosphorus balance in rats consuming the brown variety. The addition of crayfish produced a slight nonsignificant decrease in rodents eating the brown variety. Hence, varietal BGN difference and treatment affected iron and phosphorus utilization in rats fed BGN in conjunction with sorghum and/or crayfish.

Qaku et al. (2020) studied the effect of BGN supplementation on the nutritional characteristics of *mahewu* (a sorghum-based non-alcoholic beverage product) supplemented with different proportions of BGN (80:20 and 70:30, respectively). Sorghum-BGN *mahewu* blends showed a rapid reduction in pH and significantly lower final pH (4.3) after 24 h than the control (pH 5.1) (100% Sorghum *mahewu*). With total titratable acidity (TTA), sorghum–BGN *mahewu* blends had significantly higher final TTA (0.7%) than control (0.4%). Sorghum–BGN *mahewu* blends had higher protein and crude fiber contents (up to 7.7% and 7.9%, respectively) than control (6.1% and 5.7%, respectively). In general, sorghum *mahewu* supplemented with BGN did not show a substantial increase in mineral and individual amino acid contents after fermentation than the control, possibly due to the increase in tannin content with BGN supplementation. Fermentation significantly reduced phytate content by more than 90% in sorghum *mahewu* supplemented with BGN than the control (28% reduction). Significantly, BGN supplementation improved the level of some essential amino acids such as threonine, methionine, phenylalanine, isoleucine, leucine, histidine, and lysine. Therefore, supplementation with BGN can improve the nutritional quality of sorghum *mahewu,* but a modification in the processing procedure is required to effectively improve the nutritional quality of sorghum–BGN blends.

Nwadi et al. (2020) reviewed the literature from 1993 to 2019 on the protein content and use of BGN in processed foods. In most of the products, a high quantity of BGN is used, rendering the products unacceptable for taste and flavour. Research efforts to improve the sensory quality of such BGN products include roasting, germination, and extrusion.

Quedraogo et al. (2012) described BGN seed as a complete diet, making it a good supplement to cereal-based diets. Substituting other grains/flours with BGN grain/flour have been studied. In one of such studies, substituting maize (masa) flour with BGN flour in tortillas production and measuring the thermal, antioxidant, physical (spread ration, degree of puffing and rollability), and chemical properties, degree of puffing and rollability of flour and tortillas was reported. Mashau et al. (2020) observed that tortillas produced from maize and BGN flours at the ratio of 100:0, 95:5, 90:10, 85:15 and 80:20, respectively, showed improvement on the proximate composition, increased onset temperature (To) of flour blends from 57.50 to 71.95 °C, peak temperature (Tp) from 74.94 to 76.74 °C and the end temperature (Te) from 81.72 to 91.58 °C; higher values of polyphenolic compounds and antioxidant activities, increased degree of puffing and rollability but inferior textural attributes in hardness, springiness, gumminess, cohesiveness and chewiness.

The enrichment of cereal-based food with legumes in cereal-legume food has increased due to the resulting essential amino acids of high biological value (Nnam, 2001; Mensah & Tunkins, 2003). Cereals are deficient in essential amino acids like lysine and tryptophan. Legumes are deficient in sulphur-containing amino acids like methionine and cystine, but rich in tryptophan and lysine. Fonio/acha (*Digitaria exilis*), a crop with an attractive research focus (Jideani, 1999; Ayo & Okoloiko,

2003; Zhu, 2020), has the following properties (i) a high water absorption capacity carbohydrate that gives it the potential to be utilized in baked foods, (ii) contain pentosan (Zhu, 2020; Babarinde et al., 2020) which provides it with the ability to form a gel in the presence of oxidizing agent at room temperature with high residual protein-coupled with a high level of sulphur and hydrophobic amino acid residue making it useful in baking processes (Obizoba & Egbonna, 1992) and (iii) unique property of diminishing the blood glucose level (Ayo & Okoloiko, 2003) and the subsequent reduction of the diabetic populace. The effect of added BGN flour on the quality of acha-based fura by Ayo et al. (2014), to improve the nutrient quality of fura (a semi-solid dumpling cereal meal produced mainly from moist pearl millet (*Pennisettum glaucum*) flour as investigated. BGN flour resulted in (i) increased protein (6.14–9.30%), fat (3.01–4.80%), crude fiber (1.20–4.4%) and ash (0.50–1.12%), but decreased carbohydrate (88.20–79.21%) content of the fura with an increase in quantity (0–25%), (ii) increased bulk density and foam capacity 0.57–0.71% and 5.0–12.2%, respectively on increasing addition of BGN (0–25%), (iii) increased mean score for taste, texture and general acceptability (6.4–7.55, 7.1–7.8, and 6.9–7.85, respectively) with increasing addition of BGN up to 10% and after that a decrease in these parameters, (iv) the most acceptable substitution with BGN was at 10%. Ayo et al. (2014) recommended the pretreatment of the BGN before use for improved texture, colour and water absorption capacity of the product (Table 11.1).

Bambara-fortified high protein fermented maize dough (Mbata et al., 2009) is possible, taking advantage of the essential amino acids like lysine, tryptophan and threonine more in legume than in cereal grains. BGN can be used to produce milk by soaking overnight, dry frying the seeds, homogenizing and removal of insoluble material (Brough et al., 2003). Wheat flour, known to lack certain essential amino acids like lysine, tryptophan and threonine used in biscuit, bread and cake production can be fortified with BGN (Abu-Salem & Abou-Arab, 2011).

Table 11.1 Effect of added Bambara groundnut (BGN) flour on the chemical composition(%) of acha-based fura

Acha	Bambara groundnut	Crude protein	Fat	Crude fiber	Ash	Moisture	Carbohydrate
100	0	6.14c	3.01d	1.20c	0.50c	0.99b	88.20a
95	5	7.41b	3.50c	1.31bc	0.60bc	1.00b	86.21a
90	10	7.83ab	4.25b	1.71b	0.70b	1.05ab	84.53a
85	15	8.90a	4.40ab	2.30b	0.81ab	1.10a	82.54a
80	20	9.11a	4.61a	3.90a	0.93a	1.20a	80.33a
75	25	9.30a	4.80a	4.4a	1.12a	1.22a	79.21a

Average means scores with the same alphabet in the same column are not significantly different p = 05

11.3 Application of Innovative Technology to Process Bambara Groundnut

The growing world population projected to double by 2050 will increase the importance of legumes. However, BGN utilization is hindered by the dual challenge of inherent antinutritional factors and characteristic long preparation time. To counter this challenge, various cooking and processing methods are often applied to reduce the level of the antinutritional factors and improve the utilization efficiency of legumes. These processing methods may lead to undesirable nutritional changes. The nutritional and cooking quality changes associated with several distinct traditional and modern processing techniques are presented, including the advantages and disadvantages and the use of combined techniques to obtain the best possible quality end product.

11.3.1 Reduction of Cooking Time

Bambara groundnut and cowpea take a long time to cook by the conventional boiling method. This process is time and energy-consuming, resulting in high energy requirements and greater loss of nutrients (Oyeyinka et al., 2020) with implications on the textural characteristics. Generally, all landraces of BGN dry seeds are very difficult to cook, requiring more time (45 min for fresh seeds; 1 to 3 h for dry seeds) and energy (Berchie et al., 2010) than other grain legumes. Adegoke et al., 2019 in their study on the effects of cooking conditions on the texture of cowpea recommended as a suitable alternative careful management of cooking conditions involving specific weight (1000 g), cooking time (57.14 min) and post-cooking resident time (10 min) resulting in acceptable hardness, springiness, chewiness, gumminess, energy to break, stringiness, sensory texture and overall acceptance. Similar studies have been done for cowpea by Liu et al. (1993) and on BGN (Adegoke et al. (2019). These studies highlighted constraints to the optimization of process variables to include mass, cooking time, holding/post-cooking time, hardness, springiness, adhesiveness, cohesiveness, chewiness, fracturability, gumminess, energy to break, stringiness, colour, taste, texture, and overall acceptability.

From Adegoke et al. (2019), pressure-cooking of hard-to-cook legumes is a hydrothermal process involving hydration and heating, which may occur separately or concurrently. Many studies have been conducted on ways to reduce the cooking time of legumes (Deol & Bains, 2010; Lush & Evans, 1980; Phadi, 2004; Proctor & Watts, 1987; Sefa-Dedeh et al., 1978, 1979). Hydration capacity reflects the water holding capacity of the seeds, including water of hydration and capillary water. Hydration is influenced by the seed coat and cotyledon structure (Lush & Evans, 1980). The seed coat has the first direct contact with water during soaking in the early part of the cooking process and serves as conduits for water into the cotyledon. Cooking time is one of the primary criteria to evaluate the quality of dry legumes

(Proctor & Watts, 1987). Therefore, cooking time is a crucial quality parameter for most hard-to-cook legume seeds, and it is measured relative to the softening of the cotyledon texture (Phadi, 2004). Pressure cooking time often excluded hydration time before the attainment of full pressure (Deol & Bains, 2010) after which the boiling of the seed commenced. Cooking time is discussed as the time required by the seeds to attain a softer texture rather than flavour attributes, which would include cooked aroma (Deol & Bains, 2010, Phadi, 2004; Procter & Watts, 1987). During the holding/cooling time, structural changes in the seed continue. These may have significant effects on the quality of the cooked beans depending on the length of the holding/ cooling time.

Shanono and Muhammad (2015) on the effect of cooking on proximate and antinutritional content of two BGN landraces showed that cooked seeds, significantly ($p < 0.05$) decreased in crude fiber, crude fat, total ash, total carbohydrate, levels of tannins, oxalate and phytic acid than the raw seeds and that significant increase occurred in moisture content.

11.3.2 *Reduction in Antinutritional Factors- Fermentation and Sprouting*

There is a growing interest in promoting fermented leguminous seeds as fermentation reduces antinutritional compounds, improves protein and starch digestibility, reduces allergenicity, and increases seed antioxidant capacity (Limon et al., 2015; Nakajima et al., 2005). Fermentation, an age-long process of processing cereals and legumes, modifies some physical characteristics of the grains and increases the levels of nutrients, digestibility, and bioavailability and decreases the level of antinutrients in food (Onweluzo & Nwabugwu, 2009). The fermentation process has helped in the development of several food products including biscuits with an improved flavour, texture and nutritional properties. Investigations have reported that fermentation and other processing methods improved the nutritional quality of legumes and cereals by causing significant changes in their chemical composition and eliminating antinutritional factors (Yasmine, 2002; Ochanda et al., 2010).

Fig. 11.2 Sprouted red Bambara groundnuts and red beans (Nyau et al., 2017)

Processing and end-use would depend on the varying seed coat colour. Pre-decortication treatments on BGN are soaking, steaming/cooking, boiling, roasting and sprouting. There is a growing interest in the consumption of sprouted seeds (Fig. 11.2) at a global level as it enhances the polyphenolic profiles of seeds (Martınez et al., 2012; Nyau et al., 2017). Polysaccharides degrade into oligosaccharides and monosaccharides; fats into fatty acids, protein into free amino acids (Jom et al., 2011; Marton et al., 2010) and increased *in vitro* protein digestibility.

Abba et al. (2018) on sprouting BGN observed (i) sprouting process affected nutrient and antinutrient composition of BGN (*Vigna subterranea*), (ii) a significant increase ($p < 0.05$) in the nutrient content and decreased significantly ($p < 0.05$) the antinutrient contents of two BGN landraces with the observed decrease in the calcium, sodium, potassium and vitamins, (iii) zinc and iron contents increased in sprouted Niger cream landrace and (iv) decrease in vitamins B1 and B6 after processing in one of the landraces.

Studies have shown that germination or sprouting positively affects the chemical composition of legumes (Atudorei & Codina, 2020; Boukid et al., 2019; Bueno et al., 2020; Kigel et al., 2015; Luo et al., 2013; Nyau et al., 2017; Ohanenye et al., 2020; Xu et al. 2019a, b, 2020). It is the only agro-food processing process that provides a significant increase of the nutritional value by increasing the bioavailability of vitamins, bio-elements and other biologically active compounds (Brajdes & Vizireanu, 2012).

Atudorei and Codina (2020) highlighted that (i) germinated legumes can be successfully used in various food manufacturing recipes, to improve the nutritional value of these products, (ii) sprouted seeds can be successfully used to improve wheat flour to obtain healthier bakery products, (iii) the flour obtained from germinated seeds can be used as a substitute for different chemical additives in the bakery industry due to its high content of enzymes, ascorbic acid, lecithin, etc., (iv) many publications in the international literature on the influence of the addition of different types of legume flour in the bread-making recipe exist, but there are few studies on the addition of legume flour in a germinated form on bread making, hence the need for new alternatives in improving bread, underlining how different types of legumes in a sprouted form can improve dough rheological characteristics and the final quality of bread.

Germination of BGN involves soaking the seed in water for about 12 hours until fully imbibed, draining, spreading the seeds on a moist muslin cloth and allowing to germinate, preferably in the dark at room temperature (25 ± 1°C) with intermittent water sprinkling to keep moist. The sprouts are harvested after about 8 days when 98–100% of the seeds had germinated. Seeds with root growth of about 2 mm are considered to have sprouted. Germination capacity is expressed as the percentage of germinated seeds (Martınez et al., 2012; Nyau et al., 2017).

Polysaccharides degrade into oligo and monosaccharides; fats into fatty acids and protein into free amino acids during germination (Marton et al., 2010). There is a significant improvement in nutritional quality after sprouting cowpeas at 25 °C for 24 h. Protein increased by 9–12%, vitamin C increased by 4–38 times, phytic acid decreased by 4–16 times, trypsin inhibitor activity decreased by 28–55% and *in-vitro* protein digestibility increased by 8–20% (Devi et al., 2015) and alteration in the level of fatty acids, methyl esters, free fatty acids, monosaccharides and disaccharides (Jom et al., 2011).

11.3.3 Milling Bambara Groundnut into Flour

Danbaba et al. (2020) showed that variability in the milling behaviour, flour colour, particle size distribution and chemical composition of pre-treated BGN varying in seed coat colour improved decortication efficiency and split yield and reduced broken splits. They observed that a significant quantity of all flour types was retained on the 1 mm sieve except black coated seed when pre-treated by soaking, and significant improvement in protein for pre-treated flour than raw flour. Among the pre-treatments, roasting and germination reduced the lightness index of flour. All the treatments improved protein content.

11.3.4 Micronisation of Bambara Groundnut

Ogundele et al. (2017) reported the functional properties of flours from pre-soaked and micronized (130 °C) whole and dehulled BGN seeds for 5, 10 and 15 min. Micronisation utilizes electromagnetic radiation with a wavelength of 1.8 to 3.4μm in the infrared region. The food material conditioned with moisture is exposed to infrared radiation, as the infrared wave strikes on the food surface and parts of the penetrated energy are absorbed resulting in molecular vibration by the food molecules at a frequency of 60,000–150,000 MHz. The molecular vibration of the biomolecules generates heat due to molecular friction and a rise in water vapour pressure in the food. Increasing micronization time significantly reduced the pasting viscosity. Micronisation pre-gelatinized the flour forming a low viscous paste related to protein denaturation indicated by a decrease in nitrogen solubility index. The denatured protein could be preventing starch hydration and dispersion during pasting and thus reduced viscosity. Dehulling reduced the pasting viscosity suggesting a higher effect of micronization for dehulled than whole seeds.

Pretreatment of BGN seeds (i.e. soaking, roasting and combined soaking and roasting) improved its dehulling efficiency (Mubaiwa et al., 2018b). The protein content of flour ranged from 15.6–19.6%, starch from 47.8–52.0% and sucrose from 1.9–5.0%. Pretreatment improved the protein and ash content of flour. Heat treatments increased the onset gelatinization temperature of flour. Black-eye BGN flours with higher starch content also had better gelation capacity than the red BGN flours. All pretreatment methods decreased flour emulsification capacity and stability. Dry-roasting caused a greater decline than other methods, whereas soaking had little effect on emulsion stability. Soaking increased foaming capacity, whilst a decline was observed in roasted flour. All pretreatment methods increased the oil absorption capacity of both BGN flour varieties. Overall, soaked and combined soaked and roasted flour is recommended for the production of BGN flour The BGN flours compare nutritionally to other legumes such as cowpea and chickpea. Except for protein and ash that were higher in the red variety, minor differences exist in the chemical composition of the flours. The flours from the red variety had a higher gelatinization temperature compared to the black-eye variety. The authors concluded that both flours can be used in food product development.

11.4 Bambara Groundnut Flour in Biscuits

Composite flour technology involves mixing wheat flour with cereal and legume flours for bakery products like bread and biscuits. It has been used to extend scarce supplies of wheat or corn in the bakery industry. Composite flours are advantageous in many areas: (i) it supplements the deficiencies in nutrient composition thereby reducing malnutrition in infants and adults, (ii) alleviates challenges arising from the ban on importation of wheat used for baking and confectionery industries. Many studies have enhanced utilization of composite flour blends of many crops resulting in rich protein-protein that can help fight against protein-energy malnutrition (Ojinnaka et al., 2018; Ola & Adewole, 2019; Awolu et al., 2017; Ayo & Andrew, 2016). Flours made from a combination of tubers and legumes contain high protein and caloric value (Chinma et al., 2007].

Biscuits and cakes are used as snacks in many functions. They are produced by baking butter with flour, beaten eggs with or without shortenings and baking powders. Wheat is mostly used as the cereal of choice for production snacks. Due to climatic conditions, it is not grown in tropical regions, hence products requiring wheat flour for the production of baked products could be challenging in areas that import wheat (Holt et al., 2015).

Adegunwa et al. (2017) evaluated the quality of snacks made from unripe plantain–BGN–turmeric composite flour blend. The ratio 35:60:5, respectively for unripe plantain–BGN–turmeric resulted in increased protein, fat contents and was highly acceptable. Andrew (2016) prepared biscuits from varying acha and BGN flour with Date palm fruit flour as sucrose replacer. The proximate (protein, ash, fat, moisture, fiber and carbohydrate), physical (weight, spread ratio and break strength) and sensory (taste, odour, colour, texture, general acceptability) properties of the acha–Bambara composite biscuits were analyzed. Up to 10% BGN substitution improved the protein, fat, fiber and ash content of the biscuits while carbohydrate decreased with an increase in the added Bambara groundnut.

The production or enrichment of cereal-based foods with legumes has received considerable attention with legumes being high in lysine than most cereals (Akpapunam & Darbe, 1994). Agu et al. (2020) evaluated acha, malted BGN and date palm flours in gluten-free, sugar-free biscuits. The biscuit produced with 60:10:30 acha, malted BGN and date palm, respectively was the most preferred by the consumers. There was no significant difference between the preferred biscuit and the control in colour, taste, flavour, texture except in crispiness. The physical properties for the most preferred biscuit are weight 23.0 g, height 3.0 g, spread ratio 7.7 and break strength 750 g with fat 20.1%, protein 11.9%, ash 2.6%, moisture 1.2%, crude fiber 4.3% and carbohydrate 60.0%. Hence, acha, malted BGN and date palm flours could produce consumer acceptable gluten-free, sugar-free biscuits. Also, Agu et al. (2014) reported the use of [acha (*Digitaria exilis*)], BGN and unripe plantain [*M. paradisiaca]* at different proportions to produce biscuits (Fig. 11.3) and concluded that (i) high quality and acceptable ($p < 0.05$) biscuit products resulted from acha, BGN and unripe plantain (80:10:10% and 70:20:10% proportions respectively), (ii) BGN and unripe plantain improved the physical, proximate and sensory quality of acha biscuit and (iii) shelf life of one year for the biscuits. It has also been shown that the addition of leaf power, like moringa leaf powder, can affect the colour of biscuits (Nadhifa & Nugrahani Astuti, 2018).

Ola and Adewole (2019) established the potentials of fermented (spore suspension of *Rizophous oligosporous*). BGN in biscuit production by blending wheat flour and fermented BGN flour at different ratios (100:0, 90:10, 80:20, 70:30, 60:40 and 50:50 respectively) and found the protein content was higher than 100% wheat flour biscuit, moisture content ranged from 2.87–5.40%, other components of starch and reducing sugar were in the ranges 60.29–63.93%, and 2.74–4.43% and (ii) acceptability was at up to 30% substitution with BGN with thickness, diameter and spread ratio ranged from 230.33–285.67 mm, 29.00–36.00 mm and 61.70–89.19%, respectively. Their study contributed to the existing knowledge that fermentation generally improved the nutritional quality of products like a biscuit made from composite flour. Such an increase in the protein content of composite flour biscuits at the acceptable level of substitution could be used in eradicating malnutrition problems. Studies of this nature opened possibilities of incorporating fermented BGN flour into starchy diets which invariably will lead to an expansion in the utilization of fermented BGN flour (Ola & Adewole, 2019). Nwosu (2013) reported on the production and evaluation of biscuits from blends of BGN and wheat flours without the process of fermentation.

Fig. 11.3 Acha (Digitaria exilis) based biscuit (80: 10:10 % acha, Bambara groundnut and unripe plantain flours) in a polyethylene bag and paperboard packaging (Agu et al., 2014)

11.5 Bambara Groundnut Flour in Pasta

The pasta was prepared from fractionated whole-grain wheat flour enriched with 20% BGN. The fat, protein (by 37%), ash contents, lightness and antioxidant properties of the flour and pasta increased. In contrast, carbohydrate and fiber contents decreased with a reduction in particle size from 500μm to 112μm. Potassium (246.50–249.00 mg/kg), calcium (223.50–254.00 mg/kg) and magnesium (184.50–192.00 mg/kg) were the major mineral element in the pasta samples, while zinc (1.00–2.00 mg/kg) and iron (3.50–13.00 mg/kg) are present in small quantities. Pasta from finer particles is rich in fat, protein and ash, but low in carbohydrate and fiber than pasta made from larger particles. Pasta colour, protein, fiber and antioxidant properties improved with fractionation. The optimum cooking time of pasta (average 6.55 min) from the fractionated flours was shorter than the control pasta (pasta made from unfractionated wheat flour). Still, the cooking loss was not significantly affected. Pasta from flour with particle sizes of 300 and 112μm was very similar in sensory attributes with the highest overall acceptability rating. The authors concluded that fractionation of whole-grain wheat flour is a promising technology for producing pasta with fairly good antioxidant potentials and a high level of protein and fiber (Oyeyinka et al., 2021).

11.6 Formulated Complementary Foods from Bambara Groundnut

Nutritious complementary foods, also known as weaning foods, typically cover the period from six to twenty-four months of age in most developing countries (WHO/OMS, 2000). Improved complementary foods at this period of life are essential for good nutritional status, normal growth, cognitive development, prevent malnutrition, stunting and anaemia in children (Michaelsen, 2000; Uvere et al., 2010). The lysine content of BGN is very high, therefore, good for efficiently complementing cereals where lysine is the limiting amino acid (Enwere and Hung 1996).

Some work has been done on complementary foods using BGN (Adebayo-Oyetoro et al., 2017, Ijarotimi & Keshinro, 2013, Kelechi-Ebisike et al., 2020, Oluwatofumni et al., 2015, Uvere et al., 2010, Oyeyinka et al., 2017). Ijarotimi and Keshinro (2013) using fermented popcorn-African locust bean (FPA) (70% popcorn, 30% African locust bean), fermented popcorn-BGN (FPB) (70% popcorn, 30% BGN) and fermented popcorn-African locust bean-bambara groundnut (FPAB) (70% popcorn, 20% BGN, 10% African locust bean) produced and evaluated the nutritional quality of these complementary foods and observed the following (i) protein content of FPAB (26.87 ± 1.07 g/100 g) was significantly higher than FPB (20.87 ± 1.02 g/100 g) and FPA (20.49 g/100 g) respectively; potassium had the highest value in FPA (173.75 ± 0.21 mg/100 g), FPB (157.45 ± 0.25 mg/100 g) and

FPAB (132.75 ± 0.15 mg/100 g), while copper was the least mineral (Table 11.2), (ii) predicted protein efficiency ratio (P-PER) range between 0.54 and 1.43 for FPB and FPAB respectively; while biological value (BV) was between 20.13 and 39.94%, (iii) oxalate, tannin, phytate and trypsin inhibition of popcorn-based food samples was reduced to a minimum level (Table 11.2) and (iv) FPAB had a better nutritional quality over FPA and FPB using selected nutritional indices; hence, FPAB blends could be used as complementary food for infants.

Complementary feeding in Low-and-Middle-Income-Economies (LMIEs) remains a major concern as the burden of childhood malnutrition continues. Complementary foods provide energy, macronutrients and micronutrients to bridge the gaps between daily requirements for infants and young children and the amount consumed through breastfeeding (Abeshu et al., 2016; Codex Alimentarius Commission, 2017) to reduce the burden of wasting and stunting of children under five years of age. Low energy intake leads to reduced physical activity which is important for the social development of the.

child. Legumes are protein and fat sources and are increasingly gaining interest globally because of the demand for healthier foods, especially plant foods. Kelechi-Ebisike et al. (2020) formulated complementary foods in 60:40 (cereal/tuber:legume) mix ratio for diet A: sweet potatoes + millet + groundnut (*Arachis hypogaea*) + BGN (25:35:30:10% w/w), diet B: tiger nut (*Cyperus esculentus*) + millet + groundnut + BGN (25:35:30:10% w/w) and observed that (i) energy density of diets were between 4.04 ± 1.84 kcal/g – 4.13 ± 2.84 kcal/g, crude protein (15.84 ± 0.59% – 18.28 ± 0.35%), crude fat (5.84 ± 0.4% – 8.41 ± 0.21), (ii) moisture values between 1.17 ± 0.12% – 2.46 ± 0.15% suggesting longer shelf life foods, and (iii) diets had good energy and macronutrient density, a very low cost of production and overall acceptability. With the twin problem of rising malnutrition across the globe and the economic situation, especially in developing economies, it becomes important to explore the benefits of complementary food formulated with BGN.

In animal study using thirty (30) weaning rats grouped into six of five rats each and fed with five dietary samples for 28 days, Oluwatofunmi et al. (2015) prepared complementary diets with flours from BGN and cowpea at a ratio maize-Bambara (80:20), maize-cowpea (80:20) and maize-Bambara-cowpea (60:20:20) with *Nutrend*, a vegetable-based protein commercial food manufactured by Nestle plc as control and observed that (i) the physical and biochemical parameters established no significant difference in the analytical data for formulated and control diets, (ii) the formulated diets were comparable nutritionally to control diet in supporting animal growth without any significant organ impairment as indicated in the liver and kidney function tests, (iii) the diets formulated from maize, BGN and cowpea seeds blends complementary food were well accepted as shown by the amounts consumed by the rats, and (iv) the diets were acceptable, readily available, affordable and nutritionally adequate for animal growth. In most of the work on complementary food using BGN, the authors generally recommended further investigation on nutritional and biochemical studies of the formulations to substantiate their potentials (Ijarotimi and Keshinro (2013), Oyeyinka et al., 2017).

Table 11.2 Mean (±SEM) of macronutrient composition (g/100 g dry weight matter) of fermented popcorn, African locust bean and Bambara groundnut (BGN) flour blends

Nutrient	Fermented popcorn- African locust bean blend (FPA)	BGN blend (FPB)	African locust bean & BGN blend (FPAB)	Ogi	Cerelac	*FAO/WHO (g/100 g)
Moisture	$10.70^{a} \pm 0.73$	$10.27^{ab} \pm 0.01$	$10.07^{ab} \pm 0.03$	$8.31^{b} \pm 0.57$	$11.3^{a} \pm 0.50$	<5
Protein	$20.49^{b} \pm 1.08$	$20.87^{b} \pm 1.02$	$26.87^{a} \pm 1.07$	$6.52^{d} \pm 0.31$	$15.75^{c} \pm 0.01$	>15
Fat	$12.76^{b} \pm 0.20$	$12.31^{b} \pm 0.11$	$14.59^{a} \pm 0.15$	$5.17^{d} \pm 0.11$	$10.53^{c} \pm 0.02$	10–25
Ash	$6.07^{a} \pm 1.24$	$2.96^{c}0.10$	$4.87^{b} \pm 0.03$	$1.09^{c} \pm 0.01$	$3.16^{ab} \pm 0.01$	<3
Fibre	$1.72^{a} \pm 0.27$	$2.05^{a} \pm 0.21$	$2.41^{a} \pm 0.18$	$0.85^{b} \pm 0.01$	$2.11^{a} \pm 0.01$	<5
Carbohydrate	$58.95^{cd} \pm 2.41$	$61.78^{c} \pm 1.01$	$51.27^{d} \pm 2.11$	$86.38^{a} \pm 0.21$	$68.42^{b} \pm 0.01$	64
Energy (kcal)	$432.61^{bc} \pm 4.97$	$441.48^{b} \pm 3.05$	$464.94^{a} \pm 1.22$	$418.08^{c} \pm 0.47$	$431.58^{bc} \pm 0.01$	400–425

The same superscripts in a row are not significantly different ($P > 0.05$)
Ogi is fermented maize
*FAO/WHO (1991); Source: Ijarotimi and Keshinro (2013)

Food-to-food fortification as in the case of maize–BGN complementary foods still have the challenge of deficiency in calcium, iron, zinc and vitamin A. Uvere et al. (2010), adopted pre-fermentation fortification on maize–BGN complementary foods fortified for iron, zinc, calcium and vitamin A by blending with a multi-mix (1.41:1:2.25, w/w) of processed roselle calyces (*Hibiscus sabdariffa*), cattle bones, and red palm oil in a 1:2.1 (w/w) ratio and reported (i) significant increases in calcium, iron, zinc and vitamin A contents of 3.26–4.225, 0.083–0.134 and 0.015–0.017 g kg^{-1} and 4855.3–7493.7 μgRE kg^{-1}, respectively, (ii) maize–BGN foods had calcium, iron, zinc and vitamin A contents that satisfy the proposed nutrient requirements for infants, and (iii) maize–BGN and maize–BGN malt fermented by backslopping [(MB)b and (MBm)b] containing red palm oil emulsified with *Brachystegia eurycoma* had calcium contents significantly ($P < 0.05$) higher than Nutrend, a complementary food produced by Nestle (Nigeria) PLC.

Excellent complementary foods are expected to have an easy-to-swallow, semi-liquid consistency of 1000–3000 mPa s (Nout & Ngoddy, 1997), a right balance of amino acids, and be rich in bio-available micronutrients and affordable. Food fortification with vitamin and mineral premixes and the addition of milk solids have been used in many countries to address micronutrient deficiency problems (Bowely, 2003).

11.7 Bambara Groundnut Flour in Beverages and Meat Analogue

Researchers from South Africa through multivariate product and process development studied the relevant multi-factorial relationships between formulation, process and quality attributes of BGN. Murevanhema and Jideani (2015; 2017a, b) produced BGN milk from hydrated whole BGN flour from the red, brown, black-eye, brown-eye and black varieties. The production from whole flour is beneficial for energy saving and inclusion of the pigments on the hulls with its phenolics and flavonoids that may contribute to nutritional well-being. The milk differed significantly in pH, color and antioxidant capacity resulting in two clusters. Cluster 1 with a high hue (7.47°, reddish-yellow) but less yellowness (b = 5.19) relates to milk produced from black-eye and brown-eye BGN. Cluster 2 exhibited less hue (48.4°, reddish-yellow) with high yellowness (b = 5.44) relates to milk from black, brown and red BGN flours. Taste, appearance and color were the important sensory parameters considered by the panellists in their preference for BGN milk resulting in two clusters. Cluster 1 associates milk produced from red and brown BGN flour that was moderately undesirable in taste and moderately desirable in appearance and colour. Cluster 2 consisted of brown-eye and black-eye that are moderately undesirable both in taste, appearance and colour. There was a high preference for BGN milk across all age groups indicating the high probability of acceptance if commercialized. Hardy and Jideani (2018) improved the BGN milk and produced the milk powder through spray drying using gluten-free maltodextrin as an effective carrier. Dried milk powder with 10% maltodextrin had acceptable hydration properties (wettability, water absorption index and water solubility index), low water activity (a_w), bulk density comparable to those of commercial milk powders, high in protein (7.6%) and a source of fiber (3.4%). The powder is thermally stable as the glass transition temperature (49.6 °C) is well above the storage room temperature.

Hardy and Jideani (2020) studied the functional, nutritional, and physical characteristics of foam-mat dried BGN yogurt produced from reconstituted BGN milk powder (BGNPY-R) and BGN milk (BGNPY-NR). Both BGN powdered yoghurts showed favorable functional rehydration properties for reconstitution. However, BGNPY-R was significantly higher in wettability, making it superior in maintaining contact with a liquid. The BGN powdered yoghurts are high in bulk density (0.82–0.87 g/ml), a crucial requirement for commercial food powder, beneficial during packaging and material shipping. BGNPY-R and BGNPY-NR are sources of total dietary fiber (3.9–4.2%) and protein (4.8–5.7%). Both the BGN powdered yoghurts had a glass transition temperature (48.9–61.0 °C) well above room temperature (21 ± 2 °C) signifying its capability to overcome caking and stickiness. However, after undergoing drying, only BGNPY-R could be deemed as a probiotic containing food, as the probiotic viability (7.2 log cfu/ml) was above the minimum recommended value of 6 log cfu/ml. Hence, it is possible to produce BGN powdered yogurt from reconstituted BGN milk powder using the foam-mat drying with gum Arabic (6%) and methylcellulose (0.5%) as foaming agents and drying at 50 °C for 24 h.

Bambara groundnut seeds, flour, therapeutic value, functional food value, and protein content control the value of BGN for food security programs and interventions (Jideani, 2016). Bambara groundnut flour (BGNF) suitability as a natural emulsifier/stabilizer in food emulsion systems was assessed (Gabriel et al., 2013). Gelatinized BGNF stabilized an oil-in-water emulsion. All the BGNF emulsions were macro emulsions, desirable in some food applications (Fig. 11.4). BGNF-stabilized emulsions were characterized as highly spreadable (pseudoplasticity), thixotropic- desirable in mastication and chewing (good chewability) and highly rigid, necessary for withstanding harsh conditions during development, transportation, and storage (Adeyi et al. 2014).

BGN milk supported the growth of probiotics and protected them against in vitro gastric juice and bile. The probiotic beverage will benefit consumers by maintaining the right balance and composition of intestinal flora, increasing the body's ability to resist the invasion of pathogens, thereby preserving the host's efficient nutrition (Murevanhema & Jideani, 2015).

Bambara groundnut has been demonstrated to be useful in meat-alternative products and baked goods as depicted in Figs. 11.5 and 11.6, respectively (Jideani & Jideani, 2016).

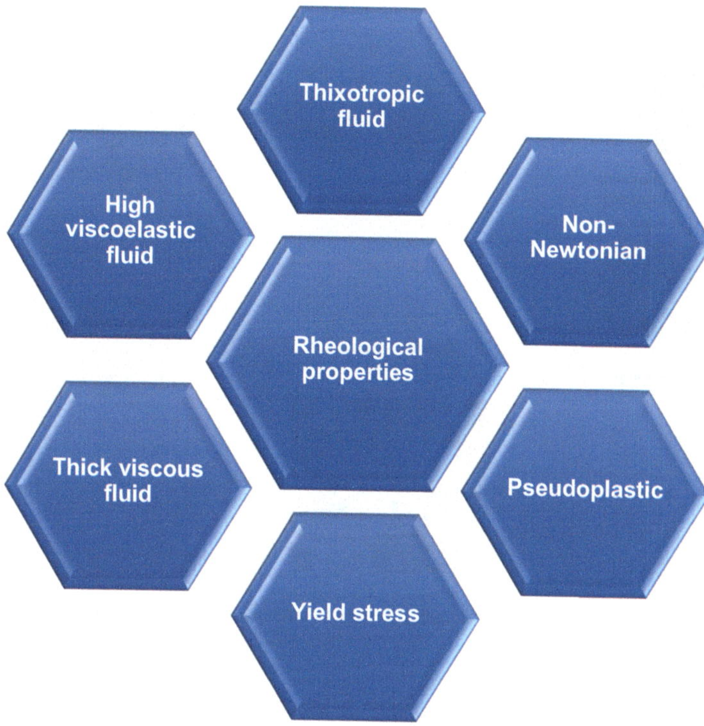

Fig. 11.4 Rheological properties of Bambara groundnut flour stabilized oil-in-water emulsion

Fig. 11.5 Prototype meat-alternatives from Bambara groundnut

11.8 Starch Blends of Bambara Groundnut with Other Plant Crops for Potential Industrial Applications

Starches from botanical sources, extracted using a conventional method, have diverse food and non-food applications. When native starch is physically or chemically modified, it increases the potential commercial application and utilization. Various starches have been previously blended like BGN starch (BGNS) with cocoyam starch (CYS) (Ashogbon, 2017); BGN starch (BGNS) with cassava starch

Fig. 11.6 Some selected gluten-free baked products from Bambara groundnut flour (Jideani & Jideani, 2016)

(Ashogbon, 2014a); cocoyam starch with wheat starch (Ashogbon, 2014b; and sweat potato starch with wheat starch (Zhu & Cooke, 2011). Ashogbon (2017) blended BGN starch (BGNS) and cocoyam starch (CYS) at three different ratios (70BGNS/30CYS, 50BGNS/50CYS and 30BGNS/70CYS) (%, w/w), evaluated compositional and some physicochemical properties and observed (i) decreases in lipid (from 0.31 to 0.06%), protein (0.12 to 0.06%) and apparent amylose (37.31 to 22.60%) contents when CYS was substituted with BGNS and vice versa when more BGNS was substituted into 100CYS, (ii) swelling power (SP) (except at 55 °C), water solubility index, bulk density (BD), dispersibility (DP), pH and pasting parameters (except peak viscosity and peak time) of the starch blends were non-additive of their individual components, (iii) bulk density, DP and pH of the starch blends ranged from 0.79 to 0.85 g/ml, 81 to 86% and 7.38 to 7.52, respectively, (iv) the breakdown viscosity of the control starches was smaller compared to that of starch blends, (v) peak time of starch blends was intermediate to that of the control starches, (vi) the 30BGNS/70CYS blend with the lowest retrogradation (120.60 RVU) could be utilized in refrigerated foods, (vii) the 50BGNS/50CYS blend with the lowest breakdown viscosity value can find application in sterilization operations in the food industry, especially in canned foods, and (viii) the 70BGNS/30CYS blend with the highest SP could find application in the pharmaceuticals as disintegrants as well as in the manufacture of noodles due to high amylose concentration. The blending of plant starches could result in blends with appropriate physicochemical properties for industrial applications. Some traditional and innovative products from BGN found in the literature are outlined in Tables 11.3 and 11.4.

Table 11.3 Proximate composition (% dwb) of formulated composite diets containing Bambara groundnut

Sample	Crude protein	Lipid	Total ash	Crude fiber	Carbohydrate	Energy (Kcal)
Maize–Bambara (80:20)	$15.46^{b} \pm 2.97$	$7.58^{a} \pm 0.29$	$1.36^{b} \pm 0.28$	$0.74^{ab} \pm 0.31$	$74.93^{a} \pm 2.87$	$429.82^{a} \pm 3.07$
Maize–cowpea (80:20)	$17.04^{ab} \pm 1.55$	$6.54^{b} \pm 0.18$	$1.61^{b} \pm 0.37$	$0.57^{ab} \pm 0.38$	$71.85^{ab} \pm 1.92$	$422.55^{b} \pm 1.56$
Maize–Bambara–cowpea (60:20:20)	$19.90^{a} \pm 0.93$	$6.88^{ab} \pm 0.70$	$1.74^{a} \pm 0.33$	$0.81^{a} \pm 0.24$	$70.18^{b} \pm 2.08$	$422.28^{b} \pm 1.90$
Nutrend	$19.11^{a} \pm 2.08$	$6.47^{b} \pm 0.21$	$2.47^{a} \pm 0.20$	$0.26^{a} \pm 0.06$	$71.29^{ab} \pm 0.54$	$419.84^{ab} \pm 0.10$
BGN starch (Ashogbon, 2017)	0.12 ± 0.01	0.31 ± 0.11a	0.05 ± 0.16	–	37.31 ± 0.20 (amylose) 62.69 ± 0.10 (amylopectin)	–

Source: Oluwatofunmi et al. (2015)

Each value represents the mean value of three replicates ±SD. Values with different superscripts in the same column are significantly different at $p < 0.05$. *BGN* Bambara groundnut, *MAB* maize–Bambara (80:20), *MAC* maize–cowpea (80:20), *MABC* maize–bambara–cowpea (60:20:20), *NUTR* nutrend – a vegetable based protein commercial food manufactured by Nestle plc. *BGNS* BGN starch (– *AM* Amylose, *AP* Amylopectin

Table 11.4 Traditional and novel products from Bambara groundnut (BGN) in the literature

Dry grain snacks	Grain and granular foods (flour)	Porridges, gruels & dumpling	Non-alcoholic beverages	Dough products	Breads	Cakes and biscuits
Puffed & popped whole BGN	Blending of precooked sorghum with roasted BGN flours + fermentation (Léopold et al., 2020) Addition of BGN in tortillas production (Mashau et al., 2020) BGN-fortified wheat flours that are nutritive (Abu-salem & Abou-Arab, 2011) Effect of added BGN flour on the quality of acha-based fura (Ayo et al., 2014). Cooked BGN snacks	Added BGN flour on the quality of acha based *fura*. (Ayo et al., 2014). Diet A – Millet: Sweet Potatoes: Groundnut: BGN. (35:25:30:10% w/w). Diet B- Millet: Tiger nut: Groundnut: BGN (Experimental) (35:25:30:10% w/w). (Kelechi-Ebisike et al., 2020)“ Complementary foods formulated from a combination of fermented popcorn, African locust and BGN seed flour (Ijarotimi & Keshinro, 2013).	Prototypes of non-dairy beverages (CPUT 2019)	BGN-fortified high protein fermented maize dough (Mbata et al., 2009).	Sudanese sorghum-based *kissra* bread fortified with BGN seed flour (Abdualrahman et al., 2019)	West Africa: Deep-fried cowpea paste – *Akara* Steamed cowpea paste – *moimoi* Steamed BGN paste – *okpa* Biscuit produced from fermented BGN & wheat flour blends. (Ola & Adewole, 2019) Quality assessment of acha-based biscuit improved with Bambara nut and unripe plantain Production & evaluation of biscuits from blends of BGN & Wheat flours (Nwosu, 2013)

References

Abba, R., Muhammad, N., Lawal, S., & Muhammad, Y. (2018). Sprouting affects nutrient and anti-nutrient composition of Bambara groundnut (*Vigna subterranea*). *Journal of Advances in Food Science & Technology, 5*(2), 34–41. http://www.ikppress.org/index.php/JAFSAT/article/view/3745

Abdualrahman, M., Ma, H., Yagoub, A., Zhou, C., Ali, A., & Yang, W. (2019). Nutritional value, protein quality and antioxidant activity of Sudanese sorghum-based kissra bread fortified with Bambara groundnut (*Voandzeia subterranea*) seed flour. *Journal of the Saudi Society of Agricultural Sciences, 18*, 32–40.

Abeshu, M. A., Lelisa, A., & Geleta, B. (2016). Complementary feeding: Review of recommendations, feeding practices, and adequacy of homemade complementary food preparations in developing countries – Lessons from Ethiopia. *Frontiers in Nutrition, 3*, 41. https://doi.org/10.3389/fnut.2016.00041

Abu-Salem, F. M., & Abou-Arab, A. A. (2011). Effect of supplementation of Bambara groundnut (*Vigna subterranean* L.) flour on the quality of biscuits. *African Journal of Food Science, 5*(7), 376–383.

Adebayo-Oyetoro, A. O., Shotunde, A. B., Adeyeye, S. A. O., & Ogundipe, O. O. (2017). Quality evaluation of millet-based fura powder supplemented with Bambara groundnut. *International Journal of Food Science, Nutrition and Dietetics, 6*(3), 358–362. https://doi.org/10.19070/2326-3350-1700063

Adegoke, B. H., Adegunwa, M. O., Akano, R. A., Alake, O. O., Ayodele, I. O., Akinsola, O. I., & Bishi, A. A. (2019). Effects of cooking conditions on the texture profile, sensory and proximate qualities pressure cooked cowpea. *Acta Scientific Nutritional Health, 3*(5), 151–167.

Adegunwa, M. O., Omolaja, N. O., Adebowale, A. A., & Bakare, H. A. (2017). Quality evaluation of snacks produced from blends of unripe plantain, Bambara groundnut and turmeric flour. *Journal of Food Processing and Preservation, 41*, 1–8.

Adeyi, O., Ikhu-Omoregbe, D., Jideani, V. A. (2014). Emulsion stability and steady shear characteristics of concentrated oil-in-water emulsion stabilized by gelatinized Bambara groundnut. *Asian Journal of Chemistry, 26*(16), 4995–5002

Agu, H. O., Ezeh, H. G. C., & Jideani, A. I. O. (2014). Quality assessment of acha-based biscuit improved with Bambara nut and unripe plantain. *African Journal of Food Science, 8*(5), 278–285.

Agu, H. O., Onuoha, G. O., Elijah, O. E., & Jideani, V. A. (2020). Consumer acceptability of acha and malted Bambara groundnut biscuits sweetened with date palm. *Heliyon, 6*(11), e05522.

Akpapunam, M. A., & Darbe, J. W. (1994). Chemical composition and functional properties of blends of maize Bambara groundnut flours for cookie production. *Plant Food for Human Nutrition, 46*, 147–155.

Andrew, J.A.A.E. (2016). Effect of added Bambara groundnut on the quality of acha-date palm-based biscuit, 4, 34–38

Anyika, J. U. (2003). Iron and phosphorus utilization in rats fed soaked and dehulled Bambara groundnut supplemented with sorghum or crayfish. *Plant Foods for Human Nutrition, 58*, 53–61.

Ashogbon, A. O. (2014a). Physicochemical properties of Bambara groundnut starch and cassava starch blends. *African Journal Food Science, 8*(6), 322–329.

Ashogbon, A. O. (2014b). Chemical and functional properties of cocoyam starch and wheat starch blends. *International Journal of Biotechnology and Food Science, 2*(5), 94–101.

Ashogbon, A. O. (2017). Evaluation of compositional and some physicochemical properties of Bambara groundnut and cocoyam starch blends for potential industrial applications. American. *Journal of Food and Nutrition, 5*(2), 62–68. https://doi.org/10.12691/ajfn-5-2-3

Atudorei, D., & Codina, G. G. (2020). Perspectives on the use of germinated legumes in the bread making process, A review. *Applied Sciences, 10*, 6244. https://doi.org/10.3390/app10186244

Awang, N. (2020). S'pore-based startup says its fresh take on instant noodles is healthy, good for the environment. *Today*, pp. 1–12.

Awobusuyi, T. D., & Siwela, M. (2019). Nutritional properties and consumer's acceptance of pro-vitamin A-biofortified *amahewu* combined with Bambara (*Vigna subterranea*) flour. *Nutrients, 11*, 1476. https://doi.org/10.3390/nu11071476

Awolu, O. O., Olorunfemi, A. M., Enujiugha, V. N., & Arowosafe, C. F. (2017). Enriched cake production and optimization of physicochemical qualities of wheat, moringa seed and African oilbean composite flours. *Acta Scientific Nutritional Health, 1*(2), 21–30.

Ayo, J. A., & Andrew, E. (2016). Effect of added Bambara groundnut on the quality of acha-date palm-based biscuit. *International Journal of Biotechnology and Food Science, 4*(3), 34–38.

Ayo, J. A., & Okoloiko, A. (2003). Effect of spray drying temperature of egg-white powder on the quality of meringue. *Journal of Arid Agriculture, 13*, 181–187.

Ayo, J. A., Popoola, C. A., Ayo, V. A., & Andrew, E. (2014). Effect of added Bambara groundnut flour on the quality of acha-based fura. *British Journal of Applied Science & Technology, 4*(1), 168–176.

Babarinde, G. O., Ebun, A. A., & Adegbola, P. I. (2020). Hepatotoxicity and biochemical evaluation of a novel breakfast food produced from the blend of fonio (*Digitaria iburua* Stapf) and pigeon pea (*Cajanus cajan* (L.) Millspaugh) in albino rats. *Bulletin of the National Research Centre, 44*, –157. https://doi.org/10.1186/s42269-020-00409-6

Barends J (2019). Do it with Bambara groundnut. https://www.cput.ac.za/newsroom/news/article/3819/do-it-with-bambara-groundnut

Berchie, J. N., Sarkodie, A. J., Adu-Dapaah, H., Agyemang, A., Addy, S., Asare, E., & Donkor, J. (2010). Yield evaluation of three early maturing Bambara groundnut (*Vigna subterranea* L. Verdc.). *Journal of Agronomy, 9*, 175–179.

Boukid, F., Zannini, E., Carini, E., & Vittadini, E. (2019). Pulses for bread fortification: A necessity or a choice? *Trends in Food Science and Technology, 88*, 416–428.

Bowely, A. (2003). Food fortification successful in Central America. *Nutriview, 3*, 2–3.

Brajdes, C., & Vizireanu, C. (2012). Sprouted buckwheat an important vegetable source of antioxidants. *The Annals of the University Dunarea de Jos of Galati Fascicle VI – Food Technology, 36*(1), 53–60.

Brough, S. H., Azam-Ali, S. N., & Taylor, A. J. (2003). The potential of Bambara groundnut in vegetable milk production and basic protein functionality systems. *Food Chemistry, 47*, 277–283.

Bueno, D. B., Silva-Júnior, S. I., Seriani Chiarotoo, A. B., Cardoso, T. M., Neto, J. A., Lopes dos Reis, G. C., Abreu Glória, M. B., & Tavano, O. L. (2020). The germination of soybeans increases the water-soluble components and could generate innovations in soy-based foods. *LWT-Food Science and Technology, 117*, 108599.

Bultosa, G., Molapisi, M., Tselaesele, N., Kobue-Lekalake, R., Haki, G. D., Makhabu, S., Sekwati-Monang, B., Seifu, E., & Nthoiwa, G. P. (2020). Plant-based traditional foods and beverages of Ramotswa Village, Botswana. *Journal of Ethnic Foods, 7*, 1. https://doi.org/10.1186/s42779-019-0041-3

Chandiposha, M., Chagonda, I., & Makuvaro, V. (2013). Utilisation of common grain crops in Zimbabwe. *African Journal of Food Science, 7*(9), 253–257. https://doi.org/10.5897/AJFS2013.1011

Chinma, C. E., Ingbian, E. K., & Akpaunam, M. (2007). Processing and acceptability of fried cassava balls (*Akara-akpu*) supplemented with melon and soybean flours. *Journal of Food Processing and Preservation, 31*(2), 143–156.

CPUT (2014). Department of Food Science and Technology, BTech Food Product Development, Cape Peninsula University of Technology.

Codex Alimentarius Commission. (2017). *Guidelines on formulated complementary foods for older infants and young children* (CAC/GL8-1991. Rev. 2013. Amended 2017). pp. 1–10.

Coltrain, D., Barton, D., Boland, M. (2000). *Value added: Opportunities and strategies*. Department of Agricultural Economics, Kansas State University, USA.

Danbaba, N., Idakwo, P. Y., Kassum, A. L., Malgwi, S. D., Igwegbe, O. A., Lawan, H. K., Maijalo, A. I., & Nkama, I. (2020). Effects of pre-decortication treatments on the milling properties and flour characteristics of Bambara groundnut (*Vigna subterranea* L.) varying in seed coat colour. *Journal of Advances In Food Science & Technology, 6*(4), 21–32. Retrieved From https://www.ikppress.org/index.php/jafsat/article/view/4932

Deol, J. K., & Bains, K. (2010). Effect of household cooking methods on nutritional and antinutritional factors in green cowpea (*Vigna unguiculata*) pods. *Journal of Food Science and Technology, 47*(5), 579–581.

Devi, C. B., Kushwaha, A., & Kumar, A. (2015). Sprouting characteristics and associated changes in the nutritional composition of cowpea *(Vigna unguiculata). Journal of Food Science and Technology, 52*(10), 6821–6827. https://doi.org/10.1007/s13197-015-1832-1

Enwere, N. J., & Hung, Y. C. (1996). Some chemical and physical properties of Bambara groundnut (*Voandzeia subterranean* Thouars) seed and products. *International Journal of Food Science and Nutrition, 47*(6), 469–475. Available: https://doi.org/10.3109/09637489609031875.

FAO/WHO. (1991). *Protein quality evaluation* (Report of Joint FAO/ WHO Expert Consultation. FAO Food and Nutrition Paper 51) (pp. 1–66). FAO/WHO.

Gabriel, E. G., Jideani, V. A., & Ikhu-omoregbe, D. I. O. (2013). Investigation of the emulsifying properties of Bambara groundnut flour and starch. *International Journal of Food Science and Engineering, 7*, 539–547.

Gregory, P.J., Mayes, S., Hui, C.H., Jahanshiri, E., Julkifle, A., Kuppusamy, G., Kuan, H.W., Lin, T.X., Massawe, F., Suhairi, T.A.S.T.M., Azam-Ali, S.N., P. J., G., S., M., Chai, H., Ebrahim, J., A., J., Giva, K., Ho Wai, K., Tan Xin, L., F., M., T. M., S. & S., A.-A. (2019). Crops For the Future (CFF): An overview of research efforts in the adoption of underutilized species. *Planta, 250*, 979–988.

Hampson, K. J., Azam-Ali, S. H., Sesay, A., Mukwaya, S., & Azam-Ali, S. N. (2000). *Assessing opportunities for increased utilization of Bambara groundnut in southern Africa*. Tropical Crops Research Unit, University of Nottingham. Internal Report. 52p.

Hardy, Z., & Jideani, V. A. (2018). Effect of spray drying compartment and maltodextrin concentration on the functional, physical, thermal, and nutritional characteristics of Bambara groundnut milk powder. *Journal of Food Processing and Preservation, 42*, 1–14.

Hardy, Z., & Jideani, V. A. (2020). Functional characteristics and microbiological viability of foam – Mat dried Bambara groundnut (*Vigna subterranea*) yogurt from reconstituted Bambara groundnut milk powder. *Food Science and Nutrition, 8*, 5238–5248.

Hillocks, R. J., Bennett, C., & Mponda, O. M. (2012). Bambara nut: A review of utilization, market potential and crop improvement. *African Crop Science Journal, 20*(1), 1–16.

Ho, B. S. (2020). Singapore Startup NamZ Disrupts Instant Noodle Market Using Climate Resistant Sustainable Crops, 1–6. https://eproofing.springer.com/books_v3/index.php?token=UHXoaS8Y6sVCEoCWnoFn9htOHS-MCX7tzLG5NK8MwYUPDIoAPTQnaw

Hoh, W. K. (2020). Starting a food revolution by making instant noodles that are good for health and the planet. *The Straits Times*, pp. 1–2.

Holt, S. D., Resurreccion, A. V. A., & Mcwatters, K. H. (1992). Formulation, evaluation and optimization of tortillas containing wheat, cowpea and peanut flours using mixture response surface methodology. *Journal of Food Science, 57*(1), 121–127. https://doi.org/10.1111/j.1365-2621.1992.tb05438.x

Holt, N., Bremner, A., Sutherland, E. D., Vliek, M., Passer, M., & Smith, R. (2015). *Psychology: The science of mind and behaviour* (UK Higher Education Psychology, 3rd edn). McGraw-Hill Education/Europe, Middle East & Africa. ISBN-13 978-0077169848.

Ijarotimi, O. S. (2008). Protein and hematological evaluations of infant formulated from cooking banana fruits (Musa spp, ABB genome) and fermented Bambara groundnut (Vigna subterranean L. Verdc) seeds. *Nutrition Research and Practice, 2*, 165.

Ijarotimi, O. S. Jr, Olopade, A. J. (2009). Determination of amino acid content and protein quality of local infant complementary food determination of amino acid content and protein quality of complementary food produced from locally available food materials in Ondo state, Nigeria. *Malaysian Journal of Nutrition, 15*(1), 87–95. PMID: 22691808. https://pubmed.ncbi.nlm.nih.gov/22691808/

Ijarotimi, O. S., & Keshinro, O. O. (2013). Determination of nutrient composition and protein quality of potential complementary foods formulated from combination of fermented popcorn, African locust and Bambara groundnut seed flour. *Polish Journal of Food and Nutrition Science, 63*(3), 155–166. https://content.sciendo.com/view/journals/pjfns/63/3/article-p155.xml

Ja'afar-Furo, M. R., Bello, K., & Sulaiman, A. (2011). Assessment of the prospects of value addition among small-scale rural enterprises in Nigeria: Evidence from North-Eastern Adamawa State. *Journal of Development and Agricultural Economics, 3*(3), 144–149.

Jideani, A. I. (1999). Traditional and possible technology uses of *Digitaria exillis* (Acha) *Digitaria ibrua*: A review. *Plant Food for Human Nutrition, 54*, 362–373.

Jideani, V. A. (2016). Utilizing Bambara groundnut in value-added products. *Food Technology, 70*, 48–52.

Jideani, V. A. (2017a). *Comparative sensory characteristics of Bambara groundnut milk with commercial milk alternative products and sensory profile for Bambara groundnut cultured beverage*. A report submitted to the Technology Innovation Agency (TIA) Seed Fund, South Africa.

Jideani, V. A. (2017b). *Consumer sentiment to Bambara groundnut milk and cultured beverage*. A report submitted to the Technology Innovation Agency (TIA) Seed Fund, South Africa.

Jideani, V. A. & Jideani, A. I. O. (2016). Versatility of Bambara groundnut (*Vigna subteranea*) for food and nutrition security. In: Pan-African Grain Legume & World Cowpea Conference. Book of Abstract 28 February to 4 March, Livingstone, Zambia.

John, S., & Babynath, V. (2011). *Corn – The versatile crop*. Science Tech Entrepreneur. National Science and Technology Entrepreneurship Development Board (NSTEDB) India.

Jom, K. N., Frank, T., & Engel, K. H. (2011). A metabolites profiling approach to follow the sprouting process of mug beans (*Vigna radiata*). *Metabolomics, 7*, 102–117. https://doi.org/10.1007/s11306-010-0236-5

Kelechi-Ebisike, V. O., Solomon, M. D., Jaryum, K. H., Ebisike, K. J., Mshelia, P., & Danjin, M. (2020). Energy/macronutrient density and sensory evaluation of formulated complementary foods for use in low-income populations in Northern Nigeria. *Acta Scientific Nutritional Health, 4*(3), 01–06.

Kigel, J., Rosenthal, L., & Fait, A. (2015). Seed physiology and germination of grain legumes, Grain legumes handbook. *Plant Breeding, 10*, 327–363.

Kiin-Kabari, D. B., & Banigo, E. B. (2015). Quality characteristics of cakes prepared from wheat and unripe plantain flour blends enriched with Bambara groundnut protein concentrate. *European Journal of Food Science and Technology, 3*(3), 1–10.

Léopold et al. (2020). Editorial: Opposites Attract at CORR®—Machine Learning and Qualitative Research Leopold, Seth S. MD; Porcher, Raphaël PhD; Gebhardt, Mark C. MD; Gioe, Terence J. MD; Manner, Paul A. MD; Rimnac, Clare M. PhD; Wongworawat, Montri D. MD, Author Information. Clinical Orthopaedics and Related Research, 478 (10), 2193-2196. doi: 10.1097/CORR.0000000000001466

Limón, R. I., Peñas, E., Torino, M. I., Martínez-Villaluenga, C., Dueñas, M., & Frias, J. (2015). Fermentation enhances the content of bioactive compounds in kidney bean extracts. *Food. Chemistry 172*, 343–352

Linnemann, A. R. (1988). Cultivation of Bambara groundnut in Northern Nigeria. *Tropical Crops and Communications, 15*, 1–14.

Linnemann, A. R., & Azam-Ali, S. (1993). Bambara groundnut (*Vigna subterranea* (L.) Verdc.). In J. T. Williams (Ed.), *Underutilized crops series 2. Vegetables and pulses* (pp. 13–58). Chapman & Hall.

Liu, K., Phillips, R. D., & McWatters, K. H. (1993). Mechanism of pectin changes during soaking and heating as related to hard-to-cook defect in cowpeas. *Journal of Agriculture and Food Chemistry, 41*(9), 1476–1480.

Luo, Y., Zie, W., Jin, X., Tao, B., Chen, Q., & Zhu, W. (2013). Impact of sprouting pretreatment on phytic acid and polyphenol level of faba bean (*Vicia faba* L.) flour. *International Food Research Journal, 20*(3), 1133–1137.

Lush, W. M., & Evans, L. T. (1980). The seed coats of cowpeas and other grain legumes: Structure in relation to function. *Field Crops Research, 3*, 267–286.

Martınez, C. J., Martınez, A. C., Ayala, A. L. M., Muzquiz, M., Pedrosa, M. M., & Davila-Ortiz, G. (2012). Changes in protein, nonnutritional factors, and antioxidant capacity during germination of *L. campestris* seeds. *International Journal of Agronomy*. https://doi.org/10.1155/2012/387407

Marton, M., Mandoki, Z. S., Csapo-Kiss, Z. S., & Csapo, J. (2010). The role of sprouts in human nutrition. A review. *Acta Universitatis Sapientiae, Alimentaria, 3*, 81–117.

Mashau, E. M., Mabodze, T., Tshiakhatho, O. J., Silungwe, H., & Ramashia, S. E. (2020). Evaluation of the content of polyphenols, antioxidant activity and physicochemical properties of tortillas added with Bambara groundnut flour. *Molecules, 25*, 3035. https://doi.org/10.3390/molecules25133035. www.mdpi.com/journal/molecules

Mbata, T. I., Ikenebomeh, M. J., & Ezeibe, S. (2009). Evaluation of mineral content and functional properties of fermented maize (generic and specific) flour blended with Bambara groundnut. *African Journal of Food Science, 3*, 107–112.

Mensah, P., & Tunkins, A. (2003). Household level technologies to improve the availability and preparation of adequate protein and safe complementary foods. *Food and Nutrition Bulletin, 24*, 104–125.

Michaelsen, K. F. (2000). Are there negative effects of an excessive protein intake? *Pediatrics, 106*(5), 1293–1294.

Misheck, C., Chagonda, I., & Makuvaro, V. (2013). Utilization of common grain crops in Zimbabwe. *African Journal of Food Science, 7*(9), 253–257. https://doi.org/10.5897/AJFS2013.1011

Mubaiwa, J., Fogliano, V., Chidewe, C., Bakker, E. J., & Linnemann, A. R. (2018a). Utilization of Bambara groundnut (*Vigna subterranea* (L.) Verdc.) for sustainable food and nutrition security in semi-arid regions of Zimbabwe. *PLOS ONE, 13*, e0204817.

Mubaiwa, J., Fogliano, V., Chidewe, C., & Linnemann, A. R. (2018b). Bambara groundnut (*Vigna subterranea* (L.) Verdc.) flour: A functional ingredient to favour the use of an unexploited sustainable protein source. *PLoS One, 13*, 1–20.

Muhammad, I., Rafii, M. Y., Ramlee, S. I., Nazli, M. H., Harun, A. R., Oladosu, Y., Ibrahim Musa, I., Arolu, F., Chukwu, S. C., Haliru, B. S., Akos, I. S., Halidu, J., & Arolu, I. W. (2020). Exploration of Bambara groundnut (*Vigna subterranea* (L.) Verdc.), an underutilized crop, to aid global food security: Varietal improvement, genetic diversity and processing. *Agronomy, 10*, 766. https://doi.org/10.3390/agronomy10060766. www.mdpi.com/journal/agronomy

Murevanhema, Y. Y., & Jideani, V. A. (2015). Production and characterization of milk produced from Bambara groundnut (*Vigna subterranea*) varieties. *Journal of Food Processing and Preservation, 39*, 1485–1498.

Nadhifa, A. A, & Nugrahani Astuti, N. (2018). Effects of added katuk leaf powder and cooking oil on the organoleptic properties of groundnut biscuits. *Advances in Social Science, Education and Humanities Research (ASSEHR), 112*, 84–88. 1st International Conference on Social, Applied Science and Technology in Home Economics (ICONHOMECS 2017) http://creativecommons.org/licenses/by-nc/4.0/

Nakajima, N., Nozaki, N., Ishihara, K., Ishikawa, A., & Tsuji, H. (2005). Analysis of isoflavone content in tempeh, a fermented soybean, and preparation of new isoflavone-enriched tempeh. *Journal of Bioscience and Bioengineering, 100*, 685–687.

Ndidi, U. S, Ndidi, C. U, Aimola, I. A., Bassa, O. Y., Mankilik, M., & Adamu, Z. (2014), Effects of processing (boiling and roasting) on the nutritional and antinutritional properties of Bambara groundnuts (*Vigna subterranea* [L.] Verdc.) from Southern Kaduna, Nigeria. *Journal of Food Processing 2014*, Article ID 472129 volume https://doi.org/10.1155/2014/472129

Nnam, N. M. (2001). Comparision of the protein nutrition value of food blends based on sorghum, Bambara groundnut and sweet potatoes. *International Journal of Food Sciences and Nutrition, 52*, 25–29.

Nout, M. J. R., & Ngoddy, P. O. (1997). Technological aspects of preparing affordable fermented complementary foods. *Food Control, 8*, 279–287.

Nwadi, O. M. M., Uchegbu, N., & Oyeyinka, S. A. (2020). Enrichment of food blends with Bambara groundnut flour: Past, present, and future trends. *Legume Science, 2*, 1–10.

Nwosu, J. N. (2013). Production and evaluation of biscuits from blends of Bambara groundnut (*Vigna subterranean*) and Wheat (*Triticum eastrum*) flours. *International Journal of Food and Nutrition Science, 2*, 21–53.

Nyau, V., Prakash, S., Rodrigues, J., & Farrant, J. (2017). Profiling of phenolic compounds in sprouted common beans and Bambara groundnuts. *Journal of Food Research, 6*(6), 74–82. ISSN 1927-0887 E-ISSN 1927-0895. https://doi.org/10.5539/jfr.v6n6p

Obizoba, I. C., & Egbonna, H. I. (1992). Effect of germination and fermentation on the nutritional quality of Bambara nut (*Vocndzeria subterranean* L. Thouars) and its product (milk). *Plant Foods for Human Nutrition, 42*, 13–23.

Ochanda, S. O., Onyango, C. A., Mwasaru, A. M., Ochieng, J. K., & Mathooko, F. M. (2010). Effects of malting and fermentation treatments on group B-vitamins of red sorghum, white sorghum and pearl millets in Kenya. *Journal of Applied Bioscience, 34*, 2128–2134.

Ogundele, O. M., Minaar, A., & Emmambux, M. N. (2017). Effects of micronisation and dehulling of pre-soaked Bambara groundnut seeds on microstructure and functionality of the resulting flours. *Food Chemistry, 214*, 655–663.

Oguntunde A. O. (1985). Development of new food products from readily available raw materials. In *Proceedings of the Nigerian Institute of Food Science and Technology Training Workshop*, Ibadan, Nigeria.

Ohanenye, I. C., Tsopmo, A., Ejike, C. E. C. C., & Udenigwe, C. C. (2020). Germination as a bioprocess for enhancing the quality and nutritional prospects of legume proteins. *Trends in Food Science and Technology, 101*, 213–222.

Ojinnaka, M. C., Shadrack, D. M., Anankware, J. P., & Odinkemere-Davidson, V. N. (2018). Formulation, proximate analysis and quality evaluation of cocoyam-wheat cake enriched with edible palm larvae (*Rhynchophorus phoenicis*). *Acta Scientific Nutritional Health, 2*(11), 19–25.

Ola, O. I., & Adewole, S. A. (2019). Nutritional quality and physical characteristics of biscuit produced from fermented Bambara nut and wheat flour blends. *Acta Scientific Microbiology, 2*(12), 99–103. (ISSN: 2581-3226) https://actascientific.com/ASMI/pdf/ASMI-02-0437.pdf

Oluwatofunmi, E. G., Ishola, I. S., & Bamidele, F. J. (2015). Formulation and nutritional evaluation of maize, Bambara groundnut and cowpea seeds blends complementary food. *American Journal of Food and Nutrition, 3*(4), 101–105. https://doi.org/10.12691/ajfn-3-4-2

Onweluzo, J. C., & Nwabugwu, C. C. (2009). Fermentation of millet (*Pennisetum americanum*) and pigeon pea (*Cajanus cajan*) seeds for flour production: Effects on composition and selected functional properties. *Pakistan Journal of Nutrition, 8*(6), 737–744.

Oyeyinka, A. T., Pillay, K., & Siwela, M. (2017). Consumer awareness and acceptability of Bambara groundnut as a protein source for use in complementary foods in rural KwaZulu-Natal. *South African Journal of Clinical Nutrition, 30*(4), 87–92.

Oyeyinka, S. A., Kayitesi, E., Adebo, O. A., Oyedeji, A. B., Ogundele, O. M., Obilana, O. A., & Njobeh, P. B. (2020). A review on the physicochemical properties and potential food applications of cowpea (*Vigna unguiculata*) starch. *International Journal of Food Science and Technology*. https://doi.org/10.1111/ijfs.14604

Oyeyinka, S. A., Adepegba, A. A., Oyetunde, T. T., Oyeyinka, A. T., Olaniran, A. F., Iranloye, Y. M., Olagunju, O. F., Manley, M., & Njobeh, P. B. (2021). Chemical, antioxidant and sensory properties of pasta from fractionated whole wheat and Bambara groundnut flour. *LWT – Food Science and Technology, 138*, 110618. https://doi.org/10.1016/j.lwt.2020.110618

Phadi, M. M. (2004). Physicochemical effects of micronization on cowpeas (*Vigna unguiculata* L. Walp). MInst Agrar Dissertation, University of Pretoria, *14*, pp. 44–56.

Proctor, J. R., & Watts, B. M. (1987). Development of a modified Mattson Bean Cooker procedure based on sensory panel cookability evaluation. *Canadian Institute of Food Science and Technology Journal, 20*(1), 9–14.

Qaku, X. W., Adetunji, A., & Dlamini, B. C. (2020). Fermentability and nutritional characteristics of sorghum *mahewu* supplemented with Bambara groundnut. *Journal of Food Science, 85*, 1661–1667.

Quedraogo, M., Zagre M'bi, B., Jorgensen, S. T., & Liu, F. (2012). Effect of mounding times on yield of Bambara groundnut (*Vigna subterranea* (L.) Verdc.) landraces in Sahel-Burkina Faso. *African Journal of Agricultural Research, 7*(32), 4505–4511.

Sefa-Dedeh, S., Stanley, D. W., & Voisey, P. W. (1978). Effects of soaking time and cooking conditions on texture and microstructure of cowpeas (*Vigna unguiculate*). *Journal of Food Science, 43*(6), 1832–1838.

Sefa-Dedeh, S., Stanley, D. W., & Voisey, P. W. (1979). Effect of storage time and conditions on the hard-to-cook defect in cowpeas (*Vigna unguiculata*). *Journal of Food Science, 44*(3), 790–796.

Shanono, I. M., & Muhammad, Y. Y. (2015). Effect of cooking on proximate and antinutritional content of Bambara groundnut. *Journal of Advances in Food Science & Technology, 2*(4), 176–180. Retrieved from http://www.ikppress.org/index.php/JAFSAT/article/view/4222

Tan, X. L., Azam-Ali, S., Goh, E. V., Mustafa, M., Chai, H. H., Ho, W. K., Mayes, S., Mabhaudhi, T., Azam-Ali, S., & Massawe, F. (2020). Bambara groundnut: An underutilized leguminous crop for global food security and nutrition. *Frontiers in Nutrition, 7*, 601496. https://doi.org/10.3389/fnut.2020.601496

Teixeira, N. C., Queiroz, V. A. V., Rocha, M. C., Amorim, C. P., Soares, T. O., Monteiro, M. A. M., & Junqueira, R. G. (2016). Resistant starch content among several sorghum (*Sorghum bicolor*) genotypes and the effect of heat treatment on resistant starch retention in two genotypes. *Food Chemistry, 197*, 291–296.

Uvere, P. O., Onyekwere, E. U., & Ngoddy, P. O. (2010). Production of maize–Bambara groundnut complementary foods fortified pre-fermentation with processed foods rich in calcium, iron, zinc and provitamin A. *Journal of Science Food and Agriculture, 90*, 566–573. www.interscience.wiley.com. https://doi.org/10.1002/jsfa.3846

WHO/OMS. (2000). *Child and adolescent health and development: Nutrition and infant feeding*. Pan American Health Organization/WHO. Guiding principles for complementary feeding of the breastfed child.

Xu, M., Jin, Z., Lan, Y., Rao, J., & Chen, B. (2019a). HS-SPME-GC-MS/olfactometry combined with chemometrics to assess the impact of germination on flavor attributes of chickpea, lentil, and yellow pea flours. *Food Chemistry, 280*, 83–95.

Xu, M., Jin, Z., Simsek, S., Hall, C., Rao, J., & Chen, B. (2019b). Effect of germination on the chemical composition, thermal, pasting, and moisture sorption properties of flours from chickpea, lentil, and yellow pea. *Food Chemistry, 295*, 579–587.

Xu, M., Jin, Z., Gu, Z., Rao, J., & Bingcan, C. (2020). Changes in odor characteristics of pulse protein isolates from germinated chickpea, lentil, and yellow pea: Role of lipoxygenase and free radicals. *Food Chemistry, 314*, 126184.

Yang, Q., Shen, Y., Foster, T., & Hort, J. (2020). Measuring consumer emotional response and acceptance to sustainable food products. *Food Research International, 131*, 108992.

Yasmine, M. (2002). Impact of small scale fermentation technology on food safety in developing countries. *International Journal of Food Microbiology, 75*, 213–229.

Zhu, F. (2020). Fonio grains: Physicochemical properties, nutritional potential, and food applications. *Comprehensive Reviews in Food Science and Food Safety, 19*, 3365–3389. https://doi.org/10.1111/1541-4337.12608

Zhu, F., & Cooke, H. (2011). Gelatinization, pasting and gelling properties of sweet potato and wheat starch blends. *Cereal Chemistry, 88*(3), 302–309.

Chapter 12
Bambara Groundnut Potential in Functional Food and Ingredients

12.1 Introduction

It is reported that the highest rates of lactose intolerance are in Africa, South America and Asia, with approximately 50% of the population affected and almost 100% in some Asian countries. Vegetarianism is on the increase either by choice or due to disease and interest in plant-based nutrition has been steadily growing. This has led the food industry, along with the consumers, to be interested in a plant-based diet because of its health benefits and environmental sustainability (Fig. 12.1). Hence, there is increasing demand from consumers for lactose-free or plant-based products. Legumes are on the increase as ingredients in food processing (Sharan et al., 2021). There is a continued trend towards healthier eating choices. BGN seeds and products will fill this gap for being plant-based and lactose-free. The usefulness of BGN is more targeted to functional food and therapeutic (Atoyebi et al., 2018). Hence, value-added products from BGN seeds and flour will provide functional foods with therapeutic benefits for food security programmes and interventions (Jideani, 2016). In this section, the rich nutritional content of BGN for techno-functional application in food for consumer health and wellness is presented.

12.2 Case for the Techno-functional Attributes of Bambara Groundnut

Despite the growing international demand for higher-value processed agro-industry products and the importance of value addition, Africa is yet to make significant progress towards adding value to its primary agricultural commodities and BGN in particular. Innovation is a driver of global competitiveness. The competitiveness of a country depends on its ability to generate sustainable and relatively higher revenues from the factors of production and high employment due to exposure to

V. A. Jideani, A. I. O. Jideani, *Bambara groundnut: Utilization and Future Prospects*, https://doi.org/10.1007/978-3-030-76077-9_12

innovation relates to the food processing part of the food chain, which is the transformation of agro-raw materials into stable, edible food products with the desired quality and functional properties through executing unit operations such as pre-treatment and separation, conversion, structure formation, stabilisation and packaging (Langelaan & Da Silva, 2013).

Pre-treatment such as washing, milling, cutting, melting and dissolving is aimed to obtain the raw materials in the desired purity and chemical and physical state to allow for proper handling in the next steps of the food manufacturing process. Separation technologies include air classification to separate husks from beans, membrane filtration for sediment removal from beer, extraction of flavour components from hops make use of differences in chemical or physical properties between the constituents of a mixture to obtain two or more distinct fractions, at least one of which is enriched in one or more of the constituents (Langelaan & Da Silva, 2013). Pre-treatment for BGN could involve grading of the seeds according to their sizes or colour with different pricing based on perceived quality. The solution to the hard-to-cook phenomenon is to reduce the seeds into smaller particular sizes to obtain flour. The flour could be further fractionated through air classification into different particle sizes to obtain different sized grits for various food applications. Crushing the BGN seeds and separating the hulls from the cotyledons will produce dehulled BGN seeds, reducing the cooking time. The hulls could then be used to extract bioactive compounds or milled to fine powders as dietary fibers in food application.

Conversion processes one or more components of a mixture into a desired intermediate or final product containing one or more chemically different components e.g. alcohol production through fermentation, flavour development during ripening of cheese or meat, enzymatic treatment and chemical reactions. Many researchers have, through the application of different conversion processes, obtained other products from BGN.

The purpose of structure formation is to change the physical properties of a food product, including its appearance. Still, it may not affect the chemical composition of the intermediate or final products. Conversion processes may also be involved in structure formation. Examples of structure formation include cooking (softening of plant tissues in vegetables or solidifying egg white and yolk) and baking (e.g., reorganising gluten and starch to give a bread product). Stabilizing processes increase the shelf-life of the processed product by fixing its biological, physical or chemical state by preventing oxidation or reducing enzymatic or microbiological activity. Such processes include drying, freezing, heat pasteurisation and sterilisation, high-pressure pasteurisation, pulsed electric fields. The purpose of packaging processes is to protect the final product against harmful environmental influences (e.g. oxygen or other chemical substances, moisture, microorganisms), maintain its physical integrity and/or bring about gas conditions that reduce microbial growth but increase the products' shelf-life. Packaging also plays an essential role in the marketing of products and can be used to inform consumers. Examples of packaging processes include bottling, foil wrapping, blistering, canning and modified atmosphere gas packaging. The previous chapters on the utilisation of BGN and the efforts made by researchers indicate that the functional properties of BGN can perform several roles in the structure and stabilizing of food products and processes as stabilizer, emulsifier, dietary fiber, fortificant, etc. BGN seeds could be repackaged as a canned product in brine or sauce to solve the hard-to-cook phenomenon.

Consequently, technologies that will turn the BGN into commercialise products can enhance the pace of large-scale adoption and farmers will be encouraged to produce more, which is sustainability (Jideani, 2019). However, duplicating the processes known for producing products from other well-known grains like rice, wheat, soybean, and maize may not be appropriate for BGN. The approach should exploit the potentially useful intrinsic qualities of BGN to produce unique and alternative value-added products. Well-designed experiments that will establish the critical variables affecting the process and product are essential. The technologies and strategies proposed for BGN product innovation include (1) holistic value chain strategy; (2) fourth industrial revolution technologies with advances in science and technology and (3) multivariate approach to product development and quality control (Jideani, 2019).

12.3 Holistic Value Chain Strategies

The line of transactions from producer to consumer constitutes the value chain. The value chain of any sector is always embedded in a market system consisting of various supporting functions and rules. Knowing how market systems work is very important for developing upgrading strategies that target systemic root causes for competitive constraints. The participatory value chain approach will facilitate the vertical integration linking to other market levels in the supply chain. In contrast, horizontal integration extends the same level of the market geographically to increase volume (Mado & Saio, 2009). Achieving a real transformation of the BGN food systems requires a holistic approach. It is engaging all stakeholders in the value chain while using various actions such as improved policy, increased investment, expanded infrastructure, business support services, access to academic and research institutions, consumer behavior change and improved resource management (Nayyar et al., 2018).

Some notable strategies that improved the food systems in Japan and other developed countries were reported (Mado & Saio, 2009) and could be adopted for the BGN value chain. One example is the One-village-One Product (OVOP), where a specialty product is identified for a locality and is developed into a business enterprise (Fig. 12.4). The farmers sell their products and value-added processed items through farmers' markets in towns or direct marketing schemes in schools, factories, offices and restaurants in the local areas. The principle of the scheme is that local products are consumed in the local areas.

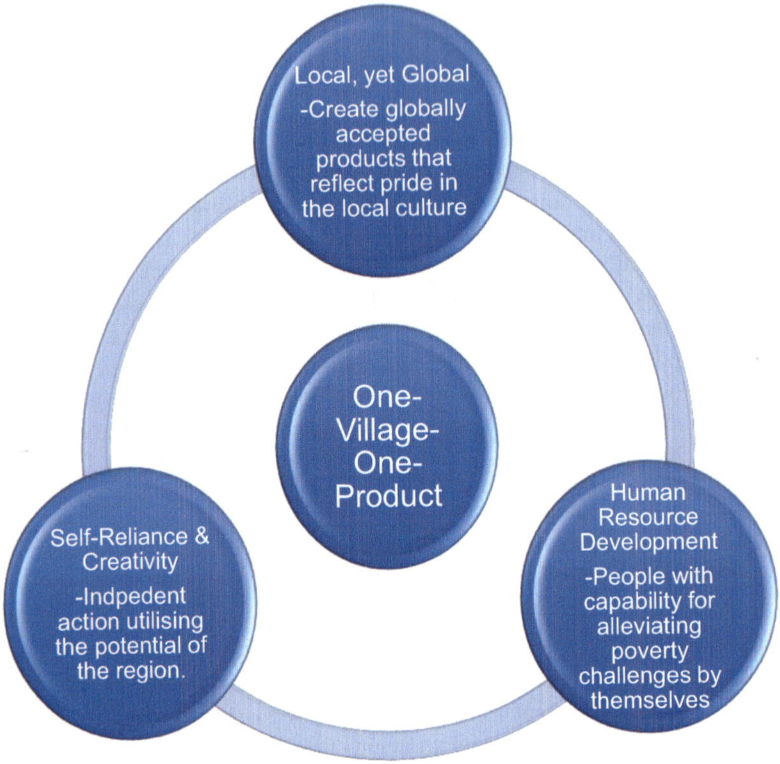

Fig. 12.4 One-village-one product strategy for Bambara groundnut adoption

12.4 Fourth Industrial Revolution Technologies with Advances in Science and Technology

Technology innovations in combination with other interventions can play important role in enabling and accelerating food systems transformation (Nayyar et al., 2018). To remain competitive in the future, the global food and drink industry must focus on technological innovations in BGN processing that support the development of (1) new and better food products to alleviate food insecurity while meeting the consumer demand for personalised nutrition in pleasure, health (natural, medical, vegetal), physical, convenience and ethical, (2) resource-efficient manufacturing processes that prevent the generation of waste and produce high value-added products from by-products, (3) integrated and transparent supply chains that connect local food production and demand with the globalisation of the food industry and retail and (4) increased innovation in the food industry by faster translation of scientific results into industrial implementation and inclusion of scientific advances from other disciplines such as nanotechnology and implementation of new processing technologies (Langelaan & Da Silva, 2013). Innovation is the key to competitiveness with special attention given to increasing small and medium enterprises' capacity to translate scientific results into industrial implementation. The use of effective, sophisticated and continuously evolving digital technologies will benefit the understanding of the potential inherent in BGN and deconstruct the apparent technological and economic constraints associated with this smart legume crop. These emerging technologies driven by the Fourth Industrial Revolution (4IR), disrupting many industries bringing rapid and large-scale change include (1) digital building blocks such as big data, the Internet of Thinks (IoT), artificial intelligence (AI) and machine learning (ML) and blockchain; (2) new physical systems such as autonomous vehicles, advanced robotics, additive manufacturing, advanced materials and nanotechnologies; (3) advances in science such as next-generation biotechnologies and genomics and new energy technologies (Nayyar et al., 2018). The 4IR technologies in conjunction with advances in science and technologies (Langelaan & Da Silva, 2013; Nayyar et al., 2018) with the potential to transform the BGN food industry is outlined in Table 12.1.

Table 12.1 4IR technologies and advances in science and technologies for Bambara groundnut

S/N	Technology	Explanation
1.	Alternative proteins	BGN plant-based proteins
2.	Food sensing technologies for food safety, quality and traceability	Embedded microscopic electronic devices such as radio-frequency identification tags and genetic markers as well as hyperspectral imaging, could be the future barcodes and used in combination with mobile phones could put information regarding the authenticity, freshness, ripeness, shelf life and nutritional content of food at consumers fingertips, quality and process control.
3.	Nutrigenetics for personalised nutrition	The ability to optimise a person's nutritional intake based on their genetics.
4.	Big data and advanced analytics	Connectivity technologies such as social networks, peer-to-peer networks and online e-commerce data could significantly influence consumption patterns and increase access to nutritious foods.
5.	Blockchain-enabled traceability	Improved value-chain efficiency driven by improved collaboration and data visibility.
6.	Sustainable packaging and refrigeration climate control, non-thermal pasteurisation and sterilisation	High-pressure processing, pulsed electric field processing, cold plasma treatment, advanced heating technologies- microwave, radio frequency, etc.
7.	Nano- and micro-technology	Microfluidic techniques, emulsions and related products ingredient fractionation and filtration.
8.	Circular economy	Increase the reuse of resources and decrease waste, pollution and carbon emissions.
9.	Information and knowledge transfer	Capacity building and linkage between academia and industry.

Adapted from Langelaan and Da Silva (2013) and Nayyar et al. (2018)

The digital technology of the 4IR, from the simplest smartphone and IoT, social media technologies like those of Twitter, etc. to AI, robotics and blockchain, digitalization, big data, autonomous vehicles, additive manufacturing, nanotechnology, biotechnology, and 3D food printing technologies and processes of the digital rich future (Aworh, 2020; Jideani et al., 2020; Jayaprakash et al., 2020) need the application to so-called orphan smart crops. Although the functional value of these technologies, like 3D printing, remains largely unrealized on a commercial scale (Jayaprakash et al., 2020), the present and approaching technologies of the future beyond blockchain hold a lot for smart crops like BGN. 3D printing (3DP) refers to an advanced production technology category characterized by layer-by-layer material deposition for creating personalized objects from pre-designed digital models (Jayaprakash et al., 2020). 3D food printing makes it easy to tailor the contents of food products to match an individual's health and activity level and their taste preferences, hence finetuning shape, nutritional content and texture of the food (Fig. 12.5). These advanced processes can help collect and process data on BGN especially in digital format, using this digital information, smart science and food chain intelligence (Selvaraj et al., 2019; Jideani et al., 2020; Marwala, 2020) to innovate healthy BGN products (Jideani, 2016).

Information is a tool to deconstruct the health benefit of BGN, hence the emerging and increasing use of various social media as a source of information for dietary market research. Aleixo et al. (2020) highlighted (i) social media has experienced rapid growth and has been subject of discussions on a wide range of topics including new lifestyles and eating habits, providing a great opportunity to obtain spontaneous consumer information, (ii) Twitter® users, can help in understanding the perception about themes of veganism and plant-based diets, (iii) social networking data mining methodology can be applied to measure the relationships, (iv) vegan group's posts are more related to the categories of recipes, trends, criticism and negative comments about veganism, being more often citing sources when compared to the other group, (v) results of the plant-based diet group are more significantly related to the impacts of nutrition, physical activity and consumer health, and (vi) Twitter® has proved to be an interesting tool for obtaining data on produced food publications on social media and their results can guide the market and the academic environment in creating new products, services and marketing strategies to answer the needs of specific consumers. Therefore, the use of social media will greatly benefit BGN. Also, integrating text mining and network analysis, Jideani et al. (2019)

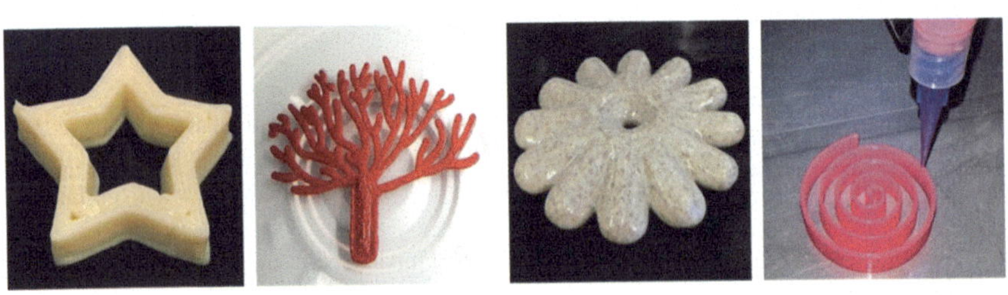

Fig. 12.5 Example food structures made using 3D food printing. (Jayaprakash et al., 2020)

integrated and profiled the ethnomedicine of BGN in South Africa as presented in Chap. 4.

A case was made by Holden et al. (2018) that food systems are not sustainable and to define the case for using technology, specifically internet technologies (hardware and software combined) to make the 'Internet of Food' to enable the transition of the food system from fossil to sustainable. The 'Internet of Food' concept (Kouma & Liu, 2011; Boulos et al., 2015), appearing in peer-reviewed literature first in 2011 described the idea of food items having an 'IP identify', which raised the question of how this might influence our eating habits, how the technology could influence food choices and predicted that by 2020 it would be possible to monitor and control food objects remotely through the Internet. This technological control of the food system will help the transition from fossil to sustainable considering the growing global population.

The technology infrastructure of high-quality data collection would help the collection of information on the BGN food system that will permit the best possible decision-making. Key requirements are standard vocabularies and food ontologies (Boulos et al., 2015) to allow integration of data sets across the Internet (Nobre & Tavares, 2017) and new analytical methods to collate, curate, analyse and utilise data across the whole BGN production system. We need an Internet of food to monitor conditions and analyse data to derive knowledge that can be combined with the means to implement control of the system to enable a step-change in BGN food systems.

An AI-based BGN disease and pest detection system using deep convolutional neural networks (DCNN) and transfer learning, successfully applied in various fields (Selvaraj et al., 2019; Howard et al., 2017), will support BGN farmers. This involves artificial intelligence (AI), deep learning, disease detection, transfer learning, convolutional neural networks, and mobile app. application. Different state-of-the-art models are found in the literature for digital crop disease detection. The use of digital detection systems like the AI (with deep learning models) for early identification of a crop disease or pest can lead to faster interventions with resulting reduced impacts on food supply chains (Selvaraj et al., 2019). Deep learning is a novel method for image processing and object detection with greater accuracy in the classification of various crop diseases (Kamilaris, 2018; Brahimi et al., 2018). Transfer learning is one such popular approach in deep learning, where pre-trained models are adapted to do a new task. Deep transfer learning (DTL) generates a fresh framework for digital image processing and predictive analytics, with greater accuracy and has huge potential in crop disease detection (Selvaraj et al., 2019). Many computer visioned approaches for automated crop disease detection and classification have been reported, but still, a detailed exploration of real-time pest and disease recognition is lagging or not there for smart crops like BGN. Earlier investigations have validated AI-based recognition of crop diseases in wheat (Siricharoen et al., 2016), cassava (Ramcharan et al., 2017) and on datasets of healthy and diseased plants (Mohanty et al., 2016; Wiesner-Hanks et al., 2018).

12.5 Multivariate Approach to Product Development and Quality Control

The technologies for assessing the quality evolution and traceability of raw materials, intermediates and final products throughout the entire food production-to-consumption chain is the quality by testing (QbT). The QbT approach tests product quality by checking it against the approved specifications at the end of the manufacturing stream, i.e. sorting out final products into good and bad samples, which does not allow for adjusting the manufacturing process to avoid products failing to meet the specifications (Langelaan & Da Silva, 2013). Consequently, leading to a lot of waste as well as an inefficient use of personnel resources. There is a great deal of unpredictability in scaling up a product from research and development to production scale and reasons for failure are generally not understood (Charoo et al., 2012). In modern control strategies, the settings of the manufacturing process are derived from the specifications of the desired final product by first defining the critical product attributes (CPA's) [e.g. concentrations, uniformity, conversion, rates, etc.], which are the most essential features of the product (Fig. 12.6).

The critical process parameters (CPP's) that directly affect the CPA's are identified during process development (Langelaan & Da Silva, 2013). Food products and processes are complex and multivariate by nature. Hence scientific understanding of the relevant multi-factorial relationships (e.g. between formulation, process and

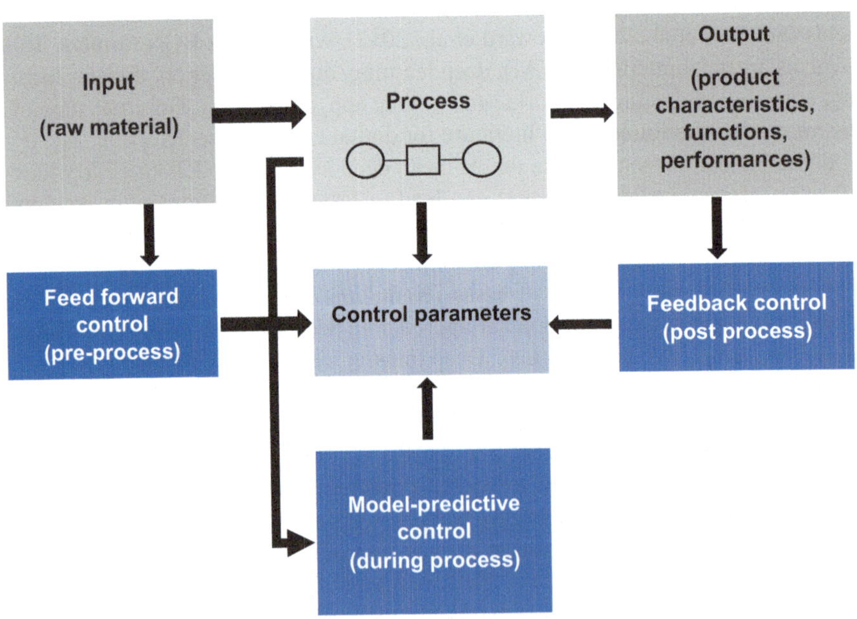

Fig. 12.6 Multivariate approach to product development and quality control. (Langelaan & Da Silva, 2013)

quality attributes) requires multivariate approaches such as the statistical design of experiments, response surface methodology, optimisation and multivariate data analysis or chemometrics (Huang et al., 2009). The combined use of experimental design, optimisation and multivariate data techniques (principal component analysis, PCA and partial least squares, PLS) provides a more powerful means to understand complex multivariate relationships in food product and process development. However, the approach is relatively few when applied to the food product and process development. When a process is implemented for routine manufacturing, sufficient knowledge should be generated on the process settings and the allowable variation (the process window). This knowledge enables the development of quality models to help predictive control, e.g. on the relation between raw materials quality and processing on the one hand and the final product quality, on the other hand. During processing, active control of the CPP's through feed-forward and feedback control loops results in a consistent output of products with the desired specifications.

A four-step process to product innovation involves (1) Product strategy development (initial screening, preliminary market assessment, detailed market research, product concept development and financial feasibility), (2) Product design and process development (prototype design, in-house testing and consumer testing); (3) Product commercialisation (trial production and market test) and (4) Product Launch and post-launch (pre-launch business analysis and production start-up). We will limit the discussion to steps one and two. At the product strategy development stage the voice of customer (VOC) is captured using methods like market research, customer surveys, customer observation, interviews, feedback reports, focus groups, conjoint analysis and KANO (maps consumer needs into basic, performance and excitement needs against the competition to determine where the product design is weak) model. Four important areas of consumer needs to be understood for product innovation are functional, cultural, sensory and aesthetic. The product strategy design research is focused on identifying and evaluating specific needs, wants, sensory properties, cultural/social values and aesthetics, so that specific product attributes can be identified in the product concept and used as a guide in product design. An important determinant of new product success is the degree of fit between the new product and consumer needs (Grunert & van Trijp, 2014). At the end of the product strategy development, the product specification and quality attributes are defined providing information on product knowledge. A business case is prepared to list compelling reasons for selecting and working on the project and the potential benefits expected to the consumers. The critical to quality (CTQ) parameters from the customer's perspective are then identified. At the product design and process development, the process parameters and controls are developed. Data on CTQ parameters to evaluate the initial capability and establish the baseline performance level using a statistically designed experiment (DOE) is collected.

The data is analysed using appropriate statistical and data analysis techniques to develop optimum parameters for the process and product design. A statistical model is developed using the design of experiments to establish the functional relationship between the input variables (X) which could influence the CTQ parameters or the response variable that needs to be optimised. DOE analysis gives optimum values of input parameters for optimising the CTQ parameter, Y. Optimisation is to develop a robust product. Robustness is the state of performance where the technology, product, or process is minimally sensitive to factors causing variability (either in the manufacturing or end-users environment) at the lowest possible cost (Alwerfalli & Lash, 2012). Full or fractional factorial designs can be used to produce an empirical model that mathematically describes the relationship between the inputs and the output(s) of interest. This mathematical prediction equation is known as a transfer function. Response surface methodology (RSM) can also be used to optimise the response parameter (Y). The product is then produced with the optimum parameters obtained through the statistical analysis and the design is then optimised using simulations. The product design is then verified to ensure it meets the customer, statutory and regulatory requirements. The prototype is developed and tested to verify that the intended objectives are met. Although the multivariate approach also known as quality-by-design is not a substitute for development expertise, they are an important supplement to maximise the benefit of the knowledge and experience the innovating team possesses and will provide more information and help avoid the "trial-and-error" approach in which a potentially endless string of "what if" scenarios are evaluated.

12.6 Recent Investigations on Techno-functionality of Bambara Groundnut

There have been various investigations on the functionality of BGN components and these range from flour as emulsifier/stabilizer in food systems (Jideani, 2015), production of provitamin A-biofortified *amahewu* combined with BGN flour with good consumer's acceptance (Awobusuyi & Siwela, 2019), the addition of BGN flour in tortillas production (Mashau et al., 2020), multi-enzyme digestibility of selected legumes (Odetokun, 2000) to foaming and gelation properties of flour (Aremu et al., 2007). Subsequent paragraphs highlight a few.

12.6.1 Bambara Groundnut Flour as Emulsifier/Stabilizer in Food Systems

Suitability of BGN flour (BGNF) as a natural emulsifier/stabilizer in food emulsion systems was assessed by Jideani (2015) and the following observations were reported: (i) gelatinized BGNF stabilised the oil-in-water emulsion, (ii) all the BGNF emulsions were macro emulsions, which are desirable in some food applications, (iii) all emulsions were stable to emulsion breakdown due to creaming and BGNF stabilized emulsions were characterized as highly spreadable (pseudoplasticity), thixotropic which is desirable in mastication and chewing, (meaning good chewability) and highly rigid, necessary for withstanding harsh conditions during development, transportation and storage.

Gabriel (2013) investigated the ability of BGN flour (BGNF) and starch (BGNS) to stabilize oil-in-water emulsions. A stable emulsion contained 9% (w/v) BGNF and 39% w/v oil compared to 5% (w/v) BGNS and 30% w/v oil. In both cases, the emulsions were stable till day 5. However, Adeyi et al. (2019a) reported a stable emulsion composition of 40% sunflower oil and 7% BGNF. Comparatively, BGNF seemed a better emulsifier with more stable emulsions than BGNS due to the functionality of other constituents other than starch such as protein and ash. Both BGNF and BGNS demonstrated emulsifying properties. The two emulsions stablized with BGNF and BGNS, respectively had only 30% and 50% growth in oil droplet diameter by day 5, compared to over 3000% in the unstable ones. In both BGNF- and BGNS-stabilized emulsions, stability improved towards neutral pH (Gabriel et al., 2013). This linearity showed that the more acidic/alkaline the emulsion medium, the more unstable they were. Generally, the stability observed in BGNF emulsions is twice that observed in BGNS, suggesting that BGNF had a higher emulsion-stabilizing capacity (Fasinu et al., 2015). The emulsion stability of BGN emulsions increased between pH 7 and 8. Emulsion stability increased with increasing NaCl concentration (65% stability at 2% NaCl compared to 82% stability at 10% NaCl). Stability however decreased sharply with increasing concentration of vinegar (45% stability at 10% vinegar compared to 81% stability at 2% vinegar). The BGNF-stabilized emulsions were more stable with or without additives than the BGNS-stabilized ones, in most cases, double those for Bambara groundnut starch. The presence of NaCl in the BGNF matrix showed a significant reduction in emulsion properties relative to emulsion without NaCl. However, the presence of NaCl at 25 mM is thought not to cause much severity to both BGNF polymer strength and emulsion formation (Adeyi et al., 2019a).

The presence of citric acid at all concentrations (0.5–6%) in the 40% BGNF and 7% oil emulsion did not significantly affect the emulsion forming properties of the resulting BGNF matrix. Compared with the BGNF matrix without citric acid, a matrix with citric acid had comparable emulsion forming ability. There was no flocculation of the oil-droplets. The phenomenon of oil-droplet aggregations was not different in both the emulsion with and without citric acid (Adeyi et al., 2019b). Vinegar (0.5–4%) at all concentrations increased the droplet size and physical instability of BGNF-stabilized emulsions. All emulsions containing vinegar possessed very large oil-droplets compared to emulsions without vinegar (Adeyi et al., 2017).

12.6.2 Bambara Groundnut Protein Concentrate and Isolate as a Functional Ingredient

Different factors, like varieties and origin, can influence the physicochemical properties of BGN cultivars. It has been shown that new BGN cultivars have the potential for industrial and household use and the possibility of shelf-stable protein products that could be a useful ingredient in food formulations. Adeleke et al. (2017) produced defatted flour and protein concentrate from newly developed BGN cultivars [Accessions No: TVSU 5 – BGN White (BGW) and TVSU 146 – BGN Brown (BGB)] and compared their nutritional, physicochemical, and functional properties with market sample [BGN commercial (BGC)] and observed the following (i) higher protein content in BGW (20.73%) and BGB (20.14%) as compared to BGC (18.50%), (ii) new varieties contain significantly higher levels of the essential fatty acids – linoleic and linolenic acids, higher levels of thiamine, riboflavin, niacin, pantothenic, ascorbic acids, pyrodoxine, alpha tocopherol, and vitamin K in the two new varieties; good sources of magnesium, calcium, iron, manganese, sodium, and potassium, (iii) defatted flour and protein concentrate of brown BGN exhibited high emulsifying activity and stability at different pH's and salt concentrations with the new varieties possessing significantly higher foaming capacity and stability than the commercial variety, and (iv) the new BGN cultivars have the potential for the industrial and household use like shelf stable protein products that could be a useful ingredient in food formulations.

BGN protein has the potential to stabilise oil-in-water (O/W) emulsions as it has emulsifying properties (Adebowale et al., 2011; Adeyi et al., 2014; Malefane et al. 2019). The stability of O/W emulsions can be described by their strength as described by hydrogen bonds, hydrophobic, electrostatic and covalent bonds amongst the components (Geremias-Andrade et al., 2016). Although there has been some research on the functionality of legume protein in food (Primozic et al., 2017) however, there is limited knowledge of the potential of brown BGN (BBGN) protein isolates and its effect on the rheological characteristics of oil-in-water (O/W) emulsions. Adebowale et al. (2011) obtained protein isolates from BGN. Malefane et al. (2019) investigated the rheological properties of O/W emulsions stabilised with BBGN protein isolates and observed that (i) different compositions: low protein emulsion (6% protein, 39% oil, 55% water), medium protein emulsion (8.14% protein, 36.63% oil 55.23% water) and high protein emulsion (15% protein, 30% oil, 55% water) exhibited shear thinning with the Hershel Bulkley model being the best fit model when compared with Power Law Model, Bingham Model and Casson Model, (ii) high protein emulsion had the longest linear viscoelastic region, (iii) increasing the frequency increased the elasticity of the O/W emulsions, and (iv) refrigeration enhanced the stability of the O/W emulsions.

Kudre and Benjakul (2012) investigating the effects of BGN protein isolate on protein degradation and gel properties of surimi from sardine (*Sardinella albella*) observed higher inhibitory activity against autolysis when isolate level increased up to 1% and that isolate effectively prevented the degradation of myosin heavy chain in a concentration-dependent manner in gels. This work demonstrated a preventive effect of BGN protein isolate on hydrolysis of myofibrillar proteins – the most important functional requirement to provide the superior quality of surimi-based products. Therefore, protein isolate from BGN contains protease inhibitors that can lower autolysis caused by endogenous proteases in surimi production and also serve as a co-gelling agent in surimi gels in the industry.

12.6.3 Bambara Groundnut Flour and Starch as a Functional Ingredient

Replacing cereal flours with BGN flour will significantly improve the nutritional profile of products and renders possible nutritional content claims for protein and dietary fiber. The replacement or addition of BGN in processed products is discussed in Chap. 10 under Innovative processing techniques". As a legume flour when BGN flour is mixed with other cereal flours, it contributes positively to the nutritional value of the products such as tortillas (Mashau, 2020), cake, bread, cookies, pasta, snacks, etc. making the flour a functional ingredient that should be exploited by the food industries. This is an area that requires investigation to ascertain the technological effect of BGN flour on established popular products thereby yielding more value-added products. Similar studies have been on tortilla added with *Moringa oleífera* flour (Páramo-Calderóna et al., 2019), tortilla added with unripe banana and cassava flour (Aparicio-Saguilán et al., 2013).

12.6.4 Bambara Groundnut Starch & Cassava Starch Blends

Starches from different botanical sources (cereal, legume, root tuber, and fruits) behave differently and can be blended to complement each other in their techno-functionality in the food and non-food industries (Lin et al., 2013; Baljeet et al., 2011; Afolabi et al. 2012, 2017). The amylose and amylopectin contents of starches determine their behavior and functionality as they control the swelling power, solubility, pasting and gelatinization of the starches (Piyarat, 2008; Oyeyinka et al., 2020). Ashogbon (2014) examined the physicochemical properties of different ratios of BGN starch (BBS) and cassava starch (CS) blends and made the following observations (i) minor components (lipids and proteins) of the starch granules were lowest for CS and highest for the blended (70BBS/30CS, 50BBS/50CS and 30BBS/70CS) starches, (ii) apparent amylose content of the starch blends was additive of their components, (iii) swelling power (SP) was highest for CS at a higher temperature and the SP of the blended starches was additive at 75 and 95 °C, while the solubility of the blended starches was additive at lower temperatures (55 and 65 °C), (iv) except peak and breakdown viscosities, all the pasting parameters of the blends were additive of their components, (v) CS had the highest peak and breakdown viscosities, but the lowest trough, final and setback viscosities, (vi) the 50BBS/50CS blend had the lowest breakdown viscosity (149.59 RVU) for thermal stability. The blending of this nature is safe, economic and advantageous when it enhances functionality thereby preventing underutilisation and chemical modification of the native starch.

12.6.5 Bambara Groundnut Flour in Beverages and Dietary Fiber

In addition to its indigenous knowledge, BGN seeds have the highest concentration of soluble fiber compared to other beans; this could contribute to the prevention of colon cancer. As an initial step to harness the rich nutrition of BGN, Jideani and Murevanhema (2013), Jideani and Diedericks (2014) and Maphosa and Jideani (2016) demonstrated that BGN value chains can contribute to national food security and obtained two patents: (1) a process for the production of BGN milk (BGNM) and BGN probiotic yoghurt; and (2) BGN dietary fiber. These products were high in polyphenols, an indication of the nutraceutical content, and may therefore contribute to a healthy diet. BGNM supported the growth of probiotics and protected them against in vitro gastric juice and bile. It is expected that the yoghurt will benefit the consumer through maintaining good balance and composition of intestinal flora, helping to increase the body's ability to resist the invasion of pathogens, thereby maintaining the host's efficient nutrition. BGN insoluble fiber (BGNIF) is high in phenolics (6.14–15.56 mg/g GAE) with higher swelling capacity compared to commercial fibers. The major neutral sugars in BGNIF include the co-eluted arabinose/galactose, xylose and mannose, with proportions of these monomers indicative of hemicellulosic polysaccharides. The introduction of BGNIF into white bread resulted in uniform crumb grain with smaller pores, higher total dietary fiber content, reduced gumminess, chewiness and softer crumbs compared with the control white bread (Fig. 12.7). The presence of galactose, xylose and mannose in BGN soluble fiber (BGNSF) suggests galactomannan properties typical of conventional hydrocolloids such as locust bean gum. All BGN fibers, especially the soluble fibers, possess higher fat absorption capacity than pea fibers. Hence, BGN fibers can be incorporated into high-fat products and emulsions. BGNSF was effective in stabilizing an orange beverage emulsion, so it could be a new ingredient for the beverage industry.

The high fiber content in BGN may be an added advantage in the development of functional foods. Although dietary fiber helps to regulate bowel movement, softens faeces and increases faecal bulk hence preventing constipation, when in excess, it may decrease appetite in infants and young children (Abeshu et al., 2016).

12.6.6 Good Water Absorption Capacity of Bambara Groundnut

The incorporation of the BGN seed flours and its protein products into aqueous food formulations, especially those involving soup and dough handling might be due to its good water absorption capacity which is the ability of flour to absorb water and swell for improved consistency in food (Adeleke et al., 2017). The high water absorption capacity of both the defatted and the concentrate of BGN suggest that it can be used as a thickener in liquid and semi-liquid foods because of its ability to

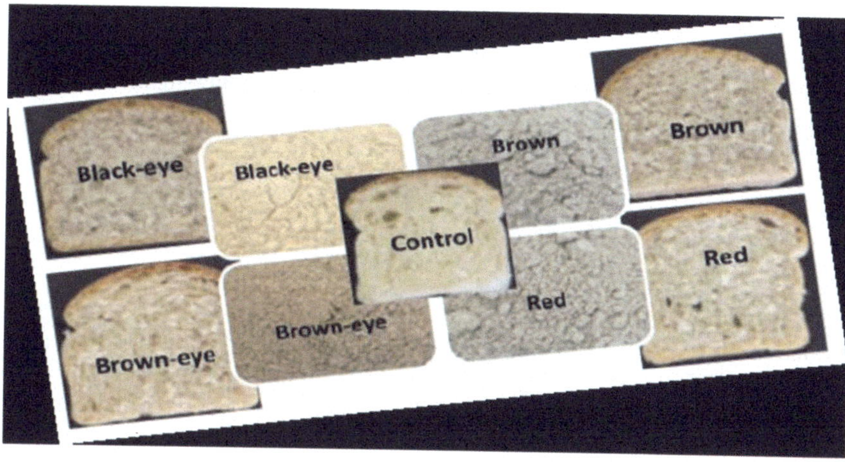

Fig. 12.7 Bambara groundnut insoluble fibers and bread baked with the fibers. (Jideani, 2015)

absorb water for improved consistency in food (Adeleke et al., 2017; Fasasi et al., 2007; Osundahunsi et al., 2003). The use of BGN flour for the formation of curd or as an additive to other gel-forming materials in food products may be an asset owing to its low gelation concentration (Akintayo et al., 2002).

12.6.7 Non-food Novel Application of the Husk/Shell of Bambara Groundnut

The rising cost of building materials has led to finding an affordable and economically available material that can be partially merged with cement in the production of blocks. Fernando et al. (2018) investigated groundnut shell ash mixed with cement on some engineering properties (density, water absorption, compressive strength and flexural strength of cement blocks) and observed the following (i) shells burnt in an open atmosphere, allowed to cool, burnt in a muffle furnace at 600 °C for 4 hours to get groundnut shell ash (GSA) revealed that 5% and 10% cement replacements were appropriate for load-bearing outer walls whereas the 20% and 25% substitution was found additional appropriate for non-load bearing indoor walls, (ii) a 15% substitution was appropriate for non-load bearing outer walls, and (iii) the findings would provide a quick reference to practicing Engineer, who will find GSA as a good partial replacement for cement, thus reducing the cost of cement block production. Groundnut shell is a secondary agricultural waste and the use of Groundnut Shell Ash (GSA) for soil stabilisation significantly scale back the value of construction and environmental hazards. GSA has been categorized beneath pozzolana given its 16–33% SiO_2, 2–6% Al_2O_3 and 0.5–5% Fe_2O_5. The SiO_2 in GSA will react with the CaOH, which liberates throughout the hardening of cement to make cementations compound (Alabadan et al., 2005; Mahmoud et al., 2012; Fernando et al., 2018). A similar investigation was conducted by Alabadan et al. (2005) using BGN shell ash (BGSA) to partially replace cement in making concrete. The compressive strength of BGN shell is higher than that of GSA, therefore BGN can produce better engineering properties. Further investigation is needed in the utilization of BGSA.

12.6.8 *By-Product Utilization of Bambara Groundnut Seed Residue*

Residue from processing BGN seeds is abundant in West Africa. Okebalama et al. (2020) applied BGN seed residue (BSR), at 0, 10, 20 and 30 t ha^{-1}, annually to an Ultisol in a two-year field experiment with cucumber and observed that (i) BSR input at 30 t ha^{-1} produced the highest yield of 256.80 g (= 0.46 t ha^{-1}) compared with 25.70 g (= 0.05 t ha^{-1}– 1) in the control, (ii) BSR had no effect on pH, the input at 20 and 30 t ha^{-1} significantly increased organic carbon, total N and K in the soil and increased macro-aggregation by 153% and 193% after the first and second year, respectively, (iii) basal soil respiration at day 3–14 showed increased and rapid labile carbon utilisation with increased BSR input rates, followed by more persistent CO_2 evolution rates during days 14–42, and significantly higher rates in the control than in the BSR treatments during days 35–42, (iv) BSR at 30 t ha^{-1}– 1 increased microbial carbon by 159%, microbial N by 205% and improved activities of β-glucosidase and tyrosine-aminopeptidase 7.5- and 3.0-fold, respectively, (v) reduced activities of S- and many carbon-cycle enzymes limited the suitability of BSR as a nutrient source for crop growth. They noted that BSR potential to enhance soil structure and organic carbon storage in the Ultisol was beneficial and the carbon pool in soil treated with 30 t ha^{-1} BSR was sufficiently large to increase nitrogen and potassium availability.

In addition to BGN by-product utilization potential, it has been used in animal feed and haulm (Adebowale & Lawal, 2004). BGN suitability for animal grazing, which is basically due to the abundance of nitrogen and phosphorus in its leaves has also been reported by Bamishaiye et al. (2011).

References

Abeshu, M. A., Lelisa, A., & Geleta, B. (2016). Complementary feeding: Review of recommendations, feeding practices, and adequacy of homemade complementary food preparations in developing countries – Lessons from Ethiopia. *Frontiers in Nutrition, 3*, 41. https://doi.org/10.3389/fnut.2016.00041

Adebowale, K. O., & Lawal, O. S. (2004). Comparative study of the functional properties of Bambara groundnut (*Voandzeia subterranean*), jack bean (*Canavalia ensiformins*) and mucuna bean (*Mucuna pruriens*) flours. *Food Research International, 37*, 355–365.

Adebowale, Y. A., Schwarzenbolz, U., & Henle, T. (2011). Protein isolates from Bambara groundnut (*Voandzeia subterranean* L.): Chemical characterization and functional properties. *International Journal of Food Properties, 14*(4), 758–775.

Adeleke, O. R., Adiamo, O. Q., & Fawale, O. S. (2017). Nutritional, physicochemical, and functional properties of protein concentrate and isolate of newly-developed Bambara groundnut (*Vigna subterrenea* L.) cultivars. *Food Science and Nutrition, 6*(1), 229–242. https://doi.org/10.1002/fsn3.552

Adeyi, O., Ikhu-Omoregbe, D., & Jideani, V. A. (2014). Emulsion stability and steady shear characteristics of concentrated oil-in-water emulsion stabilized by gelatinized Bambara groundnut. *Asian Journal of Chemistry, 26*(16), 4995–5002.

Adeyi, O., Ikhu-omoregbe, D. O., & Jideani, V. A. (2017). Effect of vinegar on the stability of sunflower oil-in-water emulsion stabilised by gelatinised Bambara groundnut flour. *Umudike Journal of Engineering and Technology, 3*(2), 69–77.

Adeyi, O., Ikhu-Omoregbe, D. I. O., & Jideani, V. A. (2019a). Physical stability characteristics of sunflower oil-in-water emulsion containing sodium chloride, stabilized by gelatinized Bambara groundnut flour. *Cogent Engineering, 6*, 1–13.

Adeyi, O., Ikhu-Omoregbe, D. I. O., & Jideani, V. A. (2019b). Effect of citric acid on physical stability of sunflower oil-in-water emulsion stabilized by gelatinized Bambara groundnut flour. *International Journal of Civil Engineering and Technology, 10*, 2260–2273.

Afolabi, T. A. (2012). Synthesis and physicochemical properties of carboxymethylated Bambara groundnut (*Voandzeia subterranean*) starch. *International Journal of Food Science and Technology, 47*, 445–451. https://doi.org/10.1111/j.1365-2621.2011.02860.x

Afolabi, A. T., Opara, A. O., Kareem, S. O., & Oladoyinbo, F. O. (2017). In vitro digestibility of hydrothermally modified Bambara groundnut (*Vigna subterranean* L.) starch and flour. *Food Science and Nutrition, 6*, 36–46. https://doi.org/10.1002/fsn3.510

Akintayo, E. T., Adebayo, E. A., & Arogundade, L. A. (2002). Chemical composition, physicochemical and functional properties of akaa pulp and seed flours. *Food Chemistry, 77*, 333–336.

Alabadan, B. A., Olutoye, M. A., Abolarin, M. S., & Zakariya, M. (2005). Partial replacement of ordinary Portland cement (OPC) with Bambara groundnut shell ash (BGSA) in concrete. *Leonardo Electronic Journal of Practices and Technologies, 4*(6), 43–48.

Aleixo, M. G. B., Sass, C. A. B., Leal, R. M., Dantas, T. M., Pagani, M. M., Pimentel, T. C., Freitas, M. Q., Cruz, A. G., Azeredo, D. R. P., & Esmerino, E. A. (2020). Using Twitter® as source of information for dietary market research: a study on veganism and plant-based diets. *International Journal of Food Science and Technology*. https://doi.org/10.1111/ijfs.14743

Alwerfalli, D., & Lash, T. (2012). *Design for six sigma (DFSS) as a proactive business process.* (pp. 2591–2600).

Aparicio-Saguilán, A., Osorio-Díaz, P., Agama-Acevedo, E., Islas-Hernández, J. J., & Bello-Pérez, L. A. (2013). Tortilla added with unripe banana and cassava flour: Chemical composition and starch digestibility. *CyTA Journal of Food, 11*, 90–95.

Aremu, M. O., Olaofe, O., & Akintayo, E. T. (2007). Functional properties of some Nigerian varieties of legume seed flours and flour concentration effect on foaming and gelation properties. *Journal of Food Technology, 5*(2), 109–115.

Ashogbon, A. O. (2014). Physicochemical properties of Bambara groundnut starch and cassava starch blends. *African Journal Food Science, 8*(6), 322–329.

Atoyebi, J. O., Osilesi, O., Abberton, M., Adebawo, O., & Oyatomi, O. (2018). Quantification of selected anti-nutrients and bioactive compounds in African Bambara groundnut (*Vigna subterranea* (L.) Verdc.). *American Journal of Food and Nutrition, 6*(3), 88–95. https://doi.org/10.12691/ajfn-6-3-5

Awobusuyi, T. D., & Siwela, M. (2019). Nutritional properties and consumer's acceptance of provitamin A-biofortified *amahewu* combined with Bambara (*Vigna subterranea*) flour. *Nutrients, 11*, 1476. https://doi.org/10.3390/nu11071476

Aworh, O. C. (2020). Food safety issues in fresh produce supply chain with particular reference to sub-Saharan Africa. *Food Control.*, ISSN 0956-7135. https://doi.org/10.1016/j.foodcont.2020.107737. http://www.sciencedirect.com/science/article/pii/S0956713520306538

Baljeet, S. Y., Ritika, B. Y., & Maheshi, K. (2011). Suitability of pigeon pea and rice starches and their blends for noodle making. *LWT – Food Science and Technology, 44*(6), 1415–1421.

Bamishaiye, O. M., Adegbola, J. A., & Bamishaiye, E. I. (2011). Bambara groundnut: An underutilized nut in Africa. *Advances in agricultural biotechnology, 1*, 60–72.

Boulos, M., Yassine, A., Shirmohammadi, S., Namahoot, C., & Brückner, M. (2015). Towards an "internet of food": Food ontologies for the internet of things. *Future Internet, 7*, 372–392.

Brahimi, M., Arsenovic, M., Laraba, S., Sladojevic, S., Boukhalfa, K., & Moussaoui, A. (2018). Deep learning for plant diseases: Detection and saliency map visualisation. In J. Zhou & F. Chen (Eds.), *Human and machine learning* (pp. 93–117). Springer.

Carlsson-Kanyama, A., Ekström, M. P., & Shanahan, H. (2003). Food and life cycle energy inputs: consequences of diet and ways to increase efficiency. *Ecological Economics 44*(2–3):293-307

Charoo, N. A., Shamsher, A. A. A., Zidan, A. S., & Rahman, Z. (2012). Quality by design approach for formulation development: A case study of dispersible tablets. *International Journal of Pharmaceutics, 423*, 167–178.

Fasasi, O. S., Adeyemi, I. A., & Fagbenro, O. A. (2007). Functional and pasting characteristics of fermented maize and Nile tilapia (*Oreochromis niloticus*) flour diet. *Pakistan Journal of Nutrition, 6*(4), 304–309. https://doi.org/10.3923/pjn.2007.304.309

Fasinu, E. G., Ikhu-Omoregbe, D. I. O., & Jideani, V. A. (2015). Influence of selected physicochemical factors on the stability of emulsions stabilized by Bambara groundnut flour and starch. *Journal of Food Science and Technology, 52*, 7048–7058.

Fernando, P. R., Hatangala, H. A. Y. N., Karunagaran, S., & Dissanayake, D. M. J. C. (2018). Evaluates some engineering properties of innovative sustainable cement blocks as a partial replacement of groundnut shell ash (GSA). *Acta Scientific Agriculture, 2*(7), 2–7.

Floros, J. D., Newsome, R., Fisher, W., Barbosa-Cánovas, G. V., Chen, H., Dunne, C. P., German, J. B., Hall, R. L., Heldman, D. R., Karwe, M. V., Knabel, S. J., Labuza, T. P., Lund, D. B., Newell-McGloughlin, M., Robinson, J. L., Sebranek, J. G., Shewfelt, R. L., Tracy, W. F., Weaver, C. M., & Ziegler, G. R. (2010). Feeding the world today and tomorrow: The importance of food science and technology. *Comprehensive Reviews in Food Science and Food Safety, 9*, 572–599.

Gabriel, E. G., Jideani, V. A., & Ikhu-omoregbe, D. I. O. (2013). Investigation of the emulsifying properties of Bambara groundnut flour and starch. *International Journal of Food Science and Engineering, 7*, 539–547.

Geremias-Andrade, I. M., Souki, N. P. B. G., Moraes, I. C. F., & Pinho, S. C. (2016). Rheology of emulsion-filled gels applied to the development of food materials. *Gels, 2*(3), 22. https://doi.org/10.3390/gels2030022

Grunert, K. G., & van Trijp, H. C. M. (2014). Consumer-oriented new product development. *Encyclopedia of Agriculture and Food Systems, 2*, 375–386.

Hatzichronoglou, T. (1996). *Globalisation and competitiveness: Relevant indicators* (OECD Science, Technology and Industry Working Papers, No. 1996/05). OECD Publishing. https://doi.org/10.1787/885511061376

Holden, N. M., White, E. P., Lange, M. C., & Oldfield, T. L. (2018). Review of the sustainability of food systems and transition using the Internet of Food. *npj Science of Food, 2*, 18. https://doi.org/10.1038/s41538-018-0027-3. www.nature.com/npjscifood

Howard, A. G., Zhu, M., Chen, B., Kalenichenko, D., Wang, W., Weyand, T., Andreetto, M., Adam, H.. (2017) MobileNets: Efficient convolutional neural networks for mobile vision applications. *arXiv:1704.04861*. https://arxiv.org/abs/1704.04861

Huang, J., Kaul, G., Cai, C., Chatlapalli, R., Hernandez-Abad, P., Ghosh, K., & Nagi, A. (2009). Quality by design case study: An integrated multivariate approach to drug product and process development. *International Journal of Pharmaceutics, 382*, 23–32.

Jayaprakash, S., Paasi, J., Pennanen, K., Ituarte, I. F., Lille, M., Partanen, J., & Sozer, N. (2020). Techno-economic prospects and desirability of 3D food printing: Perspectives of industrial experts, researchers and consumers. *Food, 9*, 1725. https://doi.org/10.3390/foods9121725. www.mdpi.com/journal/foods

Jideani, V. (2015). Should the Bambara groundnut remain underutilized? *FST Magazine* August 2015. A Publication of the South African Association of Food Science and Technology, South Africa.

Jideani, V. A. (2016). Utilizing Bambara groundnut in value-added products. *Food Technology, 70*(1), 48–52.

Jideani, V. A. (2019). Appropriate technologies in the development of novel traditional foods. In: *Traditional food processing: Prospects and challenges. 43rd conference and annual general meeting of the Nigerian Institute of Food Science and Technology (NIFST), 14-18th October 2019, at the Prof. Dora Akunyili Women Development Centre, Awka, Anambra, Nigeria* (pp. 14–18).

Jideani, V. A., & Diedericks, C. F. (2014). Nutritional, therapeutic, and prophylactic properties of *Vigna subterranea* and *Moringa oleifera*. In O. Oguntibeju (Ed.), *Antioxidant-antidiabetic agents and human health* (pp. 187–207). https://doi.org/10.5772/57338

Jideani, V. A., & Jideani, A. I. O. (2016). Versatility of Bambara Groundnut (Vigna subteranea) for Food and Nutrition Security. In: Pan-African Grain Legume & World Cowpea Conference

Jideani, V. A., & Murevanhema, Y. Y. (2013). A Bambara groundnut milk beverage and probiotic yoghurt produced therefrom and a method for the production thereof South Africa-complete patent.

Jideani, V. A., Muthoni, K. & Jideani, A. I. O. (2019). Integrating text mining and network analysis for ethnomedicinal profile of Bambara groundnut in Mpumalanga province, South Africa. Short oral presentation at International Congress on Engineering and Food – ICEF13 at Melbourne Convention Centre, South Wharf VIC 3006, 7 Convention Centre Place, Melbourne, Australia 23–26 September 2019.

Jideani, A. I. O., Mutshinyani, A., Maluleke, N., Mafukata, Z., Sithole, M., Lidovho, M., Ramatsetse, E., & Matshisevhe, M. (2020). Impact of industrial revolutions on food machinery – A review. *Journal of Food Research, 9*(5), 42–52. http://www.ccsenet.org/journal/index.php/jfr/article/view/0/43350. https://doi.org/10.5539/jfr.v9n5p42

Kamilaris, A. (2018). Prenafeta-Boldu FX. Deep learning in agriculture: A survey. *Computers and Electronics in Agriculture, 147*, 70–90.

Kouma, J.-P., & Liu, L. (2011). Internet of food. In F. Xia, Z. Chen, G. Pan, L. T. Yang, & J. Ma (Eds.), *International conference on internet of things and 4th international conference on Cyber, Physical and Social Computing*. IEEE Computer Society. https://ieeexplore.ieee.org/stamp/stamp.jsp?tp=&arnumber=6142168 , Accessed 01 Oct 2018

Kudre, T., & Benjakul, S. (2012). Effects of Bambara groundnut protein isolate on protein degradation and gel properties of surimi from sardine (*Sardinella albella*). *Journal of Food Processing and Preservation*. https://doi.org/10.1111/j.1745-4549.2012.00733.x

Langelaan, H. C. & Da Silva, F. P. (2013). *Technology options for feeding 10 billion people options for sustainable food processing*. IC STOA 2013/122, PE 513.533, t: ISBN 978-92-823-5128-4. http://www.europarl.europa.eu/stoa/

Lin, J. H., Kao, W. T., Tsai, Y. C., & Chang, Y. H. (2013). Effect of granular characteristics on pasting properties of starch blends. *Carbohydrate Polymer, 98*(2), 1553–1560.

Mado, T., & Saio, K. (2009). Value chain analysis for the development of food systems and innovative agro-based industry in Africa. SMARTPRINTWERK, Philippines.

Mahmoud, H., Belel, Z. A., & Nwakaire, C. (2012). Groundnut shell ash as a partial replacement of cement in sandcrete blocks production. *International Journal of Development and Sustainability, 1*(3), 1026–1032.

Malefane, D. G., Ikhu-Omoregbe, D., & Jideani, V. (2019). Rheological properties of oil-in-water emulsions stabilised with brown Bambara groundnut protein. *Acta Scientific Nutritional Health, 3*(12), 105–109.

Maphosa, Y., & Jideani, V. A. (2016). Physicochemical characteristics of Bambara groundnut dietary fibers extracted using wet milling. *South African Journal of Science, 112*(1–2), 1–8.

Marwala, T. (2020). *Artificial intelligence is already responding to our needs*. Retrieved: https://mg.co.za/opinion/2020-06-17-artificial-intelligence-is-already-responding-to-our-needs/

Mashau, M., Mabodze, T., Tshiakhatho, O., Silungwe, H., & Ramashia, S. E. (2020). Evaluation of the content of polyphenols, antioxidant activity and physicochemical properties of tortillas added with Bambara groundnut flour. *Molecules, 25*, 3035.

Mohanty, S. P., Hughes, D. P., & Salathe, M. (2016). Using deep learning for image-based plant disease detection. *Frontier in Plant Science, 7*, 1419.

Nayyar, S., de Cleene, S., & Dreier, L. (2018). Innovation with a purpose: The role of technology innovation. *World Economic Forum*, 1–42.
Nobre, G. C., & Tavares, E. (2017). Scientific literature analysis on big data and internet of things applications on circular economy: A bibliometric study. *Scientometrics, 111*, 463–492.
Odetokun, S. M.. (2000). Effect of fermentation on the physicochemical properties, anti-nutrients and in-vitro multi-enzyme digestibility of selected legumes. Ph.D. dissertation, Federal University of Technology, Akure, Nigeria.
Okebalama, C. B., Ibezim, C. O., Ugwu, V. U., & Marschner, B. (2020). Effects of amending an Ultisol with Bambara groundnut seed residue on soil fertility, microbial nutrient cycling and yield of cucumber. *Biological Agriculture & Horticulture, 36*(4), 249–266. https://doi.org/10.1080/01448765.2020.1808067
Osundahunsi, O. F., Fagbemi, T. N., Kesselman, E., & Shimoni, E. (2003). Comparison of the physico-chemical properties and pasting characteristics of flour and starch from red and white sweet potato cultivars. *Journal of Agriculture and Food Chemistry, 51*, 2232–2236. https://doi.org/10.1021/jf0260139
Oyeyinka, S. A., Kayitesi, E., Adebo, O. A., Oyedeji, A. B., Ogundele, O. M., Obilana, O. A., & Njobeh, P. B. (2020). A review on the physicochemical properties and potential food applications of cowpea (*Vigna unguiculata*) starch. *International Journal of Food Science and Technology*. https://doi.org/10.1111/ijfs.14604
Páramo-Calderóna, D. E., Aparicio-Saguilána, A., Aguirre-Cruza, A., Carrillo-Ahumadaa, J., Hernández-Uribeb, J. P., Acevedo-Telloc, S., & Torruco-Ucod, J. G. (2019). Tortilla added with *Moringa oleífera* flour: Physicochemical, texture properties and antioxidant capacity. *LWT-Food Science and Technology, 100*, 409–415.
Piyarat, S. (2008). Structure and functional properties of starch and flour from Bambara groundnut. *Songklanakarin Journal of Science and Technology, 30*, 51–56.
Primozic, M., Duchek, A., Nickerson, M., & Ghosh, S. (2017). Effect of lentil proteins isolate concentration on the formation, stability and rheological behavior of oil-in-water nanoemulsions. *Food Chemistry, 237*, 65–74. https://doi.org/10.1016/j.foodchem.2017.05.079
Ramcharan, A., Baranowski, K., McCloskey, P., Ahmed, B., Legg, J., & Hughes, D. P. (2017). Deep learning for image-based cassava disease detection. *Frontier in Plant Science, 8*, 1852.
Reinders, M. J., Banovic, M., & Guerrero, L. (2019). Introduction. In C. M. Galanakis (Ed.), *Innovations in traditional foods*.
Selvaraj, M. G., Vergara, A., Ruiz, H., Safari, N., Elayabalan, S., Ocimati, W., & Blomme, G. (2019). AI-powered banana diseases and pest detection. *Plant Methods, 15*, 92. https://doi.org/10.1186/s13007-019-0475-z
Sharan, S., Zanghelini, G., Zotzel, J., Bonerz, D., Aschoff, J., Saint-Eve, A., & Maillard, M. (2021). Fava bean (Vicia faba L.) for food applications: From seed to ingredient processing and its effect on functional properties, antinutritional factors, flavor, and color *Comprehensive Reviews in Food Science and Food Safety, 20*:401–428. https://doi.org/10.1111/1541-4337.12687
Siricharoen, P., Scotney, B., Morrow, P., & Parr, G. (2016). *A lightweight mobile system for crop disease diagnosis. International conference on image analysis and recognition* (pp. 783–791). Springer.
Wiesner-Hanks, T., Stewart, E. L., Kaczmar, N., DeChant, C., Wu, H., Nelson, R. J., Lipson, H., & Gore, M. A. (2018). Image set for deep learning: Field images of maize annotated with disease symptoms. *BMC Research Notes, 11*(1), 440.

Chapter 13
Miscellaneous Foods, Food Components & Consumption Trends – Marketing and Commerce

13.1 Introduction

Chaturvedi and Chakraborty (2020) in their review on potential non-dairy symbiotic beverages: a preliminary approach using legumes (i) gave brief information about the different research associating with various legume-based probiotics, prebiotics and synbiotic foods narrowing down specifically to beverages, (ii) stated that certain functional foods have gained consumer acceptance like those containing synbiotics (a blend of both prebiotics and probiotics) in different food systems, (iii) synbiotics enhance gut functionality and confer health benefits to humans; however, due to the limitations exerted by dairy-based matrices like lactose intolerance, milk allergies, limited shelf life and inclination towards veganism, consumer's interest in non-dairy foods has increased, and (iv) that integration of such information will allow researchers to explore product development using new prebiotic ingredients that have not been studied in much detail and hold an exceptional potential to be employed in the synbiotic food industry. Among the non-dairy matrices, legumes (such as chickpeas, kidney beans, lupin and BGN) are the less explored.

13.2 Consumption Trend – Legume Foods in General

China is one of the largest producers of food legumes in the world (FAOSTAT, 2015) and exports food legume products to the international markets for foreign currency or exchange with cereal grains. As documented by FAO statistics, China has remained the world's leading producer of peas, faba beans, mung beans, and adzuki beans in recent times (Li et al., 2017). Globally, whether indigenous or introduced crops, pulses have played an important role in cropping systems and made an important contribution to food resources for humans since ancient times (Brink & Belay, 2006). The demand for food legumes as healthy food will markedly increase

V. A. Jideani, A. I. O. Jideani, *Bambara groundnut: Utilization and Future Prospects*, https://doi.org/10.1007/978-3-030-76077-9_13

with the current advocacy for plant proteins. The rising consumer demand for vegetable food legumes as healthy food has led to attractive market prices and sharp production increases. In most developed economies, the future trend is such that the production of vegetable food legumes will continue to increase as the consumption continually and rapidly increase.

The large majority of consumers access food through supermarkets, which have rapidly expanded over the last 20 years. Apart from having a diverse range of food, food prices and marketing strategies, these supermarkets promote the consumption of cheap, ultra-processed foods contributing to non-communicable diseases as well as having a procurement system that marginalise small-scale farmers. The initial COVID-19 lockdown regulations in 2020 favoured supermarkets by closing informal traders and imposing stringent registration requirements for smaller retailers. The lockdown, therefore, accelerated consolidation in the food sector, disrupted livelihoods, compromised food access, and elevated transmission risks in confined, air-conditioned retail environments.

The initiation of progressive supermarkets is promoting procurement from local small-scale producers, the niche of BGN producers, by novel digital food retail platforms that are bringing producers and consumers closer while reducing COVID-19 risk. Hence the call for an urgent transition to sustainable food systems enabling diverse, small-scale food retail markets supplied by regenerative agriculture underpinned by broader stakeholder participation in food governance. Regulating and creating governance opportunities for alternative food networks, providing market access and a supportive environment to small-scale farmers will benefit BGN.

13.3 Consumption Trend – Bambara Groundnut Consumers Choices

Legumes like BGN are unique in meeting global consumer trends in the foreseeable future, including increased dietary protein consumption; heightened food safety awareness; sensitivity to environmental concerns and growing interest in ethnic-fusion cuisines. The consumption of high-protein foods and foods rich in phenolic compounds that are capable of exerting biological effects at cellular level is on the increase. BGN is increasingly known in many countries of the world due to its food composition (Hampson et al., 2000).

The overall consumption of BGN is low and the reasons need to be established, as this will help understand consumers' requirements needed by the food industry. In a study using Cragg's double hurdle model to measure consumers' choices and expenditure on selected bean varieties, Kilima and Bolle (2020) observed that gravy quality, cooking time, grain size, household size, income status of consumers, age and education of the main decision-maker are all important. Similarly, research is needed on attributes and factors on the selection of BGN varieties acceptable by consumers and the food industry. Atoyebi et al. (2017) reported on the sensory attributes of BGN (*Vigna subterranea* (L.) Verdc.) food products.

From an economic perspective and consumer act of purchase, the Cragg's Double hurdle model (Carlsson et al., 2012), Consumer Behaviour Theory (Lancaster, 1966), and Modelling food choices (Louviere et al., 2010) have all been used to determine the rational and expected outcomes in choices. Consumer behavior or decision is influenced by (i) diverse and interactive factors including intrinsic and extrinsic product characteristics as well as cultural, socioeconomic, personal, psychological, biological and situational factors (Köster, 2009), (ii) socio-economic and demographic characteristics including sex, age, education level and nutritional knowledge of the main decision-maker (Mitchell et al., 2009); (iii) factors like household size and income, price of the commodity and its attributes (Banterle et al., 2013); and food availability, culture, religious beliefs and spatial differences in lifestyles (Vui, 2008).

13.4 Commercialization: Marketing and Commerce of Bambara Groundnut and Products

Bambara groundnut [BGN] or its products have not reached marketing and commerce due to low consumption. Marketing, often done by affiliates on behalf of a company, refers to activities a company undertakes to promote the buying or selling of a product or service and includes advertising, selling, and delivering products to consumers or other businesses (CIM, 2015). Commerce, the activity of buying and selling, especially on a large scale, is the conduct of trade, the exchange of goods, services, or something of value, between businesses or entities among economic agents. There are very few of such business entities for BGN. Small business groups that buy BGN seeds from farmers and packages are beginning to emerge across countries and continents. Traces of small or individual businesses have emerged globally in few countries.

The increase in eCommerce marketing, involving the practice of using promotional tactics to drive traffic to online stores, converting that traffic into paying customers, and retaining those customers post-purchase will benefit BGN. A sound marketing strategy can help build brand awareness, drive customer loyalty, and ultimately increase online sales.

It is possible for goods not having established usage in a country to be exported throughout the world. For such to take place certain conditions must be fulfilled such as affiliates in the countries, domestic supply, foreign demand for the good, export price, etc. Every country strives to export goods for which there is strong demand, especially regions experiencing growth. For export purposes, Hatzichronoglou (1996) expressed a component of trade between a country Z and the rest of the world as:

$$X = X + A + B - C - H - E$$

Where:

X = Exports of country Z goods services, residence basis.
A = Sales by foreign affiliates from country Z to unaffiliated foreigners.
B = Exports of country Z's affiliates to their foreign affiliates.
C = Exports of country Z's firms to their affiliates abroad.
H = Exports of foreign affiliates established in-country Z to the rest of the world.
E = Local (non Z's) purchases of goods and nonfactor services by foreign affiliates of Z's companies.

Hillocks et al. (2012) indicated considerable international demand. Awang (2020) believes that demand for BGN will bring income to farmers in Malaysia and Indonesia. An international company has partnered with BGN farmers in sub-Saharan Africa to share information on BGN (Ho, 2020).

13.4.1 Global Commercialization of Bambara Groundnut Value-Added Products

13.4.1.1 Aonyx Foods Bambara Groundnut Gluten-Free Food Products

Aonyx Foods, South Africa demonstrated the versatility of BGN in gluten-free food products. Bambara groundnut flour is the starting material for its 'milk' and 'yoghurt-like or cultured beverage'. Jideani (2017a) reported the consumer sentiment of BGN milk (Bamilk) and cultured beverage (BamPro) showcased at the 2015 Innovation Showcase organised by the Cape Peninsula University of Technology, South Africa. One hundred and three consumers tasted the products and each completed a one-page questionnaire. A total of 103 people, males 54 (52.4%) and females 49 (47.6%) tasted the products and completed the product satisfaction questionnaire. Significant ($\chi^2 = 59.91$, $p < 0.05$) proportion (52.2%) of the respondents rated the Bamilk very good, 23.2% excellent and 21.7% average giving a total of 97.1%. Majority of the respondents 54.2% rated BamPro very good, 21.7% average and excellent ($\chi^2 = 78.14$, $p < 0.05$) totally 97.6%. Respondents likelihood of purchase for Bamilk ranged from probably (45.7%) to definitely (41.4%) [$\chi^2 = 65.57$, $p < 0.05$]; BamPro from probably (43.2%) to definitely (42%) [$\chi^2 = 70.42$, $p < 0.05$]. The taste, flavour, add, good, improve, market and strawberry were the dominant attributes describing both products. Sentiment analysis of the comments was done using WEKA software following J48, Random forest, Random and Naïve Bayes data mining algorithms. There was no significant ($p < 0.05$) difference between the methods in precision, recall and F-measure. The high receiver operating characteristic (ROC) curve area (> 0.70) indicated the very strong statistical relationship between the class label (negative, positive) and their descriptors. The Random Forest and Naïve Bayes will require 65.6% of the consumers to be selected in order to obtain 97.9% positive sentiments with a cost of 17 and a gain of 31. Naïve Bayes correctly classified 32(66.7%) out of 48 negative comments and 47(97.9%) out of 48 positive comments. The overall consumer sentiment for Bamilk and BamPro is positive. However, an improvement is desired for more sugar and less acidity for the BamPro. Hence, there is a market potential for Bamilk and BamPro.

Bambara groundnut alternative milk products (strawberry and pear caramel flavours) were compared with commercial variants (almond, rice and soybean) for appearance, colour, aroma, taste, mouthfeel and overall acceptability (Jideani, 2017b). The consumers perceived that the products were dissimilar in appearance and colour, although the BGN products were more similar to soybean milk than rice and almond milk. Soybean and BGN strawberry milk were similar in aroma and are dissimilar to almond, rice and BGN pear caramel milk. The BGN products were all similar to almond milk in taste. The similarity of BGN products to almond milk in taste indicates their commercial potential. BGN pear caramel is similar to almond milk in overall acceptability. The differentiation of BGN alternative milk products lies in their taste (Fig. 13.1). Both BGN beverages were liked moderately in appearance, colour aroma, taste and overall acceptability an indication of their positive commercial potential. Figure 13.2 depicts the word clouds for the panellists comments on the BGN alternative milk products and commercial variants.

Furthermore, the flour is useful for the extraction of fiber and in gluten-free baked products. The food industry had shown interest in flour; however, its absence in the marketplace is a limitation. Researchers demonstrated the isolation of dietary fibers from BGN flour and their utilization in food products (Diedericks & Jideani, 2015; Maphosa & Jideani, 2016; Maphosa et al., 2017). Consumer acceptable products were produced using the dietary fibers, BGN flour and flour mix as indicated in Fig. 13.3.

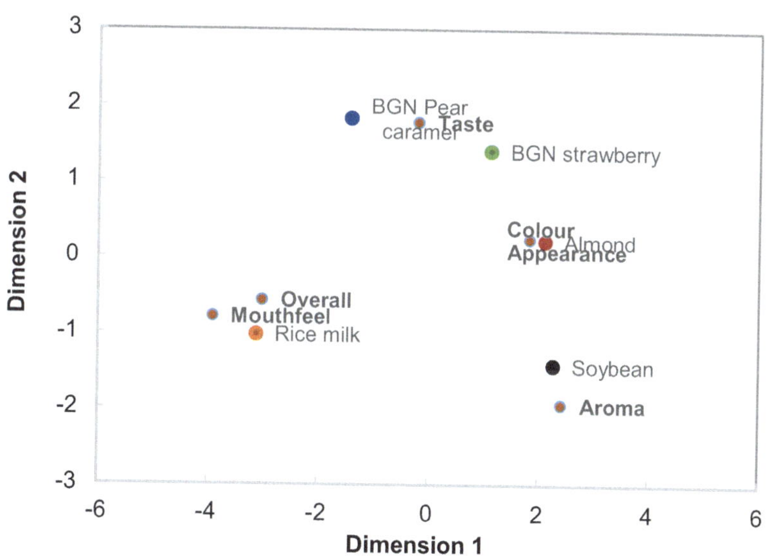

Fig. 13.1 Similarities proximities for Bambara groundnut (BGN) alternative milk compared to commercial brands

Fig. 13.2 Word cloud from panellists comments on Bambara groundnut alternative milk products and commercial variants

13.4.1.2 Singapore-Based Food Science Company NamZ Bambara Groundnut Healthy and Good for the Environment Instant Noodles

Crops For the Future (CFF) in Malaysia has established a broad range of research activities to promote the improvement and adoption of BGN with emphasis on the short-day photoperiod requirement for pod-filling and the hard-to-cook phenomena. The photoperiod requirement restricts the range of latitudes/timing of planting of the crop, while the latter increases fuel demand on households. Through selective breeding lines that are less susceptible to photoperiod were developed (Gregory et al., 2019).

NamZ, a Singapore-based food science company developed noodles using the underutilized "future-fit crops" (Ho, 2020). These crops can grow on degraded land with few resources, requires less water but produce good yields (Awang, 2020; Ho, 2020; Hoh, 2020). The noodles (Fig. 13.4) were created from BGN, lupin and moringa and other vegetables like pumpkins, tomatoes and cocoa (Ho, 2020; Hoh, 2020). The noodles are dehydrated by steaming and high-velocity air, unlike the conventional noodles dehydrated through deep-frying with palm oil. They are high in fiber, protein and plant nutrients with 70% less fat, 82% less saturated fat (Ho, 2020). The startup believes that crops like BGN will help revive the land and bring income to the farming community in Malaysia and Indonesia (Awang, 2020). The company has partnered with smallholder farmers in Ghana to share information and technical knowledge on the cultivation of BGN.

Fig. 13.3 Prototype Bambara groundnut products. (Barends, 2019)

(a) (b)

Fig. 13.4 Innovative food products from Bambara groundnut (BGN). (**a**) NamZ, BGN-based noodles (Awang, 2020); (**b**) WhatIF noodles from BGN (Anonymous, 2020)

WhatIF Food based in Singapore is the consumer brand launched by food tech company NamZ. WhatIF Foods hopes to launch its BamNuts plant-based milk in the second quarter of 2021 with original milk with nothing added, a low-GI sweetened version, and a chocolate-flavored version. The company reported that the milk is whiter than the brownish hue of almond or oat milk with a strong sensory feel. They tailored the taste to be more dairy, nuttier or more greeny, extremely creamy and similar to pasteurized dairy milk, which soy cannot compete. The unique texture is due to the starch, proteins and fat composition of BamNut, which enables it to froth like dairy. It is micronutrient-dense with its protein consisting of all the nine essential amino acids.

13.4.1.3 Bambara Groundnut – A Food Crop Seeds to Survive a Harsher Climate

BGN is on the priority list of a global search for food crop seeds that could save lives in a warming world. Over 100 scientists working across 25 countries, from the mountains of Peru to the fields of the Mediterranean island of Cyprus spend the past six years tracking down long-lost wild relatives of 28 food and forage crops that are important for world food security (Makri, 2020).

13.4.1.4 Bambara Groundnut Sustainability – The Key to Marketing Its Food Products

BamNut's sustainability attributes surpass most plant-based milk in the market because (i) it requires little to no fertilizer; (ii) able to grow on degraded land and revives the land and (iii) it can survive areas with 300 mm of rainfall and use less than 10% the water that almonds require (Neo, 2020). BGN cultivated in mining degraded soil remediated acidic soils by reducing the soils aluminium and hydrogen contents while increasing the nitrogen, phosphorus, calcium and magnesium (Zengkun, 2020).

Yang et al. (2020) investigated 100 United Kingdom consumer acceptability and emotional response to snack products containing BGN flour as an alternative sustainable ingredient by sharing BGN's sustainability and nutritional records. The participants evaluated two biscotti and two cracker products made from standard ingredients sourced commercially, and one made by replacing some of the standard flour with BGN flour. Under the blind condition, no significant differences in overall liking were observed between standard and BGN products, indicating UK consumers accept the sensory properties of products that contain BGN flour. Interestingly, when the participants received the BGN sustainability information, consumer emotional responses shifted towards more positive and less negative emotions when consuming BGN flour products. Consequently, they felt less guilt when consuming the BGN products, suggesting consumers engage with the idea of sustainable ingredients, and that this sustainable ingredient has potential for future new product development. Food neophobia status did not impact product acceptability and emotional response between BGN and standard products. However, overall, a lower emotional response was recorded for medium neophobic consumers who are more likely to evade novel products.

References

Anonymous. (2020). These instant noodles are healthy, high quality, and eco-friendly [Internet document]. THE SUPER CHARMED LIFEURL https://www.thesupercharmedlife.com/post/healthy-ecofriendly-instant-noodles. Accessed 15 Jan 2021.

Atoyebi, J. O., Osilesi, O., & Adebawo, O. (2017). Sensory evaluation of Bambara groundnut (*Vigna subterranea* (L.) Verdc.) food products in Ibadan, Nigeria. *International Journal of Agriculture Innovations and Research, 5*(5), 2319–1473.

Awang, N. (2020). S' pore-based startup says its fresh take on instant noodles is healthy, good for the environment. *Today*, 1–12.

Banterle, A., Cavaliere, A., & Ricci, E. C. (2013). Food labelled information: An empirical analysis of consumer preferences. *International Journal of Food System Dynamics, 3*(2), 156–170.

Barends, J. (2019). Do it with Bambara groundnut. Technology Transfer Office and Food Science and Technology Department, Cape Peninsula University of Technology, South Africa. https://www.cput.ac.za/newsroom/news/article/3819/do-it-with-bambara-groundnut.

Brink, M., & Belay, G. (Eds.). (2006). *Plant resources of tropical Africa* (Cereals and pulses) (p. 214). PROTA Foundation/Backhuys Publishers/CTA.

Carlsson, F., Mørkbak, M. R., & Olsen, S. B. (2012). The first time is the hardest: A test of ordering effects in choice experiments. *Journal of Choice Modelling, 5*(2), 19–37.

Chaturvedi, S., & Chakraborty, S. (2020). Review on potential non-dairy symbiotic beverages: A preliminary approach using legumes. *International Journal of Food Science and Technology.* https://doi.org/10.1111/ijfs.14779

CIM. (2015). Marketing and the 7Ps – © CIM 20 - A brief summary of marketing and how it works. The Chartered Institute of Marketing (CIM). Berkshire, United Kingdom. https://www.cim.co.uk/media/4772/7ps.pdf; Visited 26-01-2021.

Diedericks, C. F., & Jideani, V. A. (2015). Physicochemical and functional properties of insoluble dietary Fiber isolated from Bambara groundnut (*Vigna subterranea* [L.] Verdc.). *Journal of Food Science, 80*, C1933–C1944.

FAOSTAT. (2015). http://faostat3.fao.org/download/Q/QC/E

Gregory, P. J., Mayes, S., Hui, C. H., Jahanshiri, E., Julkifle, A., Kuppusamy, G., Kuan, H. W., Lin, T. X., Massawe, F., Suhairi, T. A. S. T. M., & Azam-Ali, S. N. (2019). Crops For the Future (CFF): An overview of research efforts in the adoption of underutilized species. *Planta, 250*, 979–988.

Hampson, K. J., Azam-Ali, S. H., Sesay, A., Mukwaya, S., & Azam-Ali, S. N. (2000). Assessing opportunities for increased utilisation of Bambara groundnut in southern Africa. Tropical Crops Research Unit, University of Nottingham. Internal Report. 52p.

Hatzichronoglou, T. (1996). *Globalisation and competitiveness: Relevant indicators*. OECD Science, Technology and Industry Working Papers 1996/05, OECD Publishing, Paris, https://doi.org/10.1787/885511061376.

Hillocks, R. J., Bennett, C., & Mponda, O. M. (2012). Bambara nut: A review of utilisation, market potential and crop improvement. *African Crop Science Journal, 20*(1), 1–16.

Ho, B.S. (2020). Singapore startup NamZ disrupts instant noodle market using climate resistant sustainable crops, 1–6.

Hoh, W. K. (2020). Starting a food revolution by making instant noodles that are good for health and the planet. *The Straits Times*, 1–2.

Jideani, V. A. (2017a). Consumer sentiment to Bambara groundnut milk and cultured beverage. A report submitted to the technology innovation agency (TIA) seed fund, South Africa.
Jideani, V. A. (2017b). Comparative sensory characteristics of Bambara groundnut milk with commercial milk alternative products and sensory profile for Bambara groundnut cultured beverage. A report submitted to the technology innovation agency (TIA) seed fund, South Africa.
Kilima, F. T. M., & Bolle, E. M. (2020). Factors underlying consumers' choices of dry common beans in Dar Es Salaam, Tanzania. *African Journal Food Agriculture Nutrition Development, 20*(2), 15619–15636. https://doi.org/10.18697/ajfand.90.18350
Köster, E. P. (2009). Diversity in the determinants of food choice: A psychological perspective. *Food Quality Preference, 20*(2), 70–82.
Lancaster, K. A. (1966). New approach to consumer theory. *Journal of Political Economy, 74*, 132–157.
Li, L., Yanga, T., Liu, R., Redden, B., Maalouf, F., & Zonga, X. (2017). Food legume production in China. *The Crop Journal, 5*, 115–126. https://doi.org/10.1016/j.cj.2016.06.001
Louviere, J. J., Flynn, T. N., & Carson, R. T. (2010). Discrete choice experiments are not conjoint analysis. *Journal of Choice Modelling, 3*(3), 57–72.
Makri, A. (2020). We're turning to offbeat foods to survive a harsher climate. *National Geographic*, 10–13.
Maphosa, Y., & Jideani, V. A. (2016). Physicochemical characteristics of Bambara groundnut dietary fibres extracted using wet milling. *South African Journal of Science, 112*, 72–79.
Maphosa, Y., Jideani, V. A., & Adeyi, O. (2017). Effect of soluble dietary fibres from Bambara groundnut varieties on the stability of orange oil beverage emulsion. *African Journal of Science, Technology, Innovation and Development, 9*, 69–76.
Mitchell, D. C., Lawrence, F. R., Hartman, T. J., & Curran, J. M. (2009). Consumption of dry beans, peas, and lentils could improve diet quality in the US population. *Journal of American Dietary Association, 109*(5), 909–913.
Neo, P. (2020). 'Sustainable groundnut': WhatIF foods targets ASEAN Bambara supply chain amid 2021 plant-based milk launch. Food navigator, 1–3.
Vui, H. (2008) Essays on the economics of food production and consumption in Vietnam. Dissertation for award of Ph.D. degree at the University of Minnesota, United States of America. 267.
Yang, Q., Shen, Y., Foster, T., & Hort, J. (2020). Measuring consumer emotional response and acceptance to sustainable food products. *Food Research International, 131*, 108992.
Zengkun, F. (2020). How 'future-fit' crops can protect the environment and provide for the future, 1–9. Sponsored by nutritional paradox. https://www.eco-business.com/news/how-future-fit-crops-can-protect-the-environment-and-provide-for-the-future/ Viewed 10-01-2021. https://www.nutritionalparadox.com/

Chapter 14
Bambara Groundnut Value Chain

14.1 Introduction

Agribusiness refers to commercialisation and value-addition in the agricultural sector with a focus on pre- and post-product enterprises and building linkages among enterprises. The agro-processing refers to the subset of manufacturing (agribusiness) that processes raw materials and intermediate products derived from the agricultural sector. Promotion of agro-enterprise development can provide positive impacts on employment in both rural and urban areas, offer market access to smallholder agriculture, create business linkages to Small & Medium Enterprises (SMEs), and enhance food security by reducing post-harvest losses and extending the shelf-life of food and fibers for the rapidly increasing urban poor. The combined effects of employment gains and food security through improved agro-industry competitiveness are an important strategy for reducing overall poverty. It is important to take advantage of Bambara groundnut [BGN] flour in the value chain to value-added products for the health and wellness of the community. The access to these markets will capture the growing international demand for higher-value processing agro-industry. This will address the challenges of faster and more inclusive economic growth, with a focus on implementation for small, micro and medium enterprises (SMMEs), poverty/inequality, employment, agriculture and livelihood strategies. The outcome will include economic growth through the creation of competitive industries and businesses in BGN (raw material) sectors of the economy; development of productive firms, especially small-to-medium-sized, and their inclusion in local and global value chains. Hence, it is at the level of economic sustainability, self-reliance and cost-effectiveness that the BGN value chain will prove its viability and strength. At this level, BGN products will be part of the second economy dominated by small-scale producers, manufacturers and bankers being part of the most vibrant sectors of economies.

The line of transactions from producer to consumer constitutes the value chain which describes the range of activities required to bring a product or service from

V. A. Jideani, A. I. O. Jideani, *Bambara groundnut: Utilization and Future Prospects*, https://doi.org/10.1007/978-3-030-76077-9_14

conception, through intermediary phases of production to final consumers and disposal after use (ILO, 2015). The activities include design, production, marketing, distribution and support services up to the final consumer (ILO, 2015). The value chain of any sector is always embedded into a market system consisting of various supporting functions and rules. From an agronomic perspective, yields of major crops are expected to decrease in the face of climate change. A shift of attention away from major crops and towards lesser-known or orphan crops and documented minor crops such as BGN could prove a key ingredient in the development of sustainable food and nutrition security. The knowledge of how market systems work is very important for developing upgrading strategies that target systemic root causes for competitive constraints. The objective of understanding the value chain of BGN is to move poor individuals and households out of saturated, low-return activities and into higher-return, growing markets.

14.2 Bambara Groundnut Value Chain for Sustainable Household Income

Mure et al. (2014) investigated the structure of the Bambara groundnut market and livelihood to the households involved in the production, processing and marketing in Nigeria. A multistage sampling technique was used to select 317 farmers and 151 processors and retailers of okpa (a steamed BGN flour slurry) from 56 communities. Cultivation of BGN, processing and marketing of okpa is dominated by women (86%). Most (92%) of the harvested BGN are sold at the local market while 96% of okpa produced are sold by women in community markets and roadsides. Processing is the most concentrated link in the chain relative to margin realized and marketing cost. Income from sale of okpa represented about 9.2%, 26.1% and 39.1% of total households' income for the farmers, processors and retailers, respectively. The allocation of the income shows that 44.3% and 62.7% of the income is utilized in meeting household' food needs for the retailers and processors, respectively. Among the processors and retailers, the income represents a major component of personal savings and investment in children's education. Age, formal educational level and household size of retailers and processors as well as ownership and size of farm holdings were among the factors that significantly influenced the households' dependence on income from okpa. BGN and products provide sustainable livelihood and households' food security to the value chain actors in sub-Saharan Africa. Women constitute a significant player in the BGN value chain as producers, processors and retailers. The value chain of BGN is fragmented due to a lack of product diversification. Lack of capital and liquidity, trust and formal contracts and high transaction cost were identified as problems that limit the efficiency of BGN value chain. The authors recommend the need to build the capacity of chain actors in input sourcing and management, market targeting as well as finance access.

Adzawla et al. (2016) reported that farmers and marketers in northern Ghana judged BGN positively (i) it can be processed into many products; (ii) requires a small amount of fertilizer, (iii) is drought-tolerant; (iv) has high protein content and nutritious; (v) tastes better than other legumes and (vi) more profitable. More than 50% of the respondents are less than 45 years. Consequently, the producers of BGN include young adults who are in the prime age of economic productivity, a positive impact on the sustainability of BGN production (Adzawla et al., 2016). The educational levels of BGN value chain actors are generally low, typical of most orphan crops.

BGN is sold dry without the pods; utilization involves basic processing of boiling the fresh pods and roasting the dry seeds (Mbosso et al., 2020). Most of the traders (91%) in Mali are retailers and very few semi-wholesalers (9%). Semi-wholesalers sold larger amounts to individual consumers and small businesses. Producers described BGN as an early crop that is useful in times of famine. It is used as a mid-morning snack, acts as a meat replacer as children who eat BGN are not malnourished. BGN is made into croquettes, galettes (grilled flatbreads mixed with okra powder) and tô (crushed seeds mixed with millet or sorghum flour and water into a stiff porridge) (Mbosso et al., 2020). No wholesaler of BGN was identified in the study (Mbosso et al., 2020). However, Adzawla et al. (2016) reported that 80% of the marketers in northern Ghana obtained BGN from the wholesale market. The majority of the respondents (81.7%) believe there was an adequate supply of BGN in the wholesale market (Adzawla et al., 2016). The majority (87%) of the traders in Mali are women (Mbosso et al., 2020). A similar high proportion of women in BGN value chain is reported in Nigeria (Mure et al., 2014), Ghana (Berchie et al., 2010) and western Niger (Ibrahim et al., 2018). Notably, in Ghana BGN is grown by more men in areas where it was a cash crop and by more women in areas where it is primarily a subsistence crop (Berchie et al., 2010). Involving an equal gender ratio in the BGN value chain may enable more women to assess the potentials of BGN in production and livelihood systems (Berchie et al., 2010).

Sidibé et al. (2020) utilized a community biodiversity management approach to reinforce the cultivation of BGN in six communities in the Ségou and Sikasso regions of Mali. The research involved multiple varieties of BGN (8–12) established in diversity fields in each site over 2 years where farmers engaged in experiential learning over the crop cycle. Farmers' access to intraspecific diversity was increased and capacities of community institutions for their management were developed. Significant adoption of BGN was detected in several study sites. The drivers of adoption are likely increased seed access and awareness through the diversity field settings, seed fairs and community seed banks. No significant yield advantage was detected for any of the varieties in the diversity fields, which showed variable performance by site and year. The number of varieties registered and managed by community seed banks in each site increased from 1–5 varieties of each crop to 11–12 varieties following the interventions. The number of BGN varieties cultivated in farmers' fields also increased in some communities. The community biodiversity management approach revitalised BGN cultivation and promoted diffusion and use by improving access to seed, diverse varieties and awareness of the values of these crops in diversified and resilient production systems. This approach is promising for strengthening the seed system of BGN by building on the strengths of both formal and informal systems. In the northern study sites, the activities were relevant in providing access to different varieties and building knowledge of cultivation practices, as BGN had not previously received much support from agricultural research and extension despite their important livelihood roles. In the southern study sites, a clear increase in the cultivation occurred, which could support climate change adaptation in this region where climate and soil degradation are increasingly challenging maize and cotton production.

14.3 Bambara Groundnut Value Chain Constraints

Mbosso et al. (2020) using a Rapid Market Appraisal method surveyed southern Mali traders, producers, processors and consumers in 2017 and 2018 to assess needs and opportunities for improving BGN value chains. Although BGN is consumed by the local populations in southern Mali, poor processing facilities and lack of market deter their wider socioeconomic potential (Mbosso et al., 2020).

The importance of BGN as a crop produced and marketed by women is that it provides an independent source of income for livelihood. It is sold within village boundaries in small volumes. The women considered that there was "no good market", "not market with a good price" and "no specific market" for BGN (Mbosso et al., 2020). BGN is sold at the farm gate and local market by farmers and processors, while the retailers sell at roadsides and markets (Mure et al., 2014). Ibrahin and Ogunwusi (2016) reported long cooking time, lack of best agronomic practices, lack of available and market access, lack of government support, urbanisation and inadequate funding as constraints to the BGN value chain.

Existing BGN products are not well promoted in the local or international markets and new products are needed that highlight its inherent nutritional and culinary advantages (Hillocks et al., 2012). Some projects on BGN, involving several countries in sub-Saharan Africa (SSA) since the 1980s, did not stimulate a sustainable increase in BGN crop production due to the absence of functioning value chains (Hillocks et al., 2012). Accessible market outlets might provide the required incentive for smallholder households to obtain improved seeds and invest more of their land and labour in the crop. There is little documented evidence of the trade in BGN but circumstantial evidence indicates considerable international demand. More attention should be given, therefore, to market research and development, with crop improvement programmes being more market-led, if BGN is to make a greater contribution to household income and rural development in SSA (Hillocks et al., 2012).

Hillocks et al. (2012) suggested three potential markets for BGN, (i) market among diaspora consumers: the market is relatively insensitive to prices but wants the product all year round rather than seasonal, suggesting that fresh-frozen might be necessary; (ii) a competitor to fresh legumes and soybean: while BGN might be more expensive due to under-production, it has potential as a novel alternative to peas, certain beans and soybeans in the fresh vegetable and salad markets and (iii)

BGN snacks: snacks from fresh BGN may have potential in the fast-growing developed world snacks food market which is always looking for novelty. BGN has sufficient qualities of nut size, taste and high protein content as a new international snack. The authors outlined five potential markets (i) fresh salad ingredients, (ii) fresh vegetables for home cooking and cooked snacks (e.g. boiled or roasted nuts), (iii) preserved BGN (frozen, canned, etc) example BGN beans, semi-processed and preserved, (iv) dried BGN for stew ingredient, and (v) specialist milled product for use as a food ingredient, example BGN flour).

Despite BGN's excellent nutritional profile, resilience and adaptability in use, the cultivation and trade of BGN remain below their potential throughout the world. The low potential is related to the weak BGN value chain (Mbosso et al., 2020). Trade is often confined to the immediate village boundaries, with little commercialization or processing (Hillocks et al., 2012; Mure et al., 2014; Mbosso et al., 2020; Sidibé et al., 2020), and negative traditional beliefs associated with the crop in parts of Africa (Forsythe et al., 2015). Figure 14.1 gives general information on BGN value chain actors at each level of value-adding activity, namely, production, primary processing, secondary process, distribution and consumption. The major interest is in the flow of different types of products provided by the chain through domestic and export channels. The line of transactions from the producer of BGN to the consumer constitutes the BGN value chain. Any sector's value chain is always embedded into a market system consisting of various supporting functions and rules.

Table 14.1 details the BGN value chain challenges. The informal nature of BGN economies in emerging markets means that relations between large companies and small enterprises remain unstable due to the lack of contracts and legal enforcement. Consequently, there is high production costs and problems in quality consistency; small enterprises are unable to access many business services such as credits, training and research facilities, government support, etc.; buyers are unable to trace products to their origins; workers and employees are deprived of their rights for social protection and labour standards (Herr & Muzira, 2009). The knowledge of how market systems work is very important for developing upgrading strategies that target systemic root causes for competitive constraints.

Cultivation of BGN is done by subsistence and smallholder farmers. Their contribution is hardly recognised despite the massive number of farmers involved and the potential that exist for BGN. The lack of recognition is primarily due to several production constraints including lack of capability and existing marketing infrastructure for BGN (Table 14.1).

Poverty is the result of diverse weaknesses and limitations at many levels such as individual (e.g. production know-how, market orientation, risk adversity); collective (e.g. information and knowledge sharing, joint action); institutional (e.g. access to non-financial and financial services); political and administrative (e.g. policies, legislation, land tenure rights and access to infrastructure) and socio-cultural (e.g. mistrust, networks and exclusion). Hence, an integrated poor-centric approach is required to address the key challenges and to make BGN value chain interventions work for the poor. The outcome will enhance the ability of the poor to participate in, contribute to and benefit from economic growth, thereby eradicating extreme poverty and hunger.

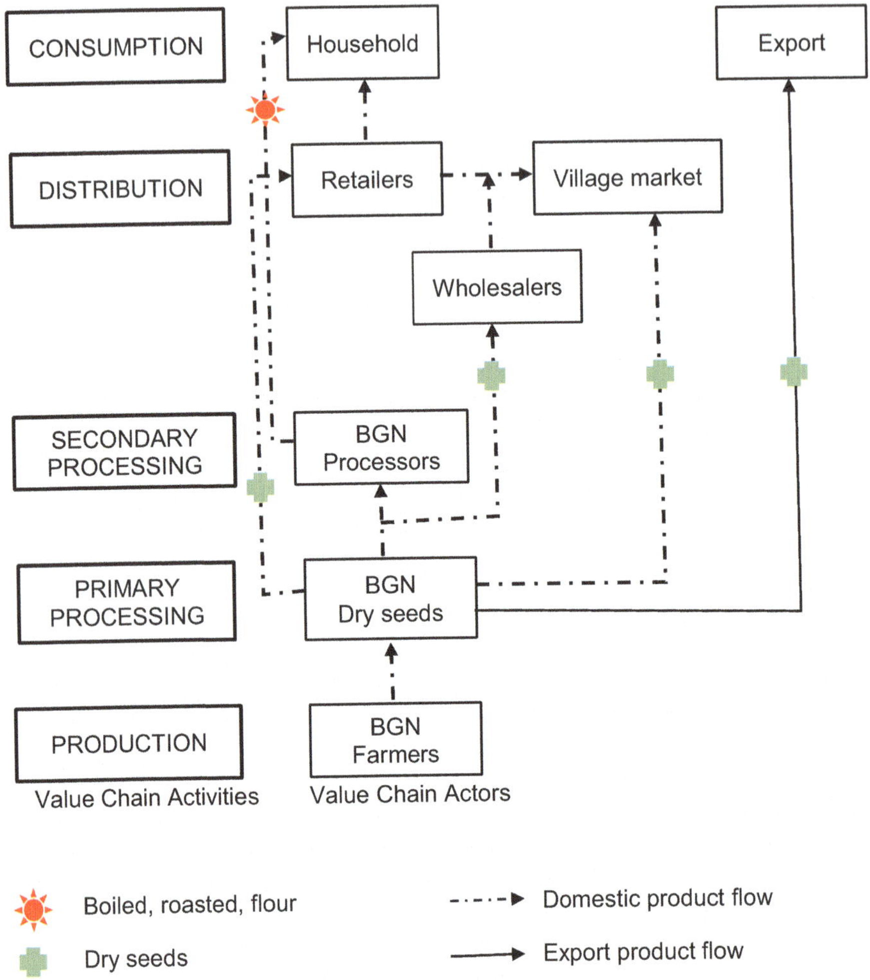

Fig. 14.1 Bambara groundnut value chain activities and actors

The objective of understanding the value chain of BGN is to move poor individuals and households out of saturated, low-return activities and into higher-return, growing markets. This will be done by linking poor producers of BGN to private sector actors who have access to growing markets, and who have a clear business interest in partnering with poorer producers as part of their supply system. Hence, bringing existing businesses together provides opportunities to share the risk and costs of developing market channels and end markets. Product traceability is a global issue and it is only possible through improved cooperation amongst value chain stakeholders and coordination of their activities. The better stakeholders cooperate along the value chain and coordinate their activities, the harder it will get for competitors to copy the product and the production process because it is not just the product they need to copy, but the entire system. This is a form of barrier to entry to the BGN value chain.

Table 14.1 Challenges within various actor in Bambara groundnut (BGN) value chain

Value chain actor	Challenge
Traders and processors	Low capital and liquidity, lack of customers, high buying price for grain Greater volume of product needed to meet demand and reduce prices Existing BGN products are not well marketed or promoted in the local market or internationally. New BGN products are needed that highlight its inherent advantages, e.g. nutrition, taste and nutrition.
Producers	Lack of processing facilities General lack of promotion and interest in the crop seen as 'petty' trade Climate constraints (not enough rain, too much at harvest time). Inadequate technical know-how. Lack of access to mechanized equipment. Lack of access to improved genotypes/varieties.
Consumers	High price during lean season. Poor quality of processed product. Takes time to cook. Bloating, constipation and flatulence.

Source: Berchie et al. (2010), Hillocks et al. (2012), Mure et al. (2014), Adzawla et al. (2016), Ibrahim et al. (2018), Mbosso et al. (2020) and Agyeman et al. (2021)

Table 14.2 Pathways to Bambara groundnut value chain development

Pathway	Action
Greater value chain integration	Market actors-large and small- working together more effectively.
	Develop and improve essential linkages along the value chain- vertical integration.
	Producers to form associations or cooperatives to enhance their bargaining power or perform bulking functions to provide the volumes required by large buyers- horizontal integration.
Greater market access	Develop linkages to new markets as a powerful path to growth. Meet new quality standards.
Better performing supporting functions	BGN market actors taking on new or improved roles
	Improved training, testing and certification services and informing enterprises about these services.
	Improved public-private dialogue for more relevant training.
	Better roads and access to a reliable source of electricity.
Improved rules of the game	Simplify relevant regulations and procedures to reduce obstacles for growth.
	Adopting and applying food safety standards to enhance the competitiveness of the BGN value chain on local markets (with international imports) and export markets.
	Environmental legislation to enhance the acceptance of products in international markets.
	More and better information on legislation and other rules to enable more market actors to benefit from them.

Adapted from ILO (2015)

14.4 Bambara Groundnut Value Chain Upgrading

Development of the BGN value chain requires attention to four main areas outlined in Table 14.2. BGN value chains are not chains because links between enterprises are weak or non-existent. There are no collection points or storage facilities for BGN resulting in processors finding it difficult to obtain inputs from producers. Producers export unprocessed BGN because there are few or no local processors. Consequently, a greater value chain for BGN integration is required as proposed in Fig. 14.2.

BGN value chain is limited to local, traditional markets because of a lack of information about other opportunities, or the actor cannot respond to these opportunities. Access to new markets involves meeting new quality standards or producing large quantities and may therefore generate innovation and be an incentive for greater value chain integration. Also acting as an incentive to improve working conditions in the BGN value chain is the consumers' demand for social and environmental standards.

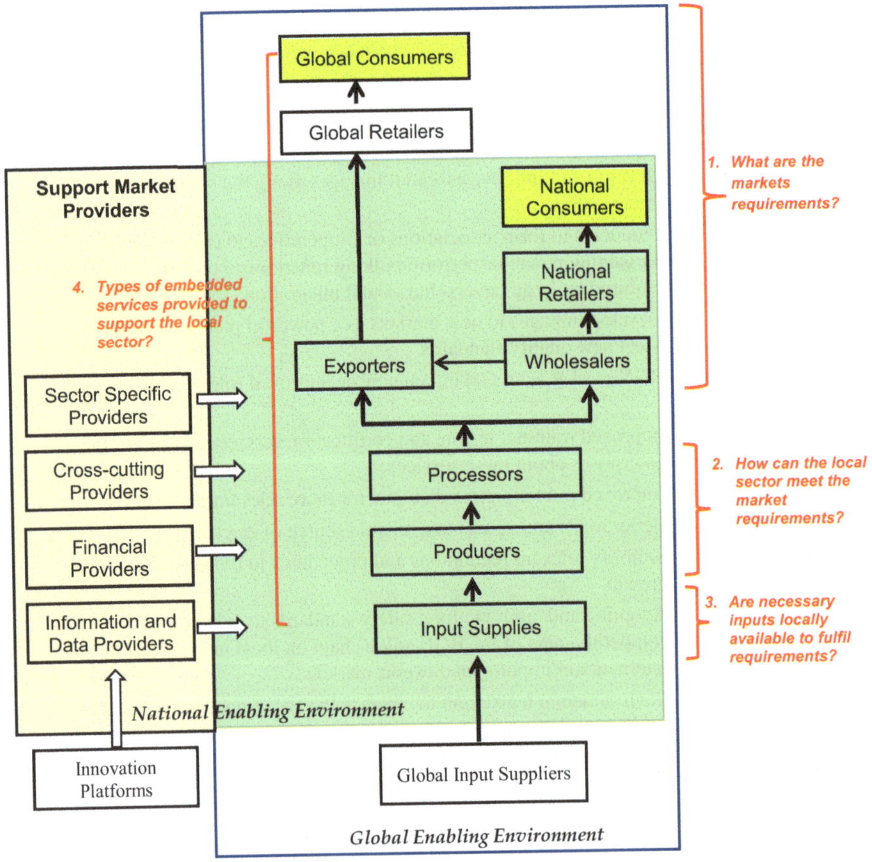

Fig. 14.2 Proposed Bambara groundnut value chain development strategy

Systemic constraints in the BGN value chain lie in the supporting functions as market actors outside the value chain are absent. The solution is that the market actors should take on new or improved roles. New workplace improvement services may help BGN value actors to enhance productivity.

Improving BGN value chain 'rule' and their application is crucial to enhancing the performance of the value chain and a powerful incentive for value chain integration and innovation. For instance, the BGN value chain is constrained by the complexities of access to land, for construction of facilities, or production. Simplifying relevant regulations and procedures can reduce such obstacles. Successful implementation of the pathways outlined in Tables 14.2 and 14.3 will result in an upgraded BGN value chain with the support market providers, national and global enabling environments as detailed in Fig. 14.2.

Agyeman et al. (2021) detailed some key attributes that farmers considered in an improved BGN variety to include ease of cooking, ease to harvest mechanically, drought tolerance, disease tolerance, tolerance to waterlogged fields, ability to fix

Table 14.3 Other strategies for Bambara groundnut (BGN) value chain development

Focus	Detail	References
Increased and sustained demand	Marketing, promotional campaigns, advertising and information dissemination	Mbosso et al. (2020)
Increased value chain efficiency	Quality affordable products Organisation of farmer groups or cooperatives to reduce intermediaries' bargaining power. Development of processing units. Better communication between different value chain actors. Decent infrastructure and transport networks to minimise transaction costs. Credit and modern inputs to scale up production Reducing cooking time.	Mbosso et al. (2020) and Adzawla et al. (2016)
Promote quality-rewarding mechanisms	Specifying necessary quality attributes that a product must have to be commercialised through certification labels. Development of certification and labelling systems to support the marketing. Simple and inexpensive certification mechanisms linked to on-farm conservation of BGN diversity. Introduction of BGN nutritious food in public procurement schemes like school meal programs. Establishment of multi-stakeholder platforms involving all value chain actors.	Adzawla et al. (2016)
Creation of processing units outside the household.	Securing availability of credit and provision of processing materials. Invest in secondary-level processing technologies as processed products are less subject to price variations compared to raw ones. Cleaning and destoning.	

Table 14.4 Preferred attributes of an improved BGN variety

Ease of cooking
Ease to harvest mechanically
Drought tolerance
Disease tolerance
Tolerance to waterlogged fields
Ability to fix nitrogen
High yield
High protein and carbohydrates
Early maturing and white or dark brown bean colour

Adapted from Agyeman et al. (2021)

nitrogen, high yields, high protein and carbohydrates, early maturing and white or dark brown bean colour (Table 14.4). Furthermore, important variables influencing farmers' preference for seed colour, nutritional and agronomic attributes to plant orientation attributes of an improved BGN variety are gender, education, years of BGN cultivation, off-farm income and household size. These factors and attributes need careful consideration in the development of improved BGN varieties in sub-Saharan Africa.

References

Adzawla, W., Donkoh, S. A., Nyarko, G., O'Reilly, P., & Mayes, S. (2016). Use patterns and perceptions about the attributes of Bambara groundnut (*Vigna subterranea* (L.) Verdc.) in Northern Ghana. *Technology and Development, 4*, 56–71.

Agyeman, K., Asante, B. O., Berchie, J. N., Sarkodie-Addo, J., Marno, P., & Adabah, R. (2021). Farmers' perceptions, constraints and preferences for improved Bambara groundnut varieties in Ghana. *Journal of Agriculture and Food Research, 3*, 100097.

Berchie, J. N., Dapaah, H. K. A., Dankyi, A. A., Plahar, W. A., Quartey, F. N., Haleegoah, J., Agyei, J. N. A., & Addo, J. K. (2010). Practices and constraints in Bambara Groundnuts production, marketing and consumption in the Brong Ahafo and upper-east regions of Ghana. *Journal of Agronomy, 9*, 111–118.

Forsythe, L., Nyamanda, M., Mwangwela, A. M., & Bennett, B. (2015). Beliefs, taboos and minor crop value chains: The case of Bambara groundnut in Malawi. *Food, Culture and Society, 18*, 501–517.

Herr, M. L., & Muzira, T. J. (2009). Value chain development for decent work- A guide for development practitioners, government and private sector initiatives. 978-92-2-122230-9, International Labour Organization 2009.

Hillocks, R. J., Bennett, C., & Mponda, O. M. (2012). Bambara nut: A review of utilisation, market potential and crop improvement. *African Crop Science Journal, 20*, 1–16.

Ibrahim, A. R., Dansi, A., Salifou, M., Ousmane, A., Alzouma, A., & Alou, W. (2018). Farmers' practices, utilization, conservation and marketing of Bambara groundnut (*Vigna subterranea* (L.) Verdc.) in Dosso Region, Western Niger. *Genetic Resources and Crop Evolution, 65*, 1907–1914.

Ibrahin, H., & Ogunwusi, A. (2016). Industrial potentials of Bambara nut. *An International Peer-reviewed Journal, 22*, 12–18.

ILO. (2015). *Value chain development for decent work- How to create employment and improve working conditions in targeted sectors* (2nd ed.). Geneva.

Mbosso, C., Boulay, B., Padulosi, S., Meldrum, G., Mohamadou, Y., Niang, A. B., Coulibaly, H., Koreissi, Y., & Sidibé, A. (2020). Fonio and Bambara groundnut value chains in mali: Issues, needs, and opportunities for their sustainable promotion. Sustainability (Switzerland).

Mure, A., Olumuyiwa, A., & Elijah, O. (2014). Value-chain analysis of Bambara Groundnut (*Vigna subterranean*) and livelihood sustainability amongst households in derived savanna belt of Nigeria.

Sidibé, A., Meldrum, G., Coulibaly, H., Padulosi, S., Traore, I., Diawara, G., Sangaré, A. R., & Mbosso, C. (2020). Revitalizing cultivation and strengthening the seed systems of fonio and Bambara groundnut in Mali through a community biodiversity management approach. *Plant Genetic Resources: Characterisation and Utilisation, 18*, 31–48.

Chapter 15
Current and Future Bambara Groundnut Research Directions

15.1 Introduction

Bambara groundnut (BGN) has potential for industrial processing and increased use for derived products (Atoyebi et al., 2017; Mohamadou et al., 2018), as well as for use in the functional foods and nutraceutical industry, therefore further research is needed in all aspect of the crop for the optimization of benefits and valorization. There are substantial interest and scientific research currently available in the literature on BGN, focusing on the entire value chain from farm to fork. The world's food supply is concentrated with up to 60% of food supply coming from conventional cereal crops like maize, rice, and wheat (Azam-Ali, 2007), conventional legumes like soybean, cowpea and groundnut (Ashogbon, 2017; Nyau et al., 2015; Li et al., 2017). The increasing world population of 2.3 billion in 2050 may exceed agricultural production (FAO, 2017; WFP, 2016; Tsoata et al., 2017; Godfray et al., 2010). According to Sahu et al. (2020), in diversifying the global food supply, enhance agricultural productivity, and eradicate malnutrition, it is necessary to focus on crop improvement of plants that are utilized in rural societies as a local source of nutrition and sustenance. Moreover, if wild foods, especially of plant origin, contribute in feeding humanity (Asprilla-Perea & Díaz-Puente, 2019; Gras et al., 2021; Pawera et al. 2020; Smith et al. 2019), BGN has vast potential in contributing not just to human diet but also to health. Bambara groundnut with highly caloric characteristics, rich in minerals, vitamins, proteins and amino acids could meet the demand of increasing crop yields while safeguarding the environment. Hence, there is a need for the adoption of mobile applications (Okoroji et al., 2021), smartphone and information communication technologies by farmers including climate-smart agricultural practices (World Bank, 2011; Lipper & Zilberman, 2018; Kifle, 2021). Consequently, further study is necessary, considering that from FAO (2014) statistics 805 million people (11.3%) were chronically undernourished from 2012 to 2014. Many aspects of BGN landraces have not been investigated, therefore, limiting their commercialization. An overview of the challenges and opportunities of

V. A. Jideani, A. I. O. Jideani, *Bambara groundnut: Utilization and Future Prospects*, https://doi.org/10.1007/978-3-030-76077-9_15

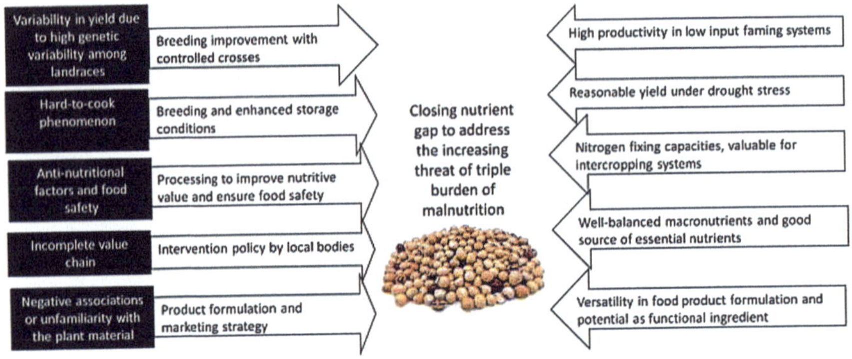

Fig. 15.1 An overview of the challenges and opportunities of utilizing Bambara groundnut in addressing food security and nutrition. (Tan et al., 2020)

utilizing BGN in addressing food security and nutrition as presented recently by Tan et al. (2020) is shown in Fig. 15.1 (Lipper & Zilberman, 2018).

To be at par on socioeconomic and scientific knowledge as conventional crops like soybean, peanut and similar legumes, there is a need for concerted investigation in the areas outlined in Fig. 15.1.

15.2 Expanded Bambara Groundnut Production

With the need to explore approaches for expanded legume production, for BGN the main thrust should be to increase the low levels of yield and to reduce the cost of production through research by addressing the following: (i) strengthening research on mechanized technology in BGN production, (ii) strengthening research on social sciences and to link technical and social science efforts more closely to accelerate the adoption of new technology, and (iii) make most effective use of the relatively limited global research resources likely to be available for research on food and feed BGN. Residue from processing BGN seeds is abundant in areas of production, but scientific knowledge on the suitability of this residue for improving soil fertility is limited (Okebalama et al., 2020).

15.3 Selective Breeding Program for Bambara Groundnut

There is limited genomic information on this economically important legume. More research on the selective breeding program using hybridization technique and other advances in breeding technologies will help understand the degree of genetic relationships between BGN accessions and hence crop improvement programmes. With the African Union's endorsement, the African Orphan Crops Consortium (AOCC, 2020), an international public-private partnership has the greater integration of orphan crops into African food systems at its heart. To support this, its goal is to focus on enhancing African orphan crops with genomics (Jamnadass et al., 2020; Sahu et al., 2020), assemble and annotate the genomes of 101 traditional African food crops, and facilitate their genetic improvement. Once genomes have been sequenced, AOCC explores genetic diversity in-depth in a 'germplasm' panel of each crop. The publication of high-quality genomes for BGN will provide information about their specific composition and the underlying genes involved in the synthesis of macro-molecules such as sugar and starch metabolism, photosynthesis, and other synthesis-related genes (Sahu et al., 2020). There is a need for the generation of assembled and high-quality annotated reference transcriptome of *Vigna subterranea*.

15.4 Effective Control Against Bambara Groundnut Viral and Fungal Diseases

Serological and molecular characterization of viruses using their corresponding polyclonal antibodies from ELISA detection, RNA extraction, RT-PCR and sequencing plus other advanced techniques are needed to establish a phylogenetic relationship with GenBank. Obtaining some effective control strategies against BGN viral diseases will mitigate the problem of farmers and food insecurity. The phylogenetic analyses explore evolutionary relationships between viral isolates and other Genbank available species. Further genetic evaluation based on the whole coat protein gene in amino acid or nucleotide with an increased number of samples will result in better characterization of the virus from genus *Potyvirus* occurring in other accessions of BGN.

In the area of fungal infestation, molecular characterisation through amplification and comparison of the partial sequence are necessary for precise identification of toxigenic strains of *Aspergillus flavus* and hence in establishing control strategies of the fungus in BGN and products. Molecular approaches will also provide great tools for use in screening BGN varieties for resistance and tolerance to *A. flavus*.

15.5 Policy Intervention for Increased Production of Bambara Groundnut

Little research has been undertaken on the contribution of BGN crops to the welfare of producers. Hillocks et al. (2012) stated that cultivating the crop could make a significant contribution to alleviating the plight of small-scale farmers by providing a nutritious alternative to animal protein and providing a source of income. Research and development interventions to promote BGN cultivation offer pathways for improving the welfare of smallholder farmers. Unfortunately, such crops continue to attract a tiny portion of global research budgets. Hence, advocating for the understanding by a wide range of potential stakeholders in the research and policy communities is needed regarding BGN crop potential in development. Policymakers and researchers also need to examine questions concerning the merits of supporting the cultivation of BGN.

15.6 Socio-scientific Knowledge of Producers and Consumers Behaviour for Bambara Groundnut

A three-pronged approach to explore BGN crop mainstreaming solutions could involve additional avenues for research to explore producers' and consumers' behaviors and preferences; firstly by the understanding of the incentives that guide the production activities by different groups of farmers (McMullin et al., 2021; Gassner et al., 2019); secondly, for BGN food products to more effectively compete with other food options in more complex market-based food systems, a better understanding is required of consumers' tastes, preferences and behaviors with respect to these foods (De Groote et al., 2018; McMullin et al., 2021; Woltering et al., 2020); and thirdly understanding how African food systems can better be diversified and made more healthful. This is an interdisciplinary process with both natural and social scientists' roles in exploring complexities and multi-scale drivers and their interactions. These areas are essential considering farm landscape, global dietary homogenisation, and the health of people and the planet (Dawson et al., 2019; Clark et al., 2019).

15.7 Unavailable Capital for Farmers

Lack of capital by some farmers was observed as a reason for not going into the production of BGN (Adzawla et al., 2016), hence the need for further research concerning the circumstances under which farmers can benefit from the wider cultivation of BGN crops. The extent and mechanism through which extension impacts the adoption of BGN require further exploration.

15.8 Bambara Groundnut Cultivars for High Protein Food Products

The development of new BGN cultivars with the potential for shelf-stable protein products could be a useful ingredient in food formulations for the food industry. Identification of high-nutrient accessions that can be utilised to eradicate the endemic burden of diseases and enhance food security. The development of new varieties of BGN by Adeleke et al. (2017) with higher protein and fat, essential fatty acids, minerals and vitamins contents and better functional properties than the commercial sample shows the potential for the industrial and household use of the new BGN cultivars into shelf-stable protein products that could be a useful ingredient in food formulations. The review of Akharume et al. (2021) on the modification of plant proteins for improved functionality highlights that though, many food industries are shifting away from a chemical modification of proteins because of the demand for clean label product and the challenge of toxicity associated with the scale-up of this technique, so physical and biological techniques are widely being adopted to produce a functional ingredient such as texturized vegetable proteins, hydrolyzed vegetable protein, clean label protein concentrates, de-flavored protein isolates, protein flour, and grits.

Protein isolate (PI) has a variety of nutritional benefits. From the report of Diedericks (2020) that BGN protein isolates showed potential as partial animal protein replacers, there is a need to improve the stability of the isolates via noncovalent interactions. Shang et al. (2021) improved the stability of whey protein isolate (WPI) using *Flammulina velutipes* polysaccharides (FVPs) for use in protein-enriched and functional beverages. The application of FVPs to develop the stability and functional properties of BGN-PI beverages in the food industry is up for investigation.

Therapeutic and scientific knowledge of the proteins and their qualities through amino acids quantification of selected accessions, leading to full utilization of its genetic resources for food and agriculture demands attention. This is not only important in breeding for genetic improvements or improved varieties of crops, but also in the utilisation of certain lines or accessions of germplasm to meet specified conditions, especially of health challenges. Available data show that BGN has the potentials to alleviate nutritional health and health-related challenges (Ademiluyi et al., 2015; Hillocks et al., 2012; Ijarotimi & Keshinro, 2012; Jideani & Diedericks, 2014; Koné et al., 2011; Murevanhema et al., 2013) like the micronutrient deficiency problems in human. There is a recent review on the nutraceutical and antimicrobial potentials of BGN (Udeh et al., 2020).

15.9 Utilization of Bambara Groundnut in Functional Ingredients and Food Products

More efforts are needed towards the development of functional food BGN products and ingredients. The feasibility of producing functional food and products from BGN is high considering research on (i) vegetable milk production (Brough et al., 1993), (ii) vegetable milk (Poulter & Caygill, 1980), and (iii) probiotic milk beverage (Murevanhema & Jideani, 2013).

15.10 Bambara Groundnut Starch Chemistry

Further investigations into BGN starch chemistry involving isolation, composition, physicochemical properties and starch modification methods in search of novel properties and the application of modified BGN starch in food systems is necessary. Exploration of the functional properties using farinograph, extensograph, amylograph and rheofermentometer devices can reveal more about BGN starches. The use of BGN flour and heat-moisture treated starch in food and food products may be an excellent way to alleviate obesity and diabetes, due to its high resistant starch content (Afolabi et al., 2017).

15.11 Development of Bambara Groundnut Functional Flour

Scientific evidence concerning the use of most climate-smart crops remains extremely limited. Hence, the development of a functional BGN-based composite flour with good nutritional, functional properties and pasting properties and with good compliance to flour properties is needed. Establishing BGN flour combination with other flours to obtain optimum rheological properties of composite dough is important.

There is high interest and an increased interest in using germinated legume seeds from lentil, chickpea, soybean, lupin, a bean in the daily diet, simply because in a germinated form, the seeds are highly improved from a nutritional point with multiple benefits for the human body (Atudorei & Codina, 2020; Boukid et al., 2019; Bueno et al., 2020; Kigel et al., 2015; Ohanenye et al., 2020; Xu et al., 2019a, b, 2020). Studies are needed on (i) the possibilities of using germinated BGN to obtain bakery products of good quality, (ii) the germination process of BGN, the benefits of the germination process on the seeds from a nutritional point of view, and the benefits of the addition of flour from germinated BGN on the rheological properties of composite flour dough, (iii) the physico–chemical and sensory characteristics of such BGN-cereal bakery products and (iv) the level and type of BGN subjected to the germination process for incorporated in different cereal flours.

15.12 Therapeutic Role of Bambara Groundnut Phytochemicals and the Use of Green Technologies in Extraction

Understanding the fate of phenolic phytochemicals in BGN during sprouting given their significant role as health-promoting constituents in human nutrition is necessary using HPLC-PDA-ESI-MS and other advanced analytical techniques. These compounds have antioxidant properties and may protect against major clinical conditions such as heart disease and cancer in which reactive oxygen species (i.e., superoxide anion, hydroxyl radicals and peroxy radicals) are involved (Duthie & Crozier, 2000; Jideani et al., 2021; Rhodes & Price, 1997). Also, dietary polyphenols or their metabolites act as signaling molecules that can increase nitrite oxide bioavailability and induce protective enzymes (Croft, 2016).

In the area of practical applications, it is essential to investigate the mechanisms of anticancer action of the phenolic compounds in BGN (*Vigna subterranea*) cultivars, processed BGN products to estimate chemoprotective potential, mechanisms (if any) involved in apoptosis pathways and proteins in modulation related to apoptotic processes. This is necessary to determine the relationship between the phenolic composition and the mechanisms of anticancer action of (i) the different assessions of BGN using an intestinal model of cancer, and (ii) as processing methods affect the phytochemical composition of such crops (Moreno-Jiménez et al., 2015). The effect of beans on the modulation of gene expression related to apoptosis, proliferation, the arrest of the cell cycle and inflammation in cellular models has been described by Campos-Vega et al. (2012) and Feregrino-Pérez et al. (2014) among others. Similar studies will go a long way in confirming that diet, including BGN (particularly darker varieties), might represent health benefits and cancer-preventive effects as in some beans varieties (Moreno-Jiménez et al., 2018). Investigation into their significant antioxidant activity and degree of polymerization may elicit prebiotic activity.

Need for the application of novel processing or and analytical technique for BGN. Some health beneficial compounds of BGN are in the literature (AOCC, 2020; Harris et al., 2018; Jideani & Diedericks, 2014; Murevanhema et al., 2013; Tsamo et al., 2018; Udeh et al., 2020) but the functionality depends on the technology of extraction and chemical structure. These functional components, biologically active compounds (BAC) have been reported (Jideani et al., 2021) together with methods of identification (Vincekovic et al., 2017; Lopez-Gutierrez et al., 2015). There is increasing global food markets demand on plant materials that are "safe", "fresh", "natural", and with "nutritional value". BACs include an extremely diverse class of compounds (polyphenols, tocopherols, organosulfur compounds, carotenoids, etc.) with different hydrophilic or lipophilic structures, a wide range of concentrations, and biological action, depending on the plant. These BACs in functional foods can be extracted by conventional extraction methods involving thermal influences on extraction yield and quality with a large expenditure of organic solvents and energy. Food production is gaining momentum on global food markets and there are industrial requirements to perfect extraction technologies for obtaining high-quality plant extracts that can be used as raw materials for its production. Advances and innovation in extraction technologies under the "green" concepts with the optimal expenditure of energy and chemicals now involve microwave-assisted extraction (MAE), ultrasound-assisted extraction (UAE), high-pressure assisted extraction (HPAE), high voltage electric discharges assisted extraction (HVED), pulsed electric fields assisted extraction (PEF), supercritical fluids extraction (SFE), and others (Gabric et al., 2018; Koubaa et al., 2017; Putnik et al., 2018; Vincekovic et al., 2017). The

Fig. 15.2 Advantages of the use of green technologies to extract bioactive compounds from plant sources. (Putnik et al., 2018)

application of these green technologies in the extraction of bioactive compounds from plant sources should be applied to BGN considering the numerous advantages as explained in Fig. 15.2 by Putnik et al. (2018). Regardless of the method used to extract BACs in BGN, there is a need for a comprehensive assessment of chemical composition using chromatographic methods and subsequently the functional effects in humans, preferably using (pre) clinical trials, aiming to achieve a dose-dependent response (Putnik et al., 2018; Ferreira et al., 2019).

15.13 Application of Big Data for Bambara Groundnut Productivity

With some recent advances and innovations such as big data, robots and more indoor farming, the digital technology of the future could lead to highly productive, and organic farming (Little, 2019). Innovative hurdle technologies involving decortication, fermentation and extrusion (Patrascu et al., 2018; Rosa-Millán et al., 2019) may have promising benefits in valorization of BGN. This is more imminent with the increasing population of vegetarians. However, innovations often look awkward at first, but they produce the food of the future. The challenge of technology diffusion lags such that it takes time for farmers in developing economies, especially small farmers, to learn about, test and adopt any new technology compare with those in advanced economies or large farmers. There is a need for farmers to reduce pesticide exposure and the associated health risks.

15.14 Flavour Profile of Bambara Groundnut

A detailed investigation and listing of the volatile compounds in the seeds of BGN and their theoretical odors are needed considering its technological attribute of strong beany flavor. Alcohols, aldehydes, alkanes, alkenes, aromatic hydrocarbons, esters, furans, ketones, organic acids, terpenes, and 3-isopropyl-2-methoxypyrazine are shown to be responsible for the green, grassy, beany, pea-like and earthy aroma notes associated with some legume seeds (Akkad et al., 2019; Ong & Liu, 2018; Mishra et al., 2017). Exploring the non-volatile taste compounds is also essential.

References

Adeleke, O. R., Adiamo, O. Q., & Fawale, O. S. (2017). Nutritional, physicochemical, and functional properties of protein concentrate and isolate of newly-developed Bambara groundnut (*Vigna subterrenea* L.) cultivars. *Food Science and Nutrition, 6*(1), 229–242. https://doi.org/10.1002/fsn3.552

Ademiluyi, A. O., Oboh, G., Boligon, A. A., & Athayde, M. L. L. (2015). Dietary supplementation with fermented legumes modulate hyperglycemia and acetylcholinesterase activities in Streptozotocin-induced diabetes. *Pathophysiology, 22*, 195–201.

Adzawla, W., Donkoh, S. A., Nyarko, G., O'Reilly, P. J., Olayide, O. E., Mayes, S., Feldman, A., & Azman, H. R. (2016). Adoption of Bambara groundnut production and its effects on farmers' welfare in Northern Ghana. *African Journal of Agricultural Research, 11*(7), 583–594. https://doi.org/10.5897/AJAR2015.10568

Afolabi, T. A., Opara, A. O., Kareem, S. O., & Oladoyinbo, F. O. (2017). In vitro digestibility of hydrothermally modified *Bambara* groundnut (*Vigna subterranean* L.) starch and flour. *Food Science and Nutrition, 6*, 36–46. https://doi.org/10.1002/fsn3.510

Akharume, F. U., Aluko, R. E., & Adedeji, A. A. (2021). Modification of plant proteins for improved functionality: A review. *Comprehensive Review in Food Science and Food Safety, 20*, 198–224. https://doi.org/10.1111/1541-4337.12688

Akkad, R., Kharraz, E., Han, J., House J. D, & Curtis, J. M. (2019). Characterisation of the volatile flavour compounds in low and high tannin faba beans (*Vicia faba* var. minor) grown in Alberta, Canada. *Food Research International, 120*, 285–294. https://doi.org/10.1016/j.foodres.2019.02.044

AOCC. (2020). *The African orphan crops consortium*. http://africanorphancrops.org/. Accessed 03 Jan 2021.

Ashogbon, A. O. (2017). Evaluation of compositional and some physicochemical properties of Bambara groundnut and cocoyam starch blends for potential industrial applications. *American. Journal of Food and Nutrition, 5*(2), 62–68. https://doi.org/10.12691/ajfn-5-2-3

Asprilla-Perea, J., & Díaz-Puente, J. M. (2019). Importance of wild foods to household food security in tropical forest areas. *Food Security, 11*, 15–22.

Atoyebi, O. J., Osilesi, O., Adebawo, O., & Abberton, M. (2017). Evaluation of nutrient parameters of selected African accessions of Bambara groundnut (*Vigna subterranea* (L.) Verdc.). *American Journal of Food and Nutrition*, (SCIEP series), *5*(3), 83–89. https://doi.org/10.12691/ajfn-5-3-1

Atudorei, D., & Codina, G. G. (2020). Perspectives on the use of germinated legumes in the bread making process, a review. *Applied Sciences, 10*, 6244. https://doi.org/10.3390/app10186244

Azam-Ali, S. (2007). Agricultural diversification: The potential for underutilized crops in Africa's changing climate. *Rivista di Bioligia, 100*(1), 27–37.

Boukid, F., Zannini, E., Carini, E., & Vittadini, E. (2019). Pulses for bread fortification: A necessity or a choice? *Trends Food Sci Tech, 88*, 416–428.

Brough, S. H., Taylor, A. J., & Azam-Ali, S. N. (1993). The potential of Bambara groundnuts (*Vigna subterranea*) in vegetable milk production and basic protein functionality systems. *Food Chemistry, 47*, 277–283.

Bueno, D. B., Silva-Júnior, S. I., Seriani Chiarotoo, A. B., Cardoso, T. M., Neto, J. A., Lopes dos Reis, G. C., Abreu Glória, M. B., & Tavano, O. L. (2020). The germination of soybeans increases the water-soluble components and could generate innovations in soy-based foods. *LWT-Food Science Technology, 117*, 108599.

Campos-Vega, R., García-Gasca, T., Guevara-González, R., Ramos-Gómez, M., Oomah, B. D., & Loarca-Piña, G. (2012). Human gut flora fermented non-digestible fraction from cooked bean (*Phaseolus vulgaris* L.) modifies protein expression associated with apoptosis, cell cycle arrest, and proliferation in human adenocarcinoma colon cancer cells. *Journal of Agricultural and Food Chemistry, 60*, 12443–12450.

Clark, M. A., Springmann, M., Hill, J., & Tilman, D. (2019). Multiple health and environmental impacts of foods. *Proceedings of the National Academy of Sciences USA, 116*, 23357–23362.

Croft, K. D. (2016). Dietary polyphenols: Antioxidants or not? *Archives of Biochemistry and Biophysics, 595*, 120–124. https://doi.org/10.1016/j.abb.2015.11.014

Dawson, I. K., Sarah, E. P., Attwood, S. J., Jamnadass, R., Powell, W., Sunderland, T., & Carsan, S. (2019). Contributions of biodiversity to the sustainable intensification of food production. *Global Food Security, 21*, 23–37.

De Groote, H., Kariuki, S. W., Traore, D., Taylor, J. R. N., Ferruzzi, M. G., & Hamaker, B. R. (2018). Measuring consumers' interest in instant fortified pearl millet products: A field experiment in Touba, Senegal. *Journal of the Science of Food and Agriculture, 98*, 2320–2331.

Diedericks, C. F. (2020) Beyond a subsistence crop: Structure and functionality of Bambara groundnut proteins Ph.D. Dissertation, Wageningen University, Wageningen. ISBN 9789463953177–159, https://library.wur.nl/WebQuery/wurpubs/563553

Duthie, G., & Crozier, A. (2000). Plant-derived phenolic antioxidants. *Current Opinion in Clinical Nutrition and Metabolic Care, 3*(6), 447–451. https://doi.org/10.1097/00075197-200011000-00006

FAO. (2014). *WFP* (The state of food insecurity in the world (2014) strengthening the enabling environment for food security and nutrition). FAO.

FAO (2017) Statistique de production (FAOSTAT). Organisation des Nation Unies pour l'Alimentation et l'Agriculture/Food and Agriculture Organization (FAO) Rome. http://www.fao.org/faostat/en/#home

Feregrino-Pérez, A. A., Piñol-Felis, C., Gómez-Arbones, X., Guevara-González, R. G., Campos-Vega, R., Acosta-Gallegos, J., & Loarca-Piña, G. (2014). A non-digestible fraction of the common bean (*Phaseolus vulgaris* L.) induces cell cycle arrest and apoptosis during early carcinogenesis. *Plant Foods for Human Nutrition, 69*, 248–254. https://doi.org/10.1007/s11130-014-0428-7

Ferreira, P. S., Victorelli, F. D., Fonseca-Santos, B., & Chorilli, M. (2019). A review of analytical methods for p-coumaric acid in plant-based products, beverages, and biological matrices. *Critical Reviews in Analytical Chemistry, 49*(1), 21–31.

Gabric, D., Barba, F., Roohinejad, S., Gharibzahedi, S. M. T., Radojcin, M., Putnik, P., & Bursac Kovacevic, D. (2018). Pulsed electric fields as an alternative to thermal processing for the preservation of nutritive and physicochemical properties of beverages: A review. *Journal of Food Process Engineering, 41*, e12638.

Gassner, A., Harris, D., Mausch, K., Terheggen, A., Lopes, C., Finlayson, R. F., & Dobie, P. (2019). Poverty eradication and food security through agriculture in Africa: Rethinking objectives and entry points. *Outlook on Agriculture, 48*, 309–315.

Godfray, H. C. J., Beddington, J. R., Crute, I. R., Haddad, L., Lawrence, D., Muir, J. F., Pretty, J., Robinson, S., Thomas, S. M., & Toulmin, C. (2010). Food security: The challenge of feeding 9 billion people. *Science, 327*, 812–818.

Gras, A., Garnatje, T., Marín, J., Parada, M., Sala, E., Talavera, M., & Vallès, J. (2021). The power of wild plants in feeding humanity: A meta-analytic ethnobotanical approach in the catalan linguistic area. *Food, 10*, 61. https://doi.org/10.3390/foods10010061

Harris, T., Jideani, V., & Le Roes-Hill, M. (2018). Flavonoids and tannin composition of Bambara groundnut (*Vigna subterranea*) of Mpumalanga. *South Africa, Heliyon, 4*, 00833.

Hillocks, R. J., Bennett, C., & Mponda, O. M. (2012). Bambara nut: A review of utilisation, market potential and crop improvement. *African Crop Science Journal, 20*(1), 1–16.

Ijarotimi, O. S., & Keshinro, O. O. (2012). Formulation and nutritional quality of infant formula produced from germinated popcorn, Bambara groundnut and African locust bean flour. *Journal of Microbiology, Biotechnology and Food Sciences, 1*(6), 1358–1388.

Jamnadass, R., Mumm, R. H., Hale, I., Hendre, P., Muchugi, A., Dawson, I. K., Powell, W., Graudal, L., Yana-Shapiro, H., Simons, A. J., & Deynze, A. V. (2020). Enhancing African orphan crops with genomics. *Nature Genetics, 52*, 356–360.

Jideani, V. A., & Diedericks, C. F. (2014). Nutritional, therapeutic, and prophylactic properties of *Vigna subterranea* and *Moringa oleifera*. In O. Oguntibeju (Ed.), *Antioxidant-antidiabetic agents and human health* (Vol. 9, pp. 187–207). Janeza Trdine. https://doi.org/10.5772/57338

Jideani, A. I. O., Silungwe, H., Takalani, T., Omolola, A. O., Udeh, H. O., & Anyasi, T. A. (2021). Antioxidant-rich natural fruit and vegetable products and human health. *International Journal of Food Properties, 24*(1), 41–67. https://doi.org/10.1080/10942912.2020.1866597

Kifle, T. (2021). Climate-smart agricultural (CSA) practices and its implications to food security in Siyadebrina Wayu District, Ethiopia. *African Journal of Agricultural, 17*(1), 92–103. https://doi.org/10.5897/AJAR2020.15100. http://www.academicjournals.org/AJAR

Kigel, J., Rosental, L., & Fait, A. (2015). Seed physiology and germination of grain legumes. In A. M. De Ron (Ed.), *Grain Legumes (Series: Handbook of plant breeding)* (Vol. 10, pp. 327–363). Springer.

Koné, M., Paice, A. G., & Touré, Y. (2011). Bambara groundnut [Vigna subterranea (L.) Verdc. (Fabaceae)] usage in human health. Chapter 22. In V. R. Preedy, R. R. Watson, & V. B. Patel (Eds.), *Nuts and seeds in health and disease prevention* (1st ed., pp. 189–196. Hardcover ISBN: 9780123756886, eBook ISBN: 9780123756893, Groote). Academic. https://doi.org/10.1016/C2009-0-01960-6

Koubaa, M., Barba, F., Bursac Kovacević, D., Putnik, P., Santos, M. S., Queirós, R. P., Moreira, S. A., Duarte, R. R., & Saraiva, J. A. (2017). Pulsed electric field processing of different fruit juices. In G. Rajauria & B. Tiwari (Eds.), *Fruit juices* (1st ed., p. 760). Academic.

Li, L., Yanga, T., Liu, R., Redden, B., Maalouf, F., & Zonga, X. (2017). Food legume production in China. *The Crop Journal, 5*, 115–126. https://doi.org/10.1016/j.cj.2016.06.001

Lipper, L., & Zilberman, D. (2018). A short history of the evolution of the climate smart agriculture approach and its links to climate change and sustainable agriculture debates. *Natural Resource Management and Policy, 52*. https://doi.org/10.1007/978-3-319-61194-5-2

Little, A. (2019). The fate of food: What we'll eat in a bigger, hotter, smarter world. : Harmony Books. 340 pp. https://www.accessagriculture.org/fate-food

Lopez-Gutierrez, N., Romero-Gonzalez, R., Plaza-Bolanos, P., & Vidal, J. L. M. (2015). Identification and quantification of phytochemicals in nutraceuticals products from green tea by UHPLC-Obitrap-MS. *Food Chemistry, 173*, 607–618.

McMullin, S., Stadlmayr, B., Mausch, K., Revoredo-Giha, C., Burnett, F., Guarino, L., Brouwer, I. D., Jamnadass, R., Graudal, L., Powell, W., & Dawson, I. K. (2021). Determining appropriate interventions to mainstream nutritious orphan crops into African food systems. *Global Food Security, 28*, 100465.

Mishra, P. K., Tripathi, J., Gupta, S., & Variyar, P. S. (2017). Effect of cooking on aroma profile of red kidney beans (*Phaseolus vulgaris*) and correlation with sensory quality. *Food Chemistry, 215*, 401–409. https://doi.org/10.1016/j.foodchem.2016.07.149

Mohamadou, M., Ngakou, A., & Tchuenguem, F.-N. F. (2018). Influence de l'interaction mycorhizes-insectes floricoles sur les rendements de *Vigna subterranea* variété rouge (Fabaceae) à Dang (Ngaoundéré, Cameroun) International Journal of Biological and Chemical Sciences, 12(4), 1897–1913. https://doi.org/10.4314/ijbcs.v12i4.30. Available online at http://www.ifgdg.org

Moreno-Jiménez, M. R., Cervantes-Cardoza, V., Gallegos-Infante, J. A., González-Laredo, R. F., Estrella, I., de García-Gasca, T. J., & Rocha-Guzmán, N. E. (2015). Phenolic composition changes of processed beans: Their antioxidant and anti-inflammatory effect in intestinal cancer cells. *Food Research International, 76*, 79–85.

Moreno-Jiménez, M. R., López-Barraza, R., Cervantes-Cardoza, V., Perez-Ramirez, I. F., Reyna-Rojas, J. A., Gallegos-Infante, J. A., Estrella, I., Rojas-Contreras, J. A., Gonzalez-Laredo, R. F., & Rocha-Guzman, N. E. (2018). Mechanisms associated to apoptosis of cancer cells by phenolic extracts from two canned common beans varieties (*Phaseolus vulgaris* L.). *Journal of Food Biochemistry, 2018*, e12680. https://doi.org/10.1111/jfbc.12680

Murevanhema, Y. Y., & Jideani, V. A. (2013). Potential of Bambara groundnut (*Vigna subterranea* (L.) Verdc) milk as a probiotic beverage. A review. *Critical Review of Food Science and Nutrition, 53*, 954–967.

Murevanhema, Y. Y., Jideani, V. A., & Oguntibeju, O. O. (2013). Review on potential of seeds and value-added products of bambara groundnut (*Vigna subterranea*): Antioxidant, anti-inflammatory, and anti-oxidative stress. *Bioactive Compounds of Medicinal Plants*, 141–188.

Nyau, V., Prakash, S., Rodrigues, J., & Farrant, J. (2015). Antioxidant activities of Bambara groundnuts as assessed by FRAP and DPPH assays. *American Journal of Food and Nutrition, 3*(1), 7–11. https://doi.org/10.12691/ajfn-3-1-2

Ohanenye, I. C., Tsopmo, A., Ejike, C. E. C. C., & Udenigwe, C. C. (2020). Germination as a bioprocess for enhancing the quality and nutritional prospects of legume proteins. *Trends in Food Science and Technology, 101*, 213–222.

Okebalama, C. B., Ibezim, C. O., Ugwu, V. U., & Marschner, B. (2020). Effects of amending an Ultisol with Bambara groundnut seed residue on soil fertility, microbial nutrient cycling and yield of cucumber. *Biological Agriculture & Horticulture*. https://doi.org/10.1080/01448765.2020.1808067

Okoroji, V., Lees, N. J., & Lucock, X. (2021). Factors affecting the adoption of mobile applications by farmers: An empirical investigation. *African Journal of Agricultural Research, 17*(1), 19–29. https://doi.org/10.5897/AJAR2020.14909. http://www.academicjournals.org/AJAR

Ong, P. K. C., & Liu, S. Q. (2018). Flavor and sensory characteristics of vegetables. In M. Siddiq & M. A. Uebersax (Eds.), *Handbook of vegetables and vegetable processing* (Vol. 1, 2nd ed., pp. 135–156). Wiley. https://doi.org/10.1002/9781119098935.ch6

Patrascu, L., Aprodu, I., Garnai, M., & Vasilean, I. (2018). Effect of germination and fermentation on the functionality of wheat-pulses flour mixtures. *Journal of Biotechnology, 280*, S52–S53.

Pawera, L., Khomsan, A., Zuhud, E. A., Hunter, D., Ickowitz, A., & Polesny, Z. (2020). Wild food plants and trends in their use: From knowledge and perceptions to drivers of change in West Sumatra, Indonesia. *Food, 9*, 1240.

Poulter, N. H., & Caygill, J. C. (1980). Vegetable milk processing and rehydration characteristics of Bambara groundnut (*Voandzeia subterranea* (L.) thou.). *Journal of the Science of Food and Agriculture, 31*(11), 1158–1163.

Putnik, P., Lorenzo, J. M., Barba, F. J., Roohinejad, S., Jambrak, A. R., Granato, D., Montesano, D., & Kovacevic, D. B. (2018). Novel food processing and extraction technologies of high-added value compounds from plant materials. *Food, 7*, 106. https://doi.org/10.3390/foods7070106Foods

Rhodes, M. J. C., & Price, K. R. (1997). Identification and analysis of plant phenolic antioxidants. *European Journal of Cancer Prevention, 6*, 5188. https://doi.org/10.1097/00008469-199712000-00005

Rosa-Millán, J., Heredia-Olea, E., Perez-Carrillo, E., Guajardo-Flores, S., & Serna-Saldívar, S. R. O. (2019). Effect of decortication, germination and extrusion on physicochemical and in vitro protein and starch digestion characteristics of black beans (*Phaseolus vulgaris* L.). *LWT-Food Science Technology, 102*, 330–333.

Sahu, S. K., Liu, M., Yssel, A., Kariba, R., Muthemba, S., Jiang, S., Song, B., Hendre, P. S., Muchugi, A., Jamnadass, R., Kao, S. M., Featherston, J., Zerega, N. J. C., Xu, X., Yang, H., Deynze, A. V., Peer, Y. V., Liu, X., & Liu, H. (2020). Draft genomes of two Artocarpus plants, jackfruit (*A. heterophyllus*) and breadfruit (*A. altilis*). *Genes, 11*(1), 27. https://doi.org/10.3390/genes11010027

Shang, J., Liao, M., Jin, R., Teng, X., Li, H., Xu, Y., Zhang, L., & Liu, N. (2021). Molecular properties of *flammulina velutipes* polysaccharide–whey protein isolate (wpi) complexes via noncovalent interactions. *Food, 10*(1). https://doi.org/10.3390/foods10010001

Smith, E., Ahmed, S., Dupuis, V., Crane, M. R., Eggers, M., Pierre, M., Flagg, K., & Shanks, C. B. (2019). Contribution of wild foods to diet, food security, and cultural values amidst climate change. *Journal of Agriculture, Food Systems and Community Development, 9*, 1–24.

Tan, X. L., Azam-Ali, S., Goh, E. V., Mustafa, M., Chai, H. H., Ho, W. K., Mayes, S., Mabhaudhi, T., Azam-Ali, S., & Massawe, F. (2020). Bambara groundnut: An underutilized leguminous crop for global food security and nutrition. *Frontiers in Nutrition, 7*, 601496. https://doi.org/10.3389/fnut.2020.601496

Tsamo, A. T., Ndibewua, P. P., & Dakora, F. D. (2018). Phytochemical profile of seeds from 21 Bambara groundnut landraces via UPLC-qTOF-MS. *Food Research International, 112*, 160–168.

Tsoata, E., Temegne, N. C., & Youmbi, E. (2017). Analysis of early biochemical criterion to screen four Fabaceae plants for their tolerance to drought stress. *International Journal of Current Research, 9*(1), 44568–44575.

Udeh, E. L., Nyila, M. A., & Kanu, S. A. (2020). Nutraceutical and antimicrobial potentials of Bambara groundnut (*Vigna subterranean*): A review. *Heliyon, 6*, e05205. https://doi.org/10.1016/j.heliyon.2020.e05205

Vincekovic, M., Viskic, M., Juric, S., Giacometti, J., Bursac Kovacević, D., Putnik, P., Donsì, F., Barba, F. J., & Režek Jambrak, A. (2017). Innovative technologies for encapsulation of Mediterranean plants extracts. *Trends Food Science and Technology, 69*, 1–12.

WFP. (2016). *Hunger statistics*. World Food Programme (WFP). http://www.worldhunger.org/2015-world-hunger-and-poverty-facts-and-statistics/

Woltering, L., Fehlenberg, K., Gerard, B., Ubels, J., & Cooley, L. (2020). Scaling – From "reaching many" to sustainable systems change at scale: A critical shift in mindset. *Agricultural Systems, 176*. https://doi.org/10.1016/j.agsy.2019.102652. article 102652.

World Bank. (2011). Climate-smart agriculture: Increased productivity and food security, enhancing resilience and reduced carbon emissions for sustainable development. Opportunities and challenges for a converging agenda: Country examples. Washington, DC. https://openknowledge.worldbank.org/handle/10986/27395

Xu, M., Jin, Z., Lan, Y., Rao, J., & Chen, B. (2019a). HS-SPME-GC-MS/olfactometry combined with chemometrics to assess the impact of germination on flavor attributes of chickpea, lentil, and yellow pea flours. *Food Chemistry, 280*, 83–95.

Xu, M., Jin, Z., Simsek, S., Hall, C., Rao, J., & Chen, B. (2019b). Effect of germination on the chemical composition, thermal, pasting, and moisture sorption properties of flours from chickpea, lentil, and yellow pea. *Food Chemistry, 295*, 579–587.

Xu, M., Jin, Z., Gu, Z., Rao, J., & Bingcan, C. (2020). Changes in odor characteristics of pulse protein isolates from germinated chickpea, lentil, and yellow pea: Role of lipoxygenase and free radicals. *Food Chemistry, 314*, 126184.

Chapter 16
Conclusion

Bambara groundnut (*Vigna subterranea* (L.) Verdc.), an underutilized crop, is a legume that can aid global food security but would require varietal improvement, genetic diversity and processing. The seed, flour, starch and other components of Bambara groundnut (BGN) have potentials. The chapters have demonstrated that BGN has the potential to contribute to food and nutrition security through its diversity of products. The use of BGN flour and heat-moisture-treated starch in food and food products may be an excellent way to alleviate obesity and diabetes, due to its high resistant starch content. Heat-moisture treatment also reduced the amylose content of the starch with a consequential decrease in its swelling power, this is a desirable property in the application of the modified starch in food and allied industries (Afolabi et al., 2017). In no distant future, BGN products will enjoy commercialization as conventional cereal and legume plant-based foods. Farmers and food industries, during the present trend of plant-based foods, should venture into this promising legume crop. The volarization and commercialization will open up markets for BGN and its products and consequently elevate its status in the global food industries.

We believe that this is one of the first books on BGN. Therefore, it is intended to: (i) Educate on the potential of BGN as a source of food and nutrition for food security and innovation, and (ii) Raise the awareness of BGN products for commercialization and elevate its status from underutilized to a grain of choice in the food industries. The chapter on "*Current and future BGN research directions*" critically highlights areas of importance requiring focus by all stakeholders ranging from the natural to the social scientists.

BGN is likely to benefit from the rapid scientific and technological advances in food science and technology considering the unprecedented technologies of the fourth industrial revolution. As stated by McClements (2019), food architecture, functional foods, nutraceuticals/probiotics, gene editing, nanotechnology and artificial intelligence, 3D food printing all put together are transforming the way our foods are produced and consumed. The application of these advances to BGN could lead to healthier, tastier, more sustainable foods. Healthiness in terms of combating

V. A. Jideani, A. I. O. Jideani, *Bambara groundnut: Utilization and Future Prospects*, https://doi.org/10.1007/978-3-030-76077-9_16

chronic diseases such as obesity, cancer, diabetes, stroke, and heart disease is more the focus in addressing modern food challenges such as feeding the growing global population, reducing greenhouse gas emissions, reducing waste, and improving sustainability (McClements, 2019). Close attention to underutilized food crops like BGN is key to tackling chronic diseases, feeding a global population, reducing food waste, and creating healthier and tastier foods.

From recent literature and emerging research on climate-smart crops, adaptive breeding and genomics, natural resource management, strategies for access to food, technology, knowledge and equity and risk governance (Campbell et al., 2016; Sarkar & vanLoon, 2019). there is much need for climate-resilient and nutritious legume crops like BGN. The effect of climate change, coupled with the COVID-19 pandemic, has been far-reaching around the world with a negative impact on food security and adaptation strategies or actions apart from taking many forms, become very challenging. To mitigate adverse situations in plant food production, the event of COVID-19, suggests that there is a need to consider broader perspectives of food security as the era of changing climate and the global pandemic linger. This book attempts to elucidate the present knowledge on BGN, highlighting different ways of processing, the possible functionality of its macromolecules as ingredients, and the need for further research to better understand how the food industry could improve their utilization in the market.

References

Afolabi, T. A., Opara, A. O., Kareem, S. O., & Oladoyinbo, F. O. (2017). In vitro digestibility of hydrothermally modified Bambara groundnut (*Vigna subterranean* L.) starch and flour. *Food Science & Nutrition, 6*, 36–46. https://doi.org/10.1002/fsn3.510

Campbell, B. M., Vermeulen, S. J., Aggarwal, P. K., Corner-Dolloff, C., Girvetz, E., Loboguerrero, A. M., Ramirez-Villegas, J., Rosenstock, T., Sebastian, L., Thornton, P. K., & Wollenberg, E. (2016). Reducing risks to food security from climate change. *Global Food Security, 11*, 34–43.

McClements, D. J. (2019). *Future foods – How modern science is transforming the way we eat.* Springer. https://doi.org/10.1007/978-3-030-12995-8_12. ISBN 978-3-030-12995-8.

Sarkar, A., & vanLoon, G. W. (2019). Chapter 12 Conclusion. In A. Sarkar, S. R. Sensarma, & G. W. vanLoon (Eds.), *Sustainable solutions for food security* (Combating climate change by adaptation). Springer. https://doi.org/10.1007/978-3-319-77878-5_25

Index

V. A. Jideani, A. I. O. Jideani, *Bambara groundnut: Utilization and Future Prospects*, https://doi.org/10.1007/978-3-030-76077-9

The manufacturer's authorised representative in the EU is Springer Nature Customer Service Centre GmbH, Europaplatz 3, 69115 Heidelberg, Germany. If you have any concerns regarding our products, please contact ProductSafety@springernature.com

Printed and bound by CPI Group (UK) Ltd, Croydon, CR0 4YY
22/09/2026
02221062-0010